T0340087

Time and Memory

The Study of Time

Founding Editor

J. T. Fraser

VOLUME XII

Time and Memory

Edited by

Jo Alyson Parker,
Michael Crawford
and
Paul Harris

BRILL
LEIDEN • BOSTON
2006

About this book series The nature of time has haunted humankind through the ages. Some conception of time has always entered into our ideas about mortality and immortality, and permanence and change, so that concepts of time are of fundamental importance in the study of religion, philosophy, literature, history, and mythology. How we experiences time physiologically, psychologically and socially enters into the research of the behavioral sciences, and time as a factor of structure and change is an essential consideration in the biological and physical sciences. On one aspect or another, the study of time cuts across all disciplines. The International Society for the Study of Time has as its goal the interdisciplinary and comparative study of time.

This book is printed on acid-free paper.

Library of Congress Cataloging-in-Publication Data

International Society for the Study of Time, Conference (12th : 2004 : Clare College, Cambridge)
    Time and memory / edited by Jo Alyson Parker, Michael Crawford, and Paul Harris.
        p. cm.—(The study of time, ISSN 0170-9704 ; v. 12)
    Includes bibliographical references and index.
    ISBN-13: 978-90-04-15427-8
    ISBN-10: 90-04-15427-2 (hardback : alk. paper)
    1. Time—Congresses. 2. Memory—Congresses. I. Parker, Jo Alyson, 1954- II. Crawford, Michael. III. Harris, Paul (Paul A.) IV. Title.

BD638.I72 2004
115—dc22

2006049032

ISSN 0170-9704
ISBN-13: 978-90-04-15427-8
ISBN-10: 90-04-15427-2

© 2006 by Koninklijke Brill NV, Leiden, The Netherlands
Koninklijke Brill NV incorporates the imprints Brill, Hotei Publishers, IDC Publishers,
Martinus Nijhoff Publishers and VSP.

PRINTED IN THE NETHERLANDS

In memory of Karen Davies, valued colleague and friend.

CONTENTS

LIST OF CONTRIBUTORS

BRIAN ALDISS is a prolific and diverse writer. His history of science fiction, *Billion Year Spree*, broke new ground in science fiction studies, and he has achieved much acclaim for his extensive publications of science fiction short stories and novels, including the trilogy the *Helliconia* novels, *Spring, Summer* and *Winter.* He has won most of the important awards in the SF field. Roger Corman filmed his novel *Frankenstein Unbound* and his short Story Supertoys *Last all Summer Long* formed the basis of the Kubrick/Spielberg movie *A.I.* His writings have become more and more diverse, with his autobiography, *Twinkling of an Eye*, his elegy for his wife, *When the Feast is Finished,* his utopia, written with Roger Penrose, *White Mars*, his poetry, plays and now a possible opera based on his recent novel, *Jocasta.* He is an actor, a lively speaker and was made a Doctor of Literature. Last year he was invested in the O.B.E. (Order of the British Empire) for services to literature.

DAWNA I. BALLARD (Ph.D., University of California, Santa Barbara, 2002) is an Assistant Professor in the Department of Communication Studies at the University of Texas at Austin, where she teaches courses in organizational communication, group communication, and temporality at work. Her research is concerned with the mutual influence between time and communication in the workplace.

ROBERT BELTON is the Dean of the Faculty of Creative and Critical Studies in the University of British Columbia Okanagan. A well-known art historian and award-winning teacher, Dr. Belton joined the faculty of the Department of Fine Arts at UBC Okanagan's predecessor, Okanagan University College, in 1992. He has published widely in his areas of interest: art theory, criticism and historiography; and modern and contemporary European and North American art and architecture. He has also acted as a curator, participated as a juror, served as an artist and/or technician, and discussed contemporary art issues in public forums not only in British Columbia but also across Canada and overseas.

MICHAL BEN-HORIN is a Lecturer in the Department of German Literature at the Hebrew University of Jerusalem. Her main research interests are theory of music and literary theory, poetics of memory in post-1945 German and

Jo Alyson Parker, Michael Crawford, Paul Harris (Eds), Time and Memory, pp. xi–xvi
© *2006 Koninklijke Brill N.V. Printed in the Netherlands.*

Austrian literature, and comparative reading of contemporary Hebrew and German literature.

EFRAT BIBERMAN is a Senior Lecturer in Hamidrasha School of Art, Beit Berl College. She teaches aesthetics at Haifa University. She has published several essays in which she explores painterly problems through psychoanalytic conceptualization.

DAVID BURROWS, Professor Emeritus of the New York University Department of Music, is a long-time Council member and former Treasurer of the International Society for the Study of Time. His papers for the Society have focused on the now.

ANN MARIE BUSH, Assistant Professor of English at Marywood University in Scranton, Pennsylvania, specializes in the Poetics of Poetry, African-American Literature, and Ethnic Literatures and has published articles on Elizabeth Bishop and African-American Literature.

MICHAEL CRAWFORD Ph.D. is an Associate Professor of Biological Sciences at the University or Windsor. His research currently focuses upon the mechanisms that underlie the "segmentation clock" that gives rise to vertebrae.

KAREN DAVIES worked at the Swedish National Institute for Working Life and was director of a national research program on social exclusion in the labor market. She was also associate professor at the Department of Sociology, University of Lund, Sweden. Since the 1980's, she published widely on the issue of time and gender.

E. R. DOUGLAS, is a Ph.D. candidate in philosophy at the University of Texas, Austin, with an emphasis in the philosophy of physics, mathematics, and science. His concentration in the natural philosophy of time, including temporal logic, the scientific modelling of time or 'temporology,' and cosmology is complemented by his interests in consciousness studies, the mind-body problem, and the foundations of artificial intelligence. He manages a time studies resource center at http://www.temporality.org, is involved in Green Party political policy development and democracy advocacy, and works as a freelance consultant in formal systems applications.

Founder of the International Society for the Study of Time, J. T. FRASER is the author of Of Time, Passion, and Knowledge, Time as Conflict, The Genesis and Evolution of Time, Time the Familiar Stranger and Time. Conflict, and Human Values. He is editor of The Voices of Time, of ten volumes of The Study of Time series and Founding Editor of KronoScope— Journal in the Study of Time. He is the author of over sixty articles on different aspects of the study of time and taught courses and conducted seminars in the integrated study of time at universities at home and abroad.

LUIS FELIPE GÓMEZ (M.A., Katholieke Universiteit Leuven, Belgium, 1995) is a doctoral candidate in the Department of Communication Studies at the University of Texas at Austin where he teaches team-based communication. His research is concerned with the effects of socially constructed time on organizational adaptation.

PAUL A. HARRIS is a Professor of English at Loyola Marymount University.

MARIE-PASCALE HUGLO is professor of French Studies at l'Université de Montréal. She has recently edited two issues of *SubStance*, one on contemporary novelist Lydie Salvayre (vol. 34, no 1, 2005), the other one, in collaboration with Johanne Villeneuve, on Memory, Media, Art (vol. 33, no 2, 2004). She has published an essay *Métamorphoses de l'insignifiant. Essai sur l'anecdote dans la modernité*, (Montréal, Balzac-Le Griot, 1997) and has co-edited, with Sarah Rocheville, *Raconter? Les enjeux de la voix narrative dans le récit contemporain* (Paris, L'Harmattan, 2004). Forthcoming in 2006 is *Le sens du récit. Lectures du roman contemporain* (Lille, Presses Universitaires du Septentrion).

HEIKE KLIPPEL is Professor of Film Studies at the Hochschule für Bildende Künste Braunschweig (Braunschweig School of Fine Arts). Her books include *Gedächtnis und Kino* (1997), and she is co-editor of the feminist film journal *Frauen und Film*. She is currently working on a book about cinema, death, and the everyday.

FRANK KRÜGER, a Ph.D. in cognitive psychology, currently holds a postdoc position in the Cognitive Neuroscience Section, National Institute of Neurological Disorders and Stroke (NINDS) at the National Institutes of Health (NIH), Bethesda, Maryland, USA. He has published articles in *Zeitschrift für Psychologie* on temporal order and pupillary response and the influence of temporal order information in general event knowledge on language comprehension.

LARS KUCHINKE is a Research Assistant in the Psychology Department (General and Neurocognitive Psychology Lab) at Free University in Berlin, Germany. He has published articles on absolute pitch and pupillary response and the incidental effects of emotional valence in single work processing. His work has appeared in such journals as *Neuroimage* and *Psychophysiology*.

DANIEL J. LEAB is Professor of History at Seton Hall University, the author of numerous books and articles including *"I Was A Communist for the FBI": the life and unhappy times of Matt Cvetic* (Penn State University Press, 2001), and *Orwell Subverted: The CIA and the Filming of* Animal Farm (forthcoming from Penn State University Press, 2007), and he is the founding editor of the journal *American Communist History*.

CARMEN LECCARDI is a Professor of Cultural Sociology in the Department of Sociology and Social Research, University of Milan-Bicocca, and publishes in the realms of time, youth and gender issues.

RÉMY LESTIENNE, Ph.D. is a Senior Researcher at CNRS (the French fundamental research agency). He has performed work in elementary particle physics, epistemology, and neuroscience. President of the ISST (1998–2004), he has published about seventy scientific articles and is the author of *The Children of Time* (University of Illinois Press, 1995) and The Creative Power of Chance (University of Illinois Press, 1998).

LINDA McKIE is Research Professor in sociology at Glasgow Caledonian University (www.gcal.ac.uk) and Associate Director at the Centre for Research on Families and Relationships (www.crfr.ac.uk). Her current work includes gender, families and violence, and organizations, work and care. Recent publications include *Families Violence and Social Change* (Open University Press, 2005) and *Families in Society: Boundaries and Relationships* (edited with Sarah Cunningham-Burley, Policy Press, 2005).

JO ALYSON PARKER is Associate Professor of English at Saint Joseph's University in Philadelphia, where she teachers courses in the eighteenth- and early nineteenth-century novel, narrative and literary theory. She is the author of *The Author's Inheritance: Henry Fielding, Jane Austen, and the*

Establishment of the Novel (Northern Illinois University Press, 1998), and she has published essays on narrative and time.

ALISON PHINNEY, RN, Ph.D. is an Assistant Professor in the School of Nursing at the University of British Columbia in Vancouver, Canada. She is also a Research Associate of the Centre for Research on Personhood in Dementia. Dr. Phinney's scholarship is focused on understanding the experiences of people living with dementia.

JEFFREY PRAGER is a Professor of Sociology at UCLA. He also has received a full clinical training in psychoanalysis and is a practicing psychoanalyst. He is the author of *Presenting the Past, Psychoanalysis and the Sociology of Misremembering* (Harvard 1998), a book that explores the influence of contemporary social and interpersonal contexts in remembering the past and the phenomenon of false memory. He has also published extensively on the relation of psychoanalysis to the social sciences, and on individual memory, collective memory and personal identity.

MARY SCHMELZER has taught in the English department at Saint Joseph's University since 1989. Her areas of specialty are critical theory and the English Renaissance. She has written about Erasmus, Robert Burton, and Michel Foucault as well as on issues of pedagogy and medical humanities. She has done a Lacanian psychoanalytic training as an offshoot of her theoretical interests and is grateful to J. T. Fraser and the ISST for the opportunities it invites for serious interdisciplinary engagement.

SHIRLEY SHARON-ZISSER is Associate Professor and Graduate Advisor at the Department of English, Tel Aviv University, and a member of the teaching staff of the Center for Lacanian Studies at Tel Aviv (under the auspices of the Department of Psychoanalysis, University of Paris VIII). Her Publications include "The Risks of Simile in Renaissance Rhetoric" (2000) and "Critical Essays on Shakespeare's A Lover's Complaint: Suffering Ecstasy" (2006, editor) and numerous essays on literature, rhetoric, philosophy, and psychoanalysis.

KATHERINE A. S. SIBLEY is Professor and chair of the History Department at Saint Joseph's University. She is currently working on a biography of Florence Kling Harding, entitled *America's First Feminist First Lady*, which will revise the prevailing portrait of Mrs. Harding as a manipulative, unhappy wife. Her work has appeared in journals including *American Communist History*,

Peace and Change, and *Diplomatic History*. Her books include the prize-winning *Loans and Legitimacy: The Evolution of Soviet-American Relations, 1919–1933* (1996); *The Cold War* (1998); and *Red Spies in America: Stolen Secrets and the Dawn of the Cold War* (2004). She is book review editor for *Intelligence and National Security* as well as Commonwealth Speaker for the Pennsylvania Humanities Council, and she serves on the U.S. State Department's Historical Advisory Committee, the Council of the Society for Historians of American Foreign Relations, and the Board of Editors of *American Communist History*.

MARLENE P. SOULSBY is Editor of *KronoScope: Journal for the Study of Time*. She has also served as editor of three volumes of the *Study of Time* series and was editor of *Times News* from 1990 to 2000. Her publications focus on the experience of time in literature, time and aging, and East-West literary comparison.

CHRISTIAN STEINECK is Senior Research Fellow at Bonn University's Center for Research on Modern Japan. He earned an M.A. degree in Japanese Studies and a Dr. Phil. in philosophy, both from Bonn University. His research has focussed on ancient and modern Japanese contributions to the understanding of body, mind, and time. He has authored books on *Fundamental Structures Of Mystical Thought, Body And Mind In Dôgen*, and *Notions Of The Human Body In Contemporary Japanese Bioethics* (forthcoming).

DIRK STRAUCH, a Ph.D. in cognitive psychology, is a psychotherapist at the Medical Center Schweriner See, in Lübsdorf, Germany. He has published on temporal relations between event concepts in the collection *Mediating between Concepts and Grammar*.

ELKE VAN DER MEER is a professor of cognitive psychology at Humboldt University Berlin, Germany. She has published articles on temporal order and pupillary response, temporal order in language comprehension, and psychological time and aging effects. Her work has appeared in journals such as *Journal of Experimental Psychology*, *Psychophysiology*, and *Zeitschrift für Psychologie* and collections such as *Mediating between Concepts and Grammar* and *Language and Cognition*.

FOREWORD

Michael Crawford, Jo Alyson Parker, Paul Harris

As sensate beings afloat on the river of time, the only faculty that we possess that can endow us with both a sense of permanence and identity is memory. Memory plants signposts along the banks of where we have been, fixes markers of our experience of the present, and helps us to chart our course into the future. Memory links experience with thought, permits reflection and planning, and helps us to form our very selves. That said, as much as memory fixes time, memory is also a labile medium of it. A philosopher of time and memory, Paul Ricoeur is careful to differentiate between the recollection of directly experienced events (memory) and the retrieval of conveyed facts (memorization). Both are prone to errors in fidelity and also to manipulation since they are potentially subject to creative and destructive impulses. Both require inscription, which also, somewhat perversely, requires forgetting and inventing. We are constantly recalling and recasting memories and in a real sense re-inventing our past and ourselves.

Memory takes many forms, from personal recollections to written histories and narratives. It can reflect an intensely personal dialog, an impersonal electronic blip, or a collaborative societal project. Memory can be fixed in architecture, DNA, silicon, or our bodies—it can be dynamic or static, reliable or ephemeral. Quite apart from serving as a medium to record and reflect the passage of time, memory also has the power alter our experience of it. Memory can dilate potent events to huge proportions, and equally, it can constrict uneventful or inconvenient experiences to the point of vanishing. Our experience of time is variable because our memory is plastic.

This volume presents selected essays from the 12th triennial conference of the International Society for the Study of Time. Inscribed memory can take many forms, so it was appropriate that the conference was housed at Clare College, Cambridge, an institution that was founded in 1326, and parts of which have been built and elaborated upon ever since. Along a similar theme, we are attempting an experiment to build and elaborate upon the arguments presented in the volume's essays: we solicited our reviewers to divest themselves of their anonymity and to provide short responses to the works that they assessed, and a number of them did so. In this manner we hope to stimulate further thought and discussion. Following the traditional lectures from the Society's Founder and from the President, the volume is divided into three sections that reflect themes apparent in the discourse of the meeting. The first two sections, "Inscribing and Forgetting", and "Inventing," are followed by "Commemoration." The essays grouped under the title "Inscribing and Forgetting" explore the mechanisms by which we remember and forget, and they compel us to consider the importance of memory for our sense of individual identity. As the essays make clear, memory and temporality are integrally related, for it is by our present accessing of past memories that we can move forward into the future. Time can distort and destroy memory—and consequently a stable sense of self. The section entitled "Inventing" features essays that address, from very different perspectives, ways in which new memory practices, concepts and effects constitute inventions in the history of human memory. Such mnemonic inventions unfold at the level of individual minds, philosophical ideas, and complex social relations alike. The processes of inventing memory reviewed here do not involve the truth or verifiability of specific memories, as the section heading could imply. Under the moniker of "Commemoration," we have grouped essays that deal with a unique and constructed form of memory. Commemoration demands collaborative inscription and subscription to a version of history that may or may not reflect reality. Commemoration serves to project past experience into the future: it is a form or memory that is often imbued with a moral or pedagogical agenda, and it can often play a profound social role. That said, as some of the essays take pains to show, when frailties rather than valorized strengths are recast to perform a commemorative function, interesting things can happen both to our sense of time as well as to our memories.

This last section, "Commemoration," is appropriate not only in a temporal sense, but in a thematic sense as well. The conference program included a beautiful and very moving memorial concert by Deborah Bradley to our colleague musicologist and composer Jon Kramer, who died unexpectedly just prior to the meeting, The volume itself is dedicated to one of its contributors, Karen Davies, who worked hard to complete her essay before her premature death in Spring 2006.

Finally, thanks are due to Thomas Weissert for creating a website that facilitated communication among the coeditors as well as for his patient and efficient support of our work. Furthermore, we must thank Tara Wessel for editorial assistance. We acknowledge our great debt to the many reviewers who helped to ensure the academic rigor of the volume and to Brill Academic Publishers, which continues to invest in our collaborative project, *The Study of Time*.

PRESIDENT'S WELCOMING REMARKS

A FEW THOUGHTS ABOUT MEMORY, COLLECTIVENESS AND AFFECTIVITY

REMY LESTIENNE

SUMMARY

The property of memory is strongly entangled with the notion of Time. Consideration of cosmological, physical, biological, and human memories shows the importance of collectiveness in its full expression. In humankind, affectivity seems to be of special importance for the transfer from short term memory to long term memory and the creation of permanent memory traces. The corresponding neurobiological mechanism in the brain implicates specific neurobiological networks, in which the hippocampus, the amygdala and the prefrontal cortex all play a part. This process seems to be largely controlled by neuromodulation networks of bioamines, in particular dopamine, the correct function of which seems to depend on the affective input from the subject, although much remain to be ascertained in this domain.

I am proud to open the 12th triennial conference of our Society in such a prestigious site as the University of Cambridge, a venue that has featured many historical events, and that has seen so many great figures. The shadows of Newton and even of Darwin are here among us. I feel also much honoured to open the conference on a theme of such an existential importance as "Time and Memory". I would like to dedicate this conference to our founder, Julius T. Fraser, who had, some fifty years ago, the marvellous idea of creating an interdisciplinary society for the study of time, and who has spent so many years accumulating experience and memories of thinking about the nature of time.

I have the delicate task of introducing the lecture of my distinguished colleague, Alex Thomson, who will momentarily present a lecture on the neurobiological mechanisms of memory. A good knowledge of this matter is essential to the further discussions that will largely reach beyond animal and human physiology, given that memories, with a plural, can also be of a physical, cosmological, biological, or societal nature. So much so, in fact, that one way of putting the question that is facing us at this conference might be formulated as follows: can we consider memory as an important attribute of time, in the sense given by Spinoza to the term attribute—an essential property of a substance—without being taxed of illegitimate anthropomorphism?

Jo Alyson Parker, Michael Crawford, Paul Harris (Eds), Time and Memory, pp. 1–6
© *2006 Koninklijke Brill N.V. Printed in the Netherlands.*

Alex Thomson will describe the mechanisms that underlie memory in the human brain. I am sure that she will introduce early in her talk the classical distinction between short term memory, now more often extended to the concept of working memory, with their properties of limitation on their capacity (classically, 5 to 7 items) and lability or erasability, and long term memories, that are much more permanent and possesses various sub-kinds such as declarative and procedural memories. One of the key questions, one of the most intriguing problems posed to neurobiologists, sociologists, psychiatrists, etc.... is the condition of passage from the short term memory to long term memory. Allow me to focus for a while on this fascinating problem.

In short, the question may be put in this way: How is it that I remember perfectly well, in complete detail, the moment of first encounter with the woman who would later become my wife, when I glanced at her in the ballroom, with her blond hair bun and her red dress. I have completely forgotten hundreds of women as attractive as she was at that moment, and that I have met before or after this first encounter. How is it that I remember perfectly well the locale, time and hour, and other circumstances in which I learnt of the murder of President John Kennedy, when so many important events of the world left only vague traces in my memory?

In opening ways to discuss this matter, I would like to stress two particular aspects of the conditions of passage from short term to long term memories. The first one is collectiveness, the opposition between individual, personal, and so to speak selfish memory, and the more open, shared, collective memories. The second is affectivity, particularly the role of positive affectivity, attractiveness, desire, love and tenderness in memorisation.

Let us pause a moment on the question of collectiveness. You know that, at the level of the material world of physics, there is no possible sense of the passage of time for an isolated particle, a single atom. A radioactive atom, for instance, always has the same probability to decay in the next second, independently of how much time has elapsed since its creation. Let us state this with a bit of anthropomorphism by saying that a single atom has no memory. By contrast, a macroscopic sample of radioactive stuff, a large collection of atoms has a collective memory, in the sense that collectively the number of decays they suffer testifies to the passage of time.

Cosmological memory of the remote events related to the origin of the universe is also evidently a kind of collective memory. A local fluctuation of the ever-encompassing background radiation, in a given narrow direction of the celestial vault, will tell nothing about the history of the universe. But the whole map of such fluctuations has been explored and exploited recently to learn lessons about the history of the universe.

At the biological level, the genome has been correctly looked at as the main repository of the history of the evolution of the considered species. In the animal kingdom, the genome is also the repository of a huge collective memory of habits, instincts, which dictate the behaviour of each individual in most circumstances of life.

But this collective aspect of memory acquires specific and even more essential characters with humankind.

Thanking in passing Mark Aultman for having shared with us the review of Juan Luis Arsuaga's book *The Neanderthal's Necklace* (2002), I will borrow from the latter a few remarks that struck me when reading through.

As it is now well known, Neanderthals are not our antecedents but our cousins. In other words, there was a time on earth, about 40 000 years ago, when two distinct species, incapable of cross-fertilisation, coexisted. Brain size, at first sight should have given the advantage to the Neanderthals, whose brain volume was in average slightly larger than that of *Homo sapiens*. Both species seemed equally aware of their mortal condition. However, the complexity of the *Homo neanderthalis* brain may have been somewhat lower than our own. One way of judging this is to compare the traces of the blood vessels on skulls of specimen of the two species. Such a comparison displays a higher blood irrigation of the *Homo sapiens* as compared to *Homo neanderthalis*, particularly in the parietal region (Saban, 1995). Arsuaga argues that there is a correlation between the complexity of primate social groups and the complexity (rather than the size) of their neocortex.

When a new ice age occurred, Neanderthals died out and Cro-Magnon flourished. One possible explanation is that, unlike Cro-Magnon, the pharynx of Neanderthal was of such a structure that he was unable to utter certain sounds, in particular most vowels, possibly enforcing a simpler social life and smaller group size relative to those of Cro-Magnon's. According to Arsuaga, the Neanderthals never attained high local population densities, and biological and cultural resources were spread too thinly for their survival. In short, their collective memory system was too simple, too short-ranged for survival in the adverse conditions that finally prevailed 40 000 years ago in Europe. By contrast, modern humans formed larger groups, with population clusters that were reproductively viable and economically self-sufficient.

I shall borrow a second reflection on the importance of collectiveness in human memorisation and development from observations of infants that were reared by animals, as illustrated by the documented cases of the "feral" or "wild" children. Is it not striking that, despite the fact that the size and organisation of their brain was similar to that of other adult humans, such infants deprived of social exchanges in their childhood never reached the state of a mature brain, capable of reasoning and socialising? An American researcher, Jaak Panksepp (1998), has devoted his entire life to study the influence of emotions on psychological development, not without some condescension from his colleagues.

In a recent review article on "Child Abuse and Neglect and the Brain", Danya Glaser (2000) enumerates the effects of child abuse and neglect in the brain: disregulation of the hypothalamic-pituitary-adrenal axis, weakening of the parasympathetic and catecholamine responses, maybe even a general reduction of the brain size. She stresses that "neglect and failure of environmental stimulation during critical periods of brain development may lead to permanent deficits in cognitive abilities". These deficits may be due, in part, to the lack of exercising the property of neuronal plasticity, which may extend well beyond the critical period of childhood; for example, recent studies on rodent have shown that neurogenesis continues throughout adult life in the dentate gyrus of the hippocampus, a very important nervous centre for memorisation.

As everyone would have guessed, the early mother-infant interaction is of particular importance in this frame. In the brain of the infant who sees the responsive mother's face, brain stem dopaminergic fibres are activated, which trigger high levels of endogenous opiates. The endorphins are bio-chemically responsible for the pleasurable aspects of social interaction and social affect and are related to attachment. Dopamine is postulated to mediate the behavioural facilitatory system, activated by rewarding (or aversive but surmountable) stimuli. An interesting psychobiological attachment theory has been expounded by Kraemer (1992), who suggests a central role for biogenic amines as mediators of secure or insecure attachment. Thus the importance of the question of the role of bioamines in the "now print" signal between short term memory and long term recording system.

While many studies have been devoted to the interaction between fear and stress and the memorisation system (where the amygdala complex seems to play a central role), very few papers are available on the role of affectiveness, loving and caring emotions, in memorisation. We know that fear messages from the amygdala arriving into the nuclei of the brain stem, and CRH (corticotropin-releasing hormone) from the hypothalamus, released by stress, stimulate the locus coeruleus and thus noradrenaline secretion in the brain (LeDoux, 1996). This demonstrates direct links between the noradrenergic and the cortisol responses to stress.

In pre-clinical studies, stress was also shown to enhance the release and metabolism of dopamine in the prefrontal cortex. Raised levels of noradrenaline and dopamine are positively associated with dysfunction of the prefrontal cortex (Arnsten, 1999), a brain area clearly involved in the planning and organising of actions using "working memory".

In part, the loving and caring emotions may act as suppressors of fear and stress responses. For example, Nachmias (1996) showed that 18-month-old children who had a secure attachment to their mother, who was present, showed no elevation of cortisol when responding fearfully to the approach of a stranger (a clown).

In the evoked passage from the working memory to the long term memorisation system, the hippocampus and its surrounding cortical areas of the medial temporal lobe are known to play an important role, which will be detailed in the forthcoming presentation. Amnesia associated with hippocampal damage is usually confined to the subsequent events, while retrograde memory is spared (Squire & Kandel, 1999), a fact that seems to point out the role of hippocampus on the "print now" mechanism. In relation to our questioning, it might be important to note that the hippocampal region, as well as the amygdala and the prefrontal cortex, are quite rich in innervation from dopaminergic nerve endings.

Probably no circumstance of life, however, is as emotionally important as death. At the very moment where one becomes conscious of the imminence of one's end, neuroregulator or neuromediator signals are very likely to hit the brain in a particularly strong way. They might even produce irreversible damage in the brain, as the fact that military combat can affect the short term memory system of adults seems to indicate (Bremner et al., 1993). On the other hand, as we have seen before, care and love emotions seems to lower or suppress the fear and stress responses. Thus, the neurochemical correlates of emotions, which can be detected and carefully studied in the brain, seem to corroborate the psychological evidence regarding the importance of social care and love accompanying people at the ultimate epoch of their lives, in sharp contradistinction to the occidental tendency to abandon them, a tendency that was clearly illustrated during the heat wave that stroke France in 2003, that made evident the loneliness of so many elderly people.

But I do not want to close my introductory remarks on such a disenchanted note. Even though little is presently known about the positive, constructive role of care and love emotions on memory, I am inclined to presume that the future progress of neuroscience will demonstrate that, well beyond their suppressive role on fear responses, they play a very beneficial role on memorisation and on the construction of everyone's mental and noetic kingdom.

REFERENCES

Arnsten, A. F. "Development of the Cerebral Cortex: XIV. Stress Impairs Prefrontal Cortical Function." *Journal of the American Academy of Child and Adolescent Psychiatry* 38, no. 2 (1999): 220–2.

Arsuaga, Juan Luis de, Andy Klatt, and Juan Carlos Sastre. *The Neanderthal's Necklace: In Search of the First Thinkers*. New York: Four Walls Eight Windows, 2002.

Bremner J. D., Scott T. M., Delaney R. C., Southwick S. M., Mason J. W., Johnson D. R., Innis R. B., McCarthy G., Charney D. S. "Deficits in short-term memory in posttraumatic stress disorder". *American Journal of Psychiatry* 150, no. 7 (1993): 1015–9.

Glaser, D. "Child Abuse and Neglect and the Brain—a Review." *Journal of Child Psychology and Psychiatry* 41, no. 1 (2000): 97–116.

Kraemer, G. W. "A Psychobiological Theory of Attachment." *Behavioral and Brain Sciences* 15, no. 3 (1992): 493–511.

LeDoux, J. E. *The Emotional Brain*, New York: Simon & Schuster, 1996.

Nachmias, M., M. Gunnar, S. Mangelsdorf, R. H. Parritz, and K. Buss. "Behavioral Inhibition and Stress Reactivity: The Moderating Role of Attachment Security." *Child Development* 67, no. 2 (1996): 508–22.

Panksepp, J. *Affective Neuroscience: The foundations of Human and Animal Emotions*, New York: Oxford Univ. Press, New York, 1998.

Saban, R. "Image of the human fossil brain: endocranial and meningeal vessels in young and adult subjects" in: *Origins of the Human Brain* (Changeux J. P. and Chavaillon J. ed.), Oxford: Clarendon, Oxford, 1995, pp. 11–39.

Squire, L. R. and Kandel, E. R. *Memory: From Mind to Molecules*, New York: Sc. Am. Library, 1999.

REFLECTIONS UPON AN EVOLVING MIRROR

J. T. Fraser

Summary

This paper suggests that the violent turmoil of our age is a symptom of an identity crisis of humankind at large, precipitated by globalization. For an understanding of that identity crisis, the evolutionary origins and uses of intent, memory and identity are sought and interpreted. This interpretation is then applied to our global laboratory in which many, incompatible needs demand fulfillment. In that perspective, the identity crisis may be seen as a struggle to decide upon whose understanding of the past, upon whose collective memories are the plans for the future of mankind to be based.

1. About Final Conflicts

The reasons that led to the founding of this Society had nothing to do with anyone's interest in the nature of time. They had to do with the puzzlement in the mind of a man of twenty-one who, in the autumn of 1944, found himself on a mountainside between two vast armadas. Behind him was the armed might of Nazi Germany, in front of him the immense masses of the Soviet Union. He knew that he was watching a struggle between two ideologies, each of which was convinced that it, and it alone, was destined to fight and win the final conflict of history. The Soviets had their creed summed up in their revolutionary anthem: "This is the final conflict / Let each stand in his place / The international party / Will be the human race...."[1] The official Nazi march said the very same thing, in different words. "This is the final bugle call to arms / Soon Hitler's flag will wave o'er every single street. / Enslavement ends / When soon we set things right."

Having been aware of both dogmas, I came to wonder whether there does exist a final conflict in history. Perhaps the buzz bombs the Nazis kept on sending over London were not the ultimate weapons they were claimed to

---

[1] "The Communist Internationale." The words are those of Eugene Pottier, a Parisian transport worker who, in 1871, wrote of "c'est la lutte final."

Jo Alyson Parker, Michael Crawford, Paul Harris (Eds), Time and Memory, pp. 7–26
© *2006 Koninklijke Brill N.V. Printed in the Netherlands.*

be. But, being hungry, cold and miserable, I did not pursue the puzzlement. All I did was to promise myself that if I ever got out of that hell alive, I would enroll in Plato's Academy and report to it about the wisdom of Robin Goodfellow, "Oh what fools these mortals be!".[2]

Nine months after I witnessed the clash of those final conflicts, I stood in an almost empty St. Peter's in Rome, in front of Michelangelo's early Pietà, a piece of Renaissance marble transfigured by human feelings. I saw two sculptures in it: a heavenly and an earthly one, joined by the two natures of the female figure: the *mater dolorosa* and the *amante dolorosa*, the grieving mother and the grieving lover.

The heavenly sculpture showed the Virgin holding the dead body of her son. In it Michelangelo asserted that the suffering of the Redeemer freed man from his earthly conflicts and opened up the way to a fulfilled, everlasting life.

I could not help but observe that the Virgin's figure was that of a woman much younger than the man whose dead body she held. I did not then know that her youth, compared to that of the man, had a veritable literature and that Michelangelo himself was asked about it.[3]

In the earthly sculpture, the youth of the female figure was no problem. She was Michelangelo's Italian model, real or imagined. She was also Dante's Beatrice, murmuring "L'Amor che muove il Sole e altre stelle" I gave the age difference an interpretation that made its way into my writings. This is from *Time, Conflict, and Human Values.*

> Just out of the havoc of World War II, I was ready to jettison all received teachings. I failed to see the Virgin holding the body of Christ. What I did see was a young woman of exquisite beauty holding the body of her man, murdered by the powers of law and order. Her face is one of infinite sadness as the irreversibility of his death permeates her unbelieving mind. Her beauty suggested to me that she was with child for I believed that women were most beautiful when they were pregnant. In a melodrama the woman of the statue would faint. In the Roman Pietà she bears up because she carries the child of the man whose body she holds.[4]

---

[2] "Midsummer-Night's Dream," iii–ii–115.

[3] Those youthful features, explained Michelangelo, were the result of her "never having entertained the slightest immodest thought which might have troubled her body." R. S. Liebert, *Michelangelo*. New Haven: Yale Universoity Press, 1983, pp. 67–8.

(EB 12–98). An entry to that dialogue may be had through the comments on this Pietà in Umberto Baldini, *The Sculpture of Michelangelo*, New York: Rizzoli, 00000, pp. 39–40.

[4] J. T. Fraser, *Time, Conflict, and Human Values.* Chicago: University of Illinois Press, 1999, pp. 151–2. Also in the introduction to *Il tempo: una presenza sconosciuta*, Milano: Feltrinelli (1991) p. 7.

During my visit, Michelangelo reassured me about the affinity between Eros and Agape. He also told me that if one is pregnant with life or with an idea, one cannot afford to faint.

A year later I was on board an American troop ship en route to the United States. On a foggy September morning I sailed by the Statue of Liberty in New York harbor.

> Give me your tired, your poor,
> Your huddled masses yearning to breath free,
> The wretched refuse of your teaming shore.
> Send these, the homeless, the tempest-tossed to me
> I lift my lamp beside the golden door.

Homeless and tempest-tossed. That was me. The words "Give me your tired, your poor" joined in my mind a much earlier invitation: "Come unto me, all ye that labor and are burdened, and I will refresh you."[5] My only concern was that the door was made of gold. First, because that earlier invitation did not say, "Come unto me all you that labor, and I will give you lots of gold." Second, because for me gold was only a dead metal, atomic weight 197 and, having survived the dictatorships of the true believers in final conflicts, I did not want to become subject to the censorship of the true believers in gold as the final arbiter of all things human.[6] I arrived in the land of my dreams: of Buffalo Bill, Thomas Alva Edison and Thomas Jefferson. I was where I wanted to live, love, die and be buried. And what a privilege it was to be among people who did not worry about final conflicts but were committed, instead, to a permanent revolution.

2. Why Study the Nature of Time

Soon after my arrival, the promise I made to myself called me to task. How was I going to tell people about that awesome stage upon which I was an insignificant walk-on? War stories were coming out in great profusion and I thought of contributing to the flood. One day, while browsing in a bookstore in New York's Greenwich Village, I came upon a comic book called, "The

---

[5] Matt. xi, 28.

[6] "Give me your energetic, your rich / Your privileged yearning to be free / The executives of your teaming shore / Send these, the achievers to us: / We lift our lamp beside the golden door." Poem titled "Invitation to New York Corporations Thinking of Moving Their Headquarters to Fairfield County." Author is E. J. Brennan, publisher, *Fairfield County Magazine*, 1976 [TAC 274].

Nazis and the Invisible Man." I did not then know the poetry of T. S. Eliot. If I had, I would have thought of "Go, go, go, said the bird: human kind / Cannot bear very much reality."[7] Even without Eliot, I gave up the idea of writing or even speaking about my war experiences.

Instead, I began to search for a vehicle that could carry "Oh what fools these mortals be" as well as Beatrice's planetary theory of love. It had to be a subject of universal interest, yet one that demanded clarity of thought and exposition, so as to protect it from the merchants of bogus scholarship and wooly science.[8]

My memory obliged. I remembered that one morning in gymnasium we learned about the mathematical pendulum and how it may be used to measure time. That evening I saw a movie in which people danced around a fire. The subtitle said that they danced to help them forget the passage of time. The next day, all across town, there were crowds, ecstatic with hatred and love. They marched in ways that looked to me like dancing around the fire.

Obviously, the pendulum was used to measure something people wanted to forget. If I could trace a connection between the swings of a pendulum and the desire to forget whatever it measured, then I could bracket both the foolishness and the greatness of the species. My theme, then, could serve as did the images on the shield of Achilles: an illustrated encyclopedia in which people could see themselves both as heavenly and earthly.

Four years after I sailed by the Statue of Liberty, while finishing my work for my first degree, I wrote a paper called "A short essay on time." It won a national humanities award for science students. This encouraged me to search the literature of time—which led me to the writings of S. G. F. Brandon, then professor of comparative religion at the University of Manchester.

He maintained that the human knowledge of time is a powerful tool in the struggle for life because, with the help of memory, it makes preparations for future contingencies possible. But, it is also the source of "an abiding sense of personal insecurity" which inspires people to seek such forms of refuge as represent their ideals of safety from all they fear and help conserve all they desire.[9]

---

[7] T. S. Eliot, "Burnt Norton." In *Four Quartets*, New York: Harcourt, Brace & World, 1971, p. 14.

[8] In a later view: from Blake's "idiot questioner." In: "Milton: Book the Second" Line 12 of the verses for Plate 42. David V. Erdman, ed. *The Complete Poetry and Prose of William Blake*, Berkeley: University of California Press, 1982, p. 142.

[9] S. G. F. Brandon, *Man and his Destiny in the Great Religions*. Manchester: Manchester University Press, 1962, p. 384. For a summary and continuation of this idea see his essay,

Brandon's lines met in my mind my memories of the war and came to be expressed in the Introduction to *The Voices of Time* (1966).

Watching the clash of cultures and the attendant release of primeval emotions stripped of their usual niceties, I could not help observing that man is only superficially a reasoning animal. Basically he is a desiring, suffering, death-conscious and hence, a time-conscious creature.[10]

I realized with a pleasant shock that the question I posed many years earlier, namely, whether there can exist a final conflict in history, was too crude to be fundamental. Namely, it is possible to imagine a world without mass murders but it is not possible to imagine humans who will not declare, in innumerably many ways, "Death, be not proud...."[11] because the conflict that gives rise to such a rhetorical command—the conflict between the knowledge of an end of the self and the desire to negate that knowledge—is at the very foundation of being human. This conflict is unresolvable because if it ceases, personhood collapses. A man or woman may well remain alive but only with impaired or absent mental identity. For this reason I came to regard that conflict as constitutive of personhood and came to see all other, overt conflicts as derivative from the fundamental one. Also, with my interest in the natural sciences, I began to wonder how such a merely human conflict fit the dynamics of nature at large?

In agreement with Brandon, I came to believe that the efforts to be able to live with that unresolvable conflict drive both the immense creativity and the frightening destructiveness of the species.

I wrote to Professor Brandon. He replied kindly and suggested that I write to Joseph Needham in Cambridge. By and by I was guided to an impressive group of British scientists and scholars and through them to their colleagues in Germany, Switzerland and France. Through these people, whose writings and letters awed and inspired me, I found my way back to my fellow countrymen, starting with David Park, a physicist at Williams College and George Kubler, an art historian at Yale.

"Time and the Destiny of Man," in J. T. Fraser. ed. *The Voices of Time*, 2nd ed. Amherst: University of Massachusetts Press, 1981, pp. 140–60.

[10] J. T. Fraser, ed. *The Voices of Time*, (1966) 2nd ed. Amherst: University of Massachusetts Press, p. xvii, 1981.

[11] John Donne, "Death be not proud..."

3. ISST

Correspondence with an increasingly larger group of scientists and scholars convinced me that the intellectual climate was ready for a collective enterprise directed to an interdisciplinary, integrated study of time. To prepare for such an undertaking, it appeared useful to publish a volume about the significance of time in the major academic disciplines. *The Voices of Time* (1966) is that survey.[12] The title had two sources. One was St Paul, "There are... so many kinds of voices in the world, and none of them is without significance."[13] The other was Simon and Garfunkel's 1964 lyric, "The Sounds of Silence."

I was ready to report to Plato's Academy about the study of time as an ideal means to explore the greatness and the foolishness of our species, but I could not find the Academy's address. I decided, therefore, to help create a Grounds, Buildings and Services Department to the Academy.

In January 1966 the New York Academy of Sciences held a conference on the theme "Interdisciplinary Perspectives of Time." Participants included Gerald Whitrow and Brian Goodwin of England, Olivier Costa de Beauregard of France and Georg Schaltenbrand of Germany. In my talk I remarked that the way to an interdisciplinary study of time must include support for "an intellectual climate where creativity common to all forms of knowledge is permitted to flourish, and aspects of reality previously separately understood are permitted to produce their synthesis by interacting through the idea and experience of time."[14]

After the conference Gerald Whitrow, then professor of the history and application of mathematics at the University of London, Satoshi Watanabe, who taught quantum theory at Yale, and I sequestered ourselves to a quiet corner and, following my proposal to form a professional group for the study of the nature of time, we declared the International Society for the Study of Time as having been founded.[15]

There was a strong personal reason for my proposal. I needed a lookout tower. Let Sam Walter Foss, a New Hampshire poet, supply the words. The title of the poem is "The House by the Side of the Road."

---

[12] J. T. Fraser, ed. *The Voices of Time* (1966) 2nd ed. Amherst: University of Massachusetts Press, 1981.

[13] 1 Corinthians 14: 10–11.

[14] New York Academy of Sciences, *Annals*. V. 138, Art. 2. "Interdisciplinary Perspectives of Time," pp. 822–48. Quote is from p. 845.

[15] G. J. Whitrow, "Foreword" in J. T. Fraser, F. C. Haber and G. H. Müller, eds. *The Study of Time*, Heidelberg, New York: Springer Verlag, 1972, p. v.

Let me live in a house by the side of the road
Where the race of men go by—
The men who are good and the men who are bad,
As good and as bad as I. [16]

I had high hopes for such a Society, provided it did not collapse into mediocrity and provincialism. The challenge of finding people who could articulate the similarities and differences among a crowbar, a candy bar and a kilobar, was still ahead.

4. TIME, THE EVOLVING MIRROR

My inquiries began to bring books and articles by the drove. The themes people judged essential for a study of time extended from the iconography of Renaissance art to information conveyed by the bees' dance, from medieval poetry to the entropic measure of human migrations. With the flow of ideas the problems of any interdisciplinary dialog became evident. Namely: different disciplines employed different jargons, had different criteria for testing for truth and maintained different, unstated assumptions about reality. Also, opinions about which field of knowledge was the most appropriate one for studying the nature of time, though widely divergent, were always accompanied by deadly parochialism.

I had the privilege of discussing the problems of interdisciplinary exchanges with Joseph Needham. He responded by giving me a copy of his Herbert Spencer lecture "Integrative levels: a revaluation of the idea of progress." [17] "See, Fraser, whether this will help," he said. It helped immensely.

The idea of integrative or organizational levels extends from Plato and Aristotle to the Christian Platonists and Aquinas, to Hegel, Marx and Bertrand Russell. It occurred to me that recognizing in nature a nested hierarchy of stable integrative levels, distinct in their complexities[18] and languages,[19] could accommodate the different epistemologies necessary for

---

[16] Sam Walter Foss, "The House by the Side of the Road," Hazel Felleman, ed. The Best Loved Poems of the American People, Garden City: Garden City Publishing, 1936, p. 105.

[17] Joseph Needham, (1937) Reprinted in his, Time, the Refreshing River, London: George Allen and Unwin, 1943, pp. 233–72.

[18] "Complexity and its measure," (constructed with the help of algorithmic information theory) in J. T. Fraser, Time, Conflict, and Human Values, Chicago, University of Illinois Press, 1999, pp. 235–42.

[19] By "languages" is meant a coherent family of signs and symbols necessary to describe the structures and interpret the functions of the stable integrative levels of nature.

dealing with the worlds of radiation, particle-waves, solid matter, life, the human mind and human society. And, for that reason, it could serve as a framework for an interdisciplinary, integrated study of time.

I did not realize until many years later that the reasons of the remarkable appropriateness of the nested hierarchical model of nature for integrating the epistemologies that an interdisciplinary study of time must accommodate, may be found in the logical structure of Gödel's incompleteness theorem.[20]

Indeed, the model could accommodate Brandon's recognition of the conflict between (1) the human knowledge of time as a weapon and (2) the "abiding personal insecurity" of our species, together with the dynamics of the life- and the physical sciences. It gave rise to the theory of time as a nested hierarchy of unresolvable, creative conflicts. I will now appeal to that theory to help us learn about the evolutionary origins and roles of intent, memory and identity.

The theory employs an operational definition of reality. Specifically, it extends the biologically based definition of reality, formulated by Jakob von Uexküll a century ago, to all forms of human knowledge: experiential, experimental and abstract. What emerges is an understanding of reality as a relationship between the knower and the known. Applying this understanding of reality to the diverse material that must enter an interdisciplinary study of time leads to the conclusion that what, in ordinary use is called "time," has a structure, that it comprises a nested hierarchy of qualitatively distinct temporalities.[21]

Let me introduce the two major dramatis personae of the theory. They are evolving causations and evolving temporalities.

First, let me attend to causation and name its evolutionary stages. They are: chaos, probability, determinism, organic intentionality, noetic intentionality and collective intentionality. Next, let me visit each separately and identify the steps in the evolution of temporalities.

The primeval chaos is without any connections among events. It supports no causation, its world is without any features that may be associated with time. Absolute chaos or pure becoming is atemporal.

The organizational level of nature above chaos, known through quantum theory, is that of particle-waves. In that world distinct instants do not yet exist, only probabilistically distributed likelihoods of instants do. Time is not yet continuous. That world is prototemporal.

---

[20] "J. T. Fraser, "Mathematics and Time," KronoScope 3-2 (2003) pp. 153-168.
[21] For its mature form, see "Perspectives on Time and Conflict," *ibid.*, pp. 21-43.

The next step in cosmic evolution was the coming about of the galaxies that form the astronomical universe of solid matter. Instants in that universe are well-defined. They are connected through deterministic relations, as embodied in both Newtonian and Einsteinian physics. That level-specific time is eotemporality. It is one of pure succession, without preferred direction. The reason why we cannot find purely deterministic processes is that, because of the nested hierarchical organization of nature, there can be no deterministic processes without probabilistic and chaotic components.

Michael Heller, physicist and Catholic theologian, has shown that the physical world is time orientable, that it allows for two directions of time, but it need not be so oriented, that it is complete and intelligible without directed time.[22] Heller's conclusions are consistent with P. C. W. Davies' assertion that "The four dimensional space-time of physics makes no provision whatever for either a 'present moment' or a 'movement' of time"[23] and that, "It is a remarkable fundamental fact of nature that all known laws of physics are invariant under time reversal."[24]

By the Principle of Parsimony, the hierarchical theory of time maintains that what physics reveals about time in the physical world is the total truth. The reason why the laws of physics do not make provisions for a present and for the flow of time is not because physical science is incomplete in that respect but because the temporalities of the physical world are incomplete when compared with what we, as living and thinking beings experience and think of as time.

For Einstein, according to his friend and biographer Paul A. Schilpp, "there was something essential about the Now which is just outside the realm of science."[25]

The parochialism of equating "science" entirely with physics, dooms an understanding of the "now" in terms of natural philosophy. But, if biology is

---

[22] Michael Heller, "The Origins of Time" in J. T. Fraser, N. Lawrence and D. Park, eds. *The Study of Time IV*, New York: Springer Verlag, 1981, pp. 90–93.

[23] P. C. W. Davies, *The Physics of Time Asymmetry*, U. Calif. Press, 1974, p. 21. The absence of time's flow and that of a present are corollaries. Namely, the flow of time must appeal to distinctions between future and past, and future and past have meaning only with reference to a present. If there are no distinctions between future and past, directed time can have no meaning. David Park even asked, "Should Physicists say that the Past Really Happened?" In J. T. Fraser, ed. *The Study of Time VI*. Madison, CT: International Universities Press, 1989. pp. 125–42.

[24] P. C. W. Davies, *The Physics of Time Asymmetry*, U. Calif. Press, 1974, p. 26.

[25] P. A. Schilpp, ed. *The Philosophy of Rudolf Carnap* (La Salle: Open Court, 1963), pp. 37–8.

included among the sciences as it must be, then, a scientific, operational basis of the "now" may be identified. To explain how life gives rise to the "now," to intent, to memory and to identity and how the flow of time acquires meaning through the life process, we have to think ourselves back to the primeval chaos of pure becoming.

David Layzer gave good reasons in support of the claim that chemical order was not present at the initial universe but was created by cosmic expansion.[26] It is necessary to add that ordering, created by the expansion of the universe, has been opposed all along by disordering, governed by the Second Law of Thermodynamics. Also, that ordering and disordering define each other.[27]

In the physical world ordering and disordering are random, uncoordinated. In sharp contrast, within the boundaries of an organism, ordering and disordering—growth and decay—go on simultaneously and are coordinated from instant to instant. The life process is identically equivalent to securing that events which must happen simultaneously do so happen, events that ought not happen simultaneously, do not. When coordination between growth and decay fails the conflict ceases, the organism dies. For this reason, the coordinated processes of growth and decay are the constitutive conflicts of life. They are necessary and sufficient to define the life process. The instant to instant coordination that maintains the life process, introduces nowness into the nowless universe of nonliving matter. With nowness it defines conditions of non-presence. Nowness is, a local phenomenon. This is the reason why, in the words of Whitrow, the Special Theory of Relativity denies the universal simultaneity of spatially separated events. Consequently, the simultaneity of events throughout the universe becomes an indeterminate concept until a frame of reference (or observer) has been specified.[28]

In Shakespeare's *As you Like It* we learn that "from hour to hour we ripe and ripe / and then from hour to hour we rot and rot."[29] The Bard had youth and age in mind. But the life process, as I suggested, is identically equivalent to simultaneous ripening and rotting. Or, one may speak of simultaneous

[26] David Layzer, *Cosmogenesis—The Growth of Order in the Universe*. Oxford: Oxford University Press, 1990, p. 144.

[27] On this, see J. T. Fraser, "From Chaos to Conflict," J. T. Fraser, M. P. Soulsby and A. Argyros, eds. *Time, order, Chaos (The Study of Time IX)*, Madison: International Universities Press, 1998, pp. 3–17.

[28] G. J. Whitrow, *The Natural Philosophy of Time*, 2nd ed. Oxford: Clarendon Press, 1980, p. 253. See also "Conventionality of Simultaneity," in *Stanford Encyclopedia of Philosophy*, http://plato,stanford.edu/entrieds/spacetime-comvensimul/.

[29] "And so from hour to hour we ripe and ripe/And then from hour to hour we rot and rot...." ("As You Like It," 2–7–26.).

and coordinated entropy decreasing and entropy increasing processes. Their conflicts, as I mentioned, are unresolvable in the sense that if they cease, the organism dies. Ripening and rotting are, as I mentioned, the constitutive conflicts of life.

That nowness has no meaning in the physical world does not mean that our experience of the present is a figment of human imagination as are leprechauns. No more so than the death of a man or woman is a figment of his or her imagination just because galaxies or crystals do not die.

Let me turn to biogenesis as the evolutionary origin of intent and memory. Living systems are thermodynamically open. They demand matter and information from the world external to them to be able to maintain their constitutive conflicts. With that demand, need is born. Need directs behavior toward need satisfaction, known as intentionality. Memory, I suggest, was selected for because of its usefulness for guiding intentionality. The subject of our conference could and should have been, "Time, Intentionality and Memory."

Into the directable but not directed temporality of the physical world intentionality and memory introduced distinctions between two non-present conditions. Imagined non-present conditions that relate to intent, driven by desire for need satisfaction, are said to be in the future. Imagined non-present conditions that suggest usefulness in the pursuit of need satisfaction are subject to classification, through the complex process of reality testing as memory or as fantasy.[30]

With future and past referred to a present, a flow of time acquired meaning. When post-mortem life became imaginable, so did a future of unending time. And with it, I would think, open-ended prepartum time as well.

It took eons for the human brain to reach the degree of coordinated complexity that could display minding and, supported by memory and fantasy, project intentions to increasingly distant futures. During those eons noetic time became established as a part of human reality. Biotemporality was already a part of that reality because thinking humans are also alive. So were the physical temporalities, because we are made of matter.

This natural history clarifies the Kantian view of time as a form of pure intuition which is also empirically real. As a pre-Darwinian thinker, Kant had difficulty reconciling the two.[31] The hierarchical theory of time

---

[30] J. T. Fraser, "Temporal Levels and Reality Testing," *International Journal of Psycho-Analysis* (1981) v. 62 pp. 3–26.
[31] On the Kantian idea of time, see Charles M. Sherover's *The Human Experience of Time*, 2nd ed., Evanston, IL, Northwestern University Press, 2000, p. 109 ff.

accommodates the empirical reality of time as having become, through evolution, a form of intuition.

In summary: to the probabilistic and deterministic causations of the physical world, life added organic intentionality. It is driven by the needs of organisms, guided by whatever forms of memory the organism possesses.

Let us step up from the biological to the noetic individual. The constitutive conflicts of personhood, as I proposed, are those between, on the one hand, a person's awareness of the end of her or his self in death and, on the other hand, her or his ceaseless efforts to escape from that ending through biological, intellectual and social offspring.

As human beings we function with the nested hierarchy of all forms of causations and in all the different temporalities. Also, our dynamics subsumes all the constitutive conflicts of matter, life and the mind. For this reason, our ideas about our experience of time, to use a poetic turn of phrase, amount to reflections upon an evolving mirror.

We may sum it up with Tennyson.

> I am a part of all that I have met;
> Yet all experience is an arch wherethrough
> Gleams that untravelled world, whose margin fades
> For ever and for ever as I move.[32]

5. The Global Laboratory

Having thus sketched the evolutionary origins of intentionality and memory and with them, the emergence of the notion and experience of time's passage, we are ready to examine the thesis of this paper. Namely, that the violent turmoil of our age is a symptom of an identity crisis of humankind and that the crisis involves a struggle to decide upon whose interpretation of the past, upon whose collective memory are the plans for the future of mankind to be based.

Let us begin by recognizing the steps in the evolution of identity.

The simplest identity, that of the biological individual, resides in the peculiar manner in which it applies its memory to serve its intentions. And, does this so as to maintain its constitutive conflicts and their phenomenal manifestations, which is the living body. The instant by instant coordination of ripening and rotting, as I proposed, defines the organic present.

---

[32] Ulysses.

The noetic identity of a thinking human resides in the peculiar manner in which it applies his or her memory to serve its intentions and apply it so as to maintain his or her constitutive conflicts and their phenomenal manifestations, which are his or her mental processes. The instant by instant coordination, as I proposed, defines the mental present.

I submit that the constitutive conflicts of a human society are those between its collective needs and the needs of its members. By a practical shorthand, these opposing trends may be described as those of gathering and scattering that is, between favoring the needs of the collective or favoring the needs of individuals.

The identity of a society, then, resides in the peculiar manner it applies its collective memories to serve its collective future so as to maintain its constitutive conflicts. The instant by instant coordination between gathering and scattering defines the social present.

How does this model of identity apply to a globalized humankind?

Globalization itself is not new. It was already evident in the 4000 mile Silk Road that connected Rome with China in the 2nd century B.C. The process of globalization acquired a theological foundation in the belief that in God's eyes all humans are equal. These foundations gave rise to ideas which assumed globalization as a natural and desirable condition of humankind. This assumption turned into political convictions. I am thinking of the mercantile capitalist desire for cheap labor and rich consumers, as well as the Communist dream for the workers of the world to unite.

What is new to 21st century globalization is that the electronic global present did not come about through evolutionary selection favoring the continuity of a global community, as did the lower order organic and social presents favoring life and nations. Instead, it came about through technological developments sui generis. Humans are now selected for by their own, self-created environment. We have to learn to live with, what Daniel Boorstin called the "global instant everywhere." Where do we stand in that selection-and-learning process?

In 1881 Walt Whitman saw America as a land of "contraltos... carpenters... duck shooters... deacons... spinning girls... freaks... patriarchs [and] opium eaters"[33] If, in this kaleidoscope, we replace the pieces with merchants of Einstein bombs, with happy rappers, with merchants of slaves, with dealers

---

[33] Walt Whitman, "Songs of Myself," No. 15 in *Complete Poetry and Selected Letters*, Emory Holloway, ed. London: The Nonesuch Press, 1938. p. 39.

in sperms and in wombs-to-let, smugglers of drugs and of stolen art, and with an increasing separation between the haves and have-nots, a globalized humanity comes into view. We even see Faust on the road, peddling his discoveries to the highest bidder.

Whitman's America succeeded better than many other regions on earth in reconciling its contraltos, carpenters and duck-shooters. It did so under the motto, "E pluribus unum" or "Out of many, one." What could possibly serve as the "unum" for a worldwide bazaar of wheelers and dealers in life and death and things and stuff?

In 1933, the philosopher-mathematician Whitehead explored human-itarian ideals which, in his view, could be shared by all people. He found them in religions. He did not say which religion it ought to be but he did maintain that, in the long run, "that religion will conquer which can render clear to popular understanding some eternal greatness incarnate in the passage of temporal fact."[34]

On the globalized earth, according to S. P. Huntington, there are three great ideologies that struggle for the mind of man or, to employ Whitehead's words, address "the passage of temporal fact." Each offers its dogmatic interpretations of life, death, history and the right rules of conduct. They are the Judeo-Christian West, Confucian-Socialist China, and the Islamic world.[35] Each of these ideologies offers a different view of man's position in the universe and, consistently, a different story about the past of humankind and of life.[36]

I believe that an amalgamation of these assessments of the past is already in the works. To find out what it is, I propose to attend to the witness of the dramatic arts.[37] because, as we learn from Claudia Clausius, drama

---

[34] Alfred North Whitehead, *Adventures of Ideas*, New York: Macmillan, 1933, p. 41.

[35] Samuel P. Huntington, *The Clash of Civilizations and the Remaking of World Order*. New York: Simon and Schuster, 1996.

[36] "The past is not a frozen landscape that may be discovered and described once and for all, but a chart of landmarks and paths which is continuously redrawn in terms of new aspirations, values and understanding." In J. T. Fraser, *Of Time, Passion, and Knowledge*. 2nd ed. Princeton: Princeton University Press, 1990, p. xv. See also Robert Robertson, "The New Global History: History in a Global Age," in Scott Lash et al. eds. *Time and Value*, Oxford: Blackwell, 1988, pp. 210–27. and J. Prager's essay, "Collective Memory, Psychology of" in N. J. Smelser and Paul B. Baltes eds. *International Encyclopedia of the Social & Behavioral Sciences*. Oxford: Pergamon, 2001, pp. 2223–2227.

[37] My inquiry has an ancestry in Thomas Ungvàri's "Time and the Modern Self," in J. T. Fraser, F. C. Haber and G. H. Müller, eds. *The Study of Time I* New York: Springer Verlag, 1972, pp. 470–478.

"enacts the cultural history of a people at the same time as it defines its own contemporary self-consciousness."[38]

Specifically, I want to point to some important changes that have taken place in the ways tragedies depict the uses of the past in the service of the future. My reasons for selecting tragic drama are these. In tragedies obligations, memories, hopes and fears are weighed with steady reflection upon future and past.[39] Also, tragedy on the social level has the same role as death-by-aging has on the organic level. The usefulness of both is to enable new generations to differ from their ancestors in a manner that is advantageous not for persons but for the community. The tragic, no less than death-by-aging, pays for social change with human suffering.

In terms of my definition of collective identity as the mode of applying collective memory to collective intent, tragic dramas articulate the identities of the communities in which they are set. I would expect that a tragic drama, appropriate for globalized humankind, would reflect humankind's identity-in-the-making.

I propose to examine three great tragedies by three great dramatists, associated with three different epochs. Then, compare the ways their protagonists employ the past in service of their future.

The first one was written in 1602. It is the story of a man who is informed about the past by a supernatural agent. This makes him confront his destiny. "The time is out of joint;" he says, "O cursed spite / That ever I was born to set it right!"[40] Yet, that is exactly what he begins to do. The vibrant sensitivity of this Shakespearean character is recognized in Pasternak's poem, "Hamlet." In it, the Prince of Denmark, an actor acting himself, talks to himself. "I stand alone. All else is swamped in Pharisaism. To live life to the end is not a childish task."[41]

With that realization, he begins to force the future. He engages a company of actors to recreate that past and takes action to repair that past through sacrifice. Hate and love converge to a denouement of "Good night, sweet prince" echoing in an otherwise empty universe. After a heartbroken farewell and military salute, the hero is laid to rest.

Hamlet is set in the court of a king. The protagonists are a small group of

[38] Claudia Clausius, "Tragedy as Forgotten Memory in Wole Soyinka's Drama," paper at this conference.

[39] For a detailed discussion see "The Tragic," in J. T. Fraser, *Time, Conflict, and Human Values.* Chicago: University of Illinois Press, 199, pp. 157–62.

[40] Hamlet 1–5–188.

[41] "Hamlet" in Boris Pasternak, *Doctor Zhivago*, New York: Pantheon, 1958, p. 523.

the privileged. The audience—during its early existence—were those who fit
in the Globe or other small theaters. The conflicts of the plot are between
the finity of human power on the one hand and the infinity of human ideals
on the other hand. I class Hamlet, together with Goethe's Faust, as tragedies
in the Greco/Western mode of the drama of redemption, leading to a
denouement of mission accomplished.

The second tragedy I have in mind is a painful register of hope lost. It
was written in 1939. Its protagonists wanted to change the world the way
they thought would make it better, but they failed. Now they are trying to
recover their personal identities in the hope of reconciling their enthusiastic
past with an uncaring present. As these efforts also falter, they drift into the
slow death of derelicts. When one of the them jumps to his death from a fire
escape, only a single voice says, "God rest his soul in peace." All the other
voices celebrate a birthday by singing and shouting in wild cacophony.[42] The
central character stares in front of himself, "oblivious to the racket."[43]

"The Iceman Cometh" is set in a bar. In it we are watching a collective
identity crisis in a petri dish. The protagonists are men and women, down
and out. The tragic tension is between the memories of the characters and
their assessments of their present. That tension dies in the empty, futureless
life of a flophouse. I class "The Iceman" together with some of Beckett's plays
as tragedies of impotence. They are dramas of worlds where there is nothing
left either to live or die for. Their moods remind me of those religious views
which, in Brandon's words, "reject... the consciousness of the self... as an
illusion of dangerous consequences."[44]

During the sixty years after "The Iceman" the world has changed immensely.
The new epoch has no patience for character development. They must be
immediately legible as are the characters of the Audio-Animatronics figures
of Disneyland. The plot must also be simple because the vast, worldwide
audience to which the mass media caters, shares only the most primitive of
human concerns, which are the spilling of blood and of semen.

In a keen and sensitive recognition of the profundity of the human drama,
Paul Harris wrote of "a sense of sweeping change in the nature of being
human, a feeling that we are reaching the end of an epoch in the history of

---

[42] One of them sings the "Carmagnole," a folk dance, danced around the Guillotine. See
J. T. Fraser, *Of Time, Passion, and Knowledge*, Princeton: Princeton University Press, (2nd ed.)
1990, pp. 392–3.
[43] *Selected Plays of Eugene O'Neill*, New York: Random House, 1940, p. 758.
[44] S. G. F. Brandon, "Time and the Destiny of Man," in J. T. Fraser, *The Voices of Time*,
Amherst: University of Massachusetts Press, 1981, pp. 140–57. Quote from p. 154.

our species."[45] In the same paper, he also wrote that, as a professor of literature, confronting a hypertext universe, he finds himself "oscillating between a kind of naïve technophilia and the frustrated rage of a luddite."

I believe that globalized humanity is confronting a hypertext world that goes much beyond the boundaries of wired communication. I have been calling it the anthill threshold. It is an incoherent community whose members have incompatible scales of values. They could coexist as long as there were distinct cultural boundaries. But now they live in a cohabitation enforced by a tight communication network and are all subject to the information rampage. They have all been thrown off balance by the interpenetration of financial and military empires where each community, in itself, is a powerless subject of an uncritical amalgamation of human values.

The consequent tensions are expressed in the conflicts between a technophilia of unrealistic hopes on the one hand and, on the other hand, a frustrated rage due to unfulfilled expectations. The crisis gives rise to a form of tragic drama that is neither in the Greco/Western redemptive category, nor in the category of impotent fading away. Instead, it is a tragic drama of globalization-in-process.

A dramatist who wrote such a tragic drama, is a woman. She is known to carry a trumpet to announce fame, a book in which she records events and a clepsydra to tell time. She is the muse of communal memory that is, of shared beliefs about the past. She is the muse of history. Her name is Clio.

Today her trumpet is the worldwide media. Her book is the Internet with its 420 billion pages of writings. The premiere of her new creation was performed on the spherical stage of a global theater. Her viewers numbered a few hundred million people, all of whom were both protagonists and audiences. At that first performance her clepsydra, calibrated in Gregorian chronology, showed 9–11–2001.[46]

In a fine essay, Anne Lévy wrote about "an ongoing American tendency to create temporal bubbles in which only the stimulating but solvable incidents rise pell-mell to the status of great meaning, while true predicaments and

---

[45] Paul A. Harris, "www.timeandglobalization.com/narrative" in *Time and Society*, 9–2/3 (2000) pp. 319–29. The conference was jointly organized by ISST and the French Association for the Advancement of Science.

[46] In the chronology of 7980 years, used by astronomers, the minute the first airplane hit, was Julian Day 2,452,163.864,583,335. This cycle was devised in the 16th century by the Dutch philologist and chronologist Josephus Scaliger, to encompass all human history known in his time.

tragedies are whisked out of sight."[47] Clio's latest plot, in which she made four airplanes stand for humankind, cannot be whisked out of sight. There are no temporal bubbles left in which anyone can hide. The denouement, which we have not seen yet, pertains to a decision about the identity and future of humankind.

The conduct of humans and gods in Greek tragedies set the tone of Greek cultural identity. The tragedy of Christ, reenacted in the Mass, helped form the cultural identity of the West. The tragic spectacle of murder in service of a savage King Ludd, using low-tech to annihilate high-tech, is appropriate in its form for a globalized humankind. For that reason, it is likely to shape the identity of the emerging community and likely to remain, at least for a while, a deed to be "acted o'er / In states unborn and accents yet unknown."[48]

The events of 9–11 involve conflicts between the Dionysian and Apollonian trends innate in history which, Nietzsche maintained, give rise to tragedy. In a different perspective, they are also conflicts between the biotemporal and nootemporal assessments of reality. The "global instant everywhere," in cahoots with social advances, have lifted the lid off the inner turmoil of people everywhere, allowing the reptilian brain to act out its desires.

I asked earlier what, for the case of globalized mankind, may serve the same role as the "unum" did in "e pluribus unum?" When, in 211 B.C. the armies of Carthage reached Rome, the cry went up in the City: "Hannibal before the gates!" The cry appropriate for humankind in globalization "Clio before the gates."

History is now at all gates, everywhere and all the time.

If, it is indeed the case that memory evolved as an aid to intent and, if one agrees with the idea that mankind's identity resides in the manner it employs its collective memory to serve its future, then, before any realistic plans for the future may be drafted, it is the past, it is an agreement on history that will have to be negotiated.

I do not mean agreement on the dates of one or another king. I mean an agreement about the origin and evolution of man, the origin and evolution of life, and the origin and evolution of the universe. But views on these issues differ, depending on whether they are based on critical scientific reasoning, on revealed religion or on mythology. Consistently, they lead to different recommendations for future actions. Also, they recognize different and

---

[47] Anne Schulenberger Lévy, "At the Limits of the Utopian Festival." *KronoScope* 4–1 (2004) pp. 75–91. Quote from p. 75.
[48] "Julius Caesar," III–I–111.

mutually incompatible needs. The chances of reaching an agreement about those needs are very slim. Yet, in my view, until we do agree on the origin and evolution of man, life and the universe, we shall have to live in Blake's "London" of 1794.

> In every cry of every Man,
> In every Infant's cry of fear,
> In every voice, in every ban,
> The mind-forg'd manacles I hear.[49]

6. THE MAGIC SHOW

The evolving mirror I spoke about is the capacity of the human brain to assess reality in categories of open-ended futures and pasts, referred to a present. That present is defined locally by life, by the mental processes and by society.

The sources of this ability of the human brain are unclear and intriguing enough to have given rise to a question by Walter Russell Brain, later Lord Brain. He was President of the British College of Physicians and one of the scientist-scholars who helped place the interdisciplinary study of time and ISST on the map.[50] If memory corresponds to a brain state, he wrote, and sense impressions to another, and since these coexist, how are they told apart?[51] The formulation of this question surely changed since 1963 but, to my knowledge, it has not been satisfactorily answered.

To memory and sense impressions listed by Lord Brain I would add a third family of brain states, namely that which corresponds to intentions, expectation and hope. Then, I would ask: how are these three categories of time told apart?

I hope that this question will be considered at this conference, with the benefit of some useful constraints. For instance, the answer cannot appeal to an unambiguous physical future, past and present in which biological, psychological and social processes unfold and to which the individual's mental faculties, somehow, adapt.

---

[49] "London." in *The Portable Blake*, New York: Viking, 1968, p. 112.
[50] Walter Russell Brain, "Time is the Essence..." *Lancet*, May 28, 1966, pp. 00–00.
[51] Walter Russell Brain, "Some Reflections on Brain and Mind." *Brain*, 86 (1963) p. 392 TPK 252 FN 38 on 485.

My talk began with reference to a young man standing between two immense armadas, sixty some years ago. It is not inappropriate to conclude by narrowing the cosmic vistas we visited to the limited boundaries of a single life.

Here is a poem by the American poet Robert Hillier; "The Wind is from the North."

> And now at sunset, ripples flecked with gold
> Leap lightly over the profounder blue;
> The wind is from the north, and days are few
> That still divide us from the winter cold.
>   O, it was easy when the dawn was new
> To make the vow that never should be old,
> But now at dusk, the words are not so bold,—
> Thus have I learned. How fares the hour with you?
>   A heron rises from the trembling sedge,
> His vigil at an end. Mine too is done.
> A late sail twinkles on the watery edge,
> And up the shore lights sparkle one by one.
> Seasons will change before tomorrow's sun,
> So speaks the dune grass on the windy ledge.[52]

---

[52] Robert Hillyer, *Collected Poems*, New York, 1961 p. 129.

RESPONSE
GLOBALIZED HUMANITY, MEMORY, AND ECOLOGY

Paul Harris

This essay presents in condensed form a response to the Founder's Lecture delivered July 26, 2005.

J. T. Fraser's approach to the conference theme of Time and Memory is informed by an evolutionary epistemology. According to this view, human memory evolved to aid planning. Memory's role in shaping human intentions becomes more pronounced and critical at times of large-scale transitions. Fraser argues that humanity finds itself situated, rather precariously, on a kind of historical cusp, desperately in need of collectively calibrating its memory in order to move cohesively towards the future.

Fraser's lecture also sketches, in very moving fashion, the memories of his life that were formative in forming the ISST. By syllogistic inference, one might infer that the ISST is in a time of transition as well, and the Founder's Lecture memorializes its origins in order to provide some grounds for shaping its "intentions, hopes, and expectations."

This response touches briefly on each of the three strands discernible in the Founder's Lecture: (1) globalized humanity and its constitutive conflicts; (2) the need for a collective, evolutionary vision of humanity in relation to its past in order to plan for a future; (3) the history and evolution of the ISST. The general vision that informs this response may be stated in the form of an irony: in order to plan well for a future, globalized humanity needs to stretch its collective memory to evolutionary timescales; yet globalized humanity lives faster and faster, shrinking the temporal horizons in which we live and think. Or, in slightly caricatured terms: at the moment when we should be slowing our minds to take in the pace of paleontological and geological temporal spans, we are busy speeding time up by saving it, and looking at what's next on the agenda in our hand-held personal planners.

Regarding globalized humanity and its constitutive conflicts, Fraser offers this succinct formulation: "globalized humanity" comprises "an incoherent community with incompatible scales of values that could and did coexist without major calamities as long as there were distinct cultural and national boundaries. But now all communities live in a cohabitation enforced by a tight communication network. They have been thrown off balance by the interpenetration of financial and military empires where each distinct community in itself is a powerless subject of an uncritical amalgamation of human values."[1] Fraser's account of globalized humanity asserts that transnational financial and military empires have overrun and outstripped national boundaries. Certainly, on one level, the multinational corporation and entities like the World Bank and World Trade Organization have displaced the nation-state as the salient bodies and forces in terms of which one can best chart the flows of power and money that shape human lives around the world.

At the same time though, the lines among political and economic divisions may become more entangled than is allowed for in Fraser's account. For instance, economic globalization and political nationalism often do not conflict with one another but actually operate in tandem. A case in point is India, where a powerful elite has embraced corporate globalization while orchestrating a fervent Hindu nationalism at the same time. Or, inevitably, we might point to the ways in which recent American military interventions serve multinational corporate interests, protected by a war machine that amounts to a private militia for big companies, a military that in turn is used as an instrument to promote isolationist nationalism. When this nationalism gets flavored with a religious fervor, we witness a nascent American form of "tribal interest cells" (Fraser's phrase) evolving. It is this kind of tribal interest appeal that allows international war to become an instrument of national election campaigns.

One important theme or area not explicitly included in Fraser's account of "globalized humanity" is ecology. Given the exponentially increasing degree to which global nature and human history are bound in a positive feedback loop, one can hardly think about "globalized humanity" without considering its ecological ramifications. Ecology could be effectively folded into Fraser's hierarchical theory in general, as I have argued in a prior ISST conference paper.[2] There I proposed that in addition to the six temporal levels in Fraser's

---

[1] More detailed treatments of this issue are found in the essay "The Time-Compact Globe," at the conclusion of *Time: The Familiar Stranger* (1987), chapter 6, "The Global Laboratory," in *Time, Conflict, and Human Values* (1999), and "Time, Globalization and the Nascent Identity of Mankind" (2000) (*Time and Society* Vol. 9(2/3): 293–302. In these texts, the time-compact globe is defined in terms of "three-cornered struggle between national governments, transnational groups, and 'tribal interest cells.'" The latter groups are bound together by ideological rather than familial or affective ties.

[2] "Ten Soundbites for the Next Millenium: Mutations in Time, Mind and Narrative." In J. T. Fraser and Marlene Soulsby, Eds. *Time: Perspectives at the Millenium.* (Westport, CT: Greenwood Publishers, 2001): 35–48.

hierarchical theory, a seventh level could be appended, a temporal umwelt comprising our planetary ecology conceived as a single system. With tongue slightly in cheek, I christened this "Gaia-temporality," after James Lovelock's Gaia hypothesis. Fraser calls the six levels in his hierarchical theory "the canonical forms of time." I would think of Gaia-temporality as an *epi-canonical* form of time, because it comprises several canonical timescales and can only be thought as a complex set of relations among them.

In Fraser's theory, each temporal level is defined by a constitutive conflict. The constitutive conflict of the Gaia-temporal could be stated, quite simply, as the conflict between the interests of ecology and economy. In terms of time, this conflict is marked by a sharp increase in ratios between forces of capitalism, development, and globalization versus the processes of ecology and evolution. Complexity theorist Brian Arthur estimated that the rate of technological innovation is about 10 million times faster than the pace of biological evolution. Global humanity acts on global nature in terms of time with remarkable greed. Globalization is driven by the timelines of development and profit; the sorts of major ecological disruptions that such development can inflict on an environment swerve, undo, or destroy balances and ecosystems that have been growing or changing slowly for eons. Economic development under globalization involves finding and exploiting new resources in order to fuel growth. Economic interests almost inevitably conflict with ecological interests. Efforts are certainly being made in response to this problem, but such efforts remain to date marginalized by the dominant forces driving the flows of global capital. The frenetic speed of economic globalization consumes and decimates centuries and geological ages of ecological change.

When we factor in the interests of global politics, we see the same greedy consumption of time. The heavy hand of Empire seems to carry with it an inevitable force of amnesia. Currently, one sees American imperialism erasing both short-term and long-term memories. By the erasure of short-term memories, I have in mind simple historical facts—drumming up support for the American-led invasion of Iraq has been predicated on forgetting that Saddam Hussein was befriended and armed by America. By the erasure of long-term memories, I mean to press the basic point that imperialism always entails the suppression and subsumption of cultural histories. How devastating it has been to witness the chaos that has overrun Baghdad, once the great city of the world where knowledge and goods from around the globe converged.

The interpenetration of global humanity and global nature, as well as the entanglements among imperialism, transnational interests, nationalism, and tribal interest cells, may be evoked by one word: petroleum. From an ecological standpoint, petroleum is a hydrocarbon fossil fuel found in very finite reserves under select parts of the earth's surface. Calculating the ratio between the amount of time it takes for petroleum to form (millenia) in relation to the estimated amount of time that globalized humanity— predominantly the USA—will take to exhaust the remaining supply of crude oil (about 40 years) is in itself a very pedagogical thought experiment. From an economic standpoint, petroleum is not a rare resource, but it is an exceptionally valuable commodity. The politics and economics involved in transforming petroleum from natural resource into consumable commodity serve to concentrate wealth into the hands of the few at the expense of the many. The huge profits controlled by oil companies and those in power in certain countries fuel a greed that, ecologically or temporally considered, is short-sighted.

One of J. T. Fraser's most prescient observations about globalized humanity is that its evolutionary path is traced on a technological, social fitness landscape rather than a strictly natural one. In analyzing "the Global Laboratory," Fraser points out that while globalization as such is not new, "what is new to twenty-first century globalization is that the electronic global present did not come about through evolutionary selection favoring the continuity of a global community.... Rather, it came about through technological developments sui generis. Humans are now selected for by their own, self-created environment. We have to learn to live with the global instant everywhere. Where do we stand in that selection-cum-learning process." There are many different implications to this statement important to explore. From one standpoint, this statement underscores the fact that "mind" in the global laboratory can no longer be thought of only in terms of human brains. Mind now operates in a distributed fashion, in communications networks that connect up different individuals in collective processes. Here, we might see a potential bridge between a neuroscientific and global learning. How can this relation be thought in terms of memory?

A compelling way to consider this question is provided by evolutionary psychologist Merlin Donald. In his book *The Origins of the Modern Mind*, Donald outlines a "cognitive ethology of human culture" and develops an idea of "cognitive architecture." Cognitive architecture involves the interplay of individual minds and cultural representational systems; his theory is marked by "its incorporation of biological and technological factors into a single evolutionary spectrum." Donald's essential contention is that "We act in cognitive collectivities, in symbiosis with external memory systems. As we develop new external symbolic configurations and modalities, we reconfigure our own mental architecture in nontrivial ways."[3] In other words, the ways in which we technologize our environment become the channels by which we install bodily regimes and rewire our brains. In this way, we are constantly establishing mappings between our minds and the technological milieu in which they operate. A question we might consider then is how to figure "external memory systems" into the essentially anthropocentric notion of memory.

Fraser's notion that this process takes us beyond evolutionary selection is significant. Most approaches to the question of global mind figure world-wide communications networks in neurobiological terms: the development of communications networks is described in terms of neural networks. This approach naturalizes technological development; it collapses cultural representation systems back onto our models of biological systems.

J. T. Fraser's lecture articulates the great challenge facing the human race in the new millennium: it must formulate a collective, evolutionary vision of itself in relation to its past, in order to have a basis on which to plan for a future. Fraser believes that humanity needs to seek out "an agreement about the origin and evolution of man, the origin and evolution of life, and the origin and evolution of our universe." As he recognizes, "the chances of reaching an agreement on these collective issues are very slim." The difficulties in resolving such agreements are readily apparent in the debates regarding public education in the United States, specifically the question of whether creation science or the "Intelligent Design" theory should be taught alongside Darwinian evolution as science.

There are many kinds of significant, inspiring work being done in many spheres and place that attempt to express a collective, evolutionary vision of human history and its future. The timepiece for such a vision could be provided by The Clock of the Long Now, a project founded by computer innovator Danny Hillis—a project that seeks to build a clock that will chime once each millennium (information on the web at www.longnow.org). A body of work that provokes humans to consider how their lives are embedded in very different natural timescales has been created by the contemporary Italian artist Giuseppe Penone. Penone, a member of the Arte Povera movement, creates work devoted to countering the anthropocentrism of a rational, technology-driven humanity that ceases to care for its environment. His work unfolds in simple gestures. To take just one example, from 1981

---

[3] Merlin Donald, *The Origins of the Modern Mind: Three Stages in the Evolution of Culure and Cognition.* Cambridge, MA: Harvard University Press, 1991, p. 382.

to 1987, he composed a series entitled "to be river." The series juxtaposes large stones taken from the River Tanaro in Italy with stones that Penone found at the source of the river and sculpted to become precise replicas of the stones from the river. The works thus induce the artist to reproduce the work that water, hydrodynamics, and time have done on stone. The slow patience required for this meticulous work is itself a meditation on the timescales and memory to be found in the landscape. Penone's work stands as an invitation to recalibrate our bodies and minds to ecology. It is this kind of reflection and practice that might contribute to molding a collective, evolutionary memory for humanity to use as a basis for planning a viable future. I say this fully aware that it bespeaks a naïve optimism, but also insisting that we have little choice but to act with determined optimism.

One final remark in response to J. T. Fraser's lecture, regarding the history and mission of the ISST. In the light of Penone's work, I would recast Fraser's story of his inspiration to undertake the work of ISST in the mold of the character depicted in Jean Giono's delightful story "The Man Who Planted Trees." Giono's story is told by a nameless narrator who, escaping the ravages of the first and second World Wars, discovers in the mountains a peasant who, untouched by those conflicts, devotes his life to planting trees. J. T. Fraser, being the multi-tasker that he is, has lived the roles of both Giono's narrator who witnesses the horrors of war and the man who planted trees.

In recent times, J. T. Fraser has increasingly concerned himself with how the legacy of ISST will live on. Fortunately, the Founder's vision of the Society has ensured that there is no need to reinvent the ISST. We need only carry on the two pillars that have defined the Society from the outset: its international and interdisciplinary character. J. T. Fraser's sweeping reflections in his lecture only underscore the importance of humans gathering in the global laboratory to engage in synthetic thinking. The spirits of internationalism and interdisciplinarity that infuse our society need to be nurtured and cared for. It is in this spirit that we might carry on our work in the years to come.

SECTION I

INSCRIBING AND FORGETTING

PREFACE TO SECTION I

INSCRIBING AND FORGETTING

Jo Alyson Parker

> If any one faculty of our nature may be called *more* wonderful than the rest, I do think it is memory. There seems something more speakingly incomprehensible in the powers, the failures, the inequalities of memory, than in any other of our intelligences. The memory is sometimes so retentive, so serviceable, so obedient—at others, so bewildered and so weak—and at others again, so tyrannic, so beyond controul!—We are to be sure a miracle every way—but our powers of recollecting and forgetting, do seem peculiarly past finding out.
> —Fanny Price in Jane Austen's *Mansfield Park*

If contemporary pop culture is any indicator, the "speakingly incomprehensible" nature of memory that prompted Fanny Price's musings continues to bemuse us. We are fascinated with "our powers of recollecting and forgetting." The 2000 film *Memento*, for example, features a protagonist, Leonard, who, due to a head injury, cannot engage in the process of consolidation whereby short-term memories are converted into long-term ones. Because no present memories can be inscribed in his long-term memory, he tattoos brief messages upon himself, literally inscribing upon his body the information that he needs to direct him toward a future course. The 2003 film *Eternal Sunshine of the Spotless Mind* fancifully conceives of a device that can selectively remove memories, generally of a traumatic nature, from the brain. When the central character Joel wishes to erase all memories of his ex-lover Clementine, the device seeks out and destroys not only all memories of her but also any memories peripherally connected to her, including Joel's happy memories of the cartoon character Huckleberry Hound singing "My Darling Clementine." Both films prompt us to consider "the powers, the failures, the inequalities of memory"—their themes resonating for us far beyond the two hours we spend in a darkened cinema.

It is no wonder that these themes resonate. After all, our sense of individual identity consists of our carrying past memories into the present to guide us into the future. As Philips Hilts notes, "Conferring with memory's ghosts, consulting its tables of facts, we project the future and what we expect it

Jo Alyson Parker, Michael Crawford, Paul Harris (Eds), Time and Memory, pp. 35–40
© *2006 Koninklijke Brill N.V. Printed in the Netherlands.*

to look like. Memory makes us, fore and aft."[1] Memory indeed serves as the bridge across time—between that self who existed years ago and that self who exists today.

We are troubled, however, by our awareness of the fluidity of memory—our awareness that something that we have long forgotten may suddenly rise to the surface while something that we once knew may sink into the depths, never to rise again. If we are what we remember and what we remember can change, then the self becomes an unstable entity. As Shelley Jackson suggests in her hypertext novel *Patchwork Girl*, a different set of memories would entail a different self: "So, within each of you there is at least one other entirely different you, made up of all you've forgotten […] and nothing you remember […]. More accurately, there are many other you's, each a different combination of memories."[2] We have all, no doubt, had the experience of someone telling a story in which we figure but of which we have no recollection whatsoever—and, as we hear it, we feels as if the story were about someone else entirely. Why did this memory inscribe itself upon the storyteller but not upon us? What difference might it make to the self had the story been part of our storehouse of memories? In what other unremembered stories do we play a part? Such questions confound our belief in the self as a continuous entity persisting across time, and they spark our interest in the mechanisms of remembering and forgetting.

These mechanisms compel us to consider the very *being* of memory. Certainly, contemporary neurobiology has made great strides in charting the mechanisms of remembering and forgetting within the human brain. We know, for example, that an injury to the hippocampal region of the brain can impair our faculty of memory consolidation, as occurred in the case of Henry M., whose sad history is recounted in Hilts's *Memory's Ghost*. We know, too, that patients with Alzheimer's Disease "have an abundance of two abnormal structures—beta amyloid plaques and neurofibrillary tangles."[3] But is there any device like that envisioned in *Eternal Sunshine*, which could extract a particular memory from the human brain, rather as one might extract a particular photograph from an album—so that the memory of Huckleberry

---

[1] Philp J. Hilts, Memory's Ghost: The Strange Tale of Mr. M and the Nature of Memory (New York: Simon & Shuster), 13.

[2] Shelley Jackson, *Patchwork Girl by Mary/Shelley & Herself*, CD Rom (Watertown, MA: Eastgate Systems, 1995). The passage come from the following sequence in the text: story / séance / she goes on.

[3] "Alzheimer's Disease: Unraveling the Mystery," *ADEAR: Alzheimer's Disease Education & Referral Center* 3 January 2006 <http://www.alzheimers.org/unraveling/06.htm>.

Hound might be removed but the memory of Yogi Bear would not? Although we may discuss physical processes that facilitate or impair memory, we are still in the dark about the nature of memories themselves.

What, too, do we make of the physical means whereby we attempt to preserve memory? The well-known discussion of writing in Plato's *Phaedrus* makes the point that attempts to preserve memory indeed encourage forgetfulness. Socrates thus deplores writing:

> If people learn from them [letters] it will make their souls forgetful through lack of exercising their memory. They'll put their trust in the external marks of writing instead of using their own external capacity for remembering on their own. You've discovered a magic potion not for memory, but for reminding, and you offer your pupils apparent, not true, wisdom. (275a)[4]

In a passage famously deconstructed by Jacques Derrida in "Plato's Pharmacy," Socrates distinguishes between the external inscriptions that one might make on a piece of paper (or papyrus or a computer screen), which he regards as inducing forgetfulness, and the internal inscriptions on the soul, which he regards as the stuff of memory:

> Can we see another kind of speech that's a legitimate brother of that one [writing] and see how it comes into being and how it is from its birth much better and more powerful than that one? [...] That which is written along with knowledge in the soul of the learner, that's able to defend itself by itself and knows to whom it ought to speak and before whom it ought to keep silent. (276a)[5]

As Derrida notes in his gloss on the discussion, "writing is essentially bad, external to memory, productive not of science but of belief, not of truth but of appearances."[6]

According to Plato, the external inscription thwarts rather than preserves memory—creates, in effect, an inauthentic self. We might come back to Leonard's tattoos in *Memento*; although they enable Leonard to act in the present so as to influence the future, these inscriptions are merely reminders— "memories" external to himself, outside the realm of his own knowledge, creating no bridge between a past self and a present one. We might consider, too, speculative fictions, such as those by Rudy Rucker or Philip K. Dick, that

---

[4] Plato, The Symposium *and* The Phaedrus: *Plato's Erotic Dialogues*, trans. William S. Cobb (Albany: State University of New York Press, 1993), 132.

[5] Plato, 133.

[6] Jacques Derrida, "Plato's Pharmacy," *Dissemination*, trans. Barbara Johnson (Chicago: University of Chicago Press, 1981), 103.

conceive of memories downloaded into computers or uploaded into selves. Do the downloaded memories make the computer a "self"? Do the uploaded memories render the self inauthentic? Yet as Derrida claims, "The outside is already *within* the work of memory. [...] Memory always therefore already needs signs in order to recall the non-present, with which it is necessarily in relation."[7] Can we really make a distinction between the memories we inscribe in, say, a journal or the memories inscribed, according to Plato's formulation, "in the soul"? Is there a tension between the external inscription and the internal memory, and which, if either, is legitimate? The very term "inscribing," which appears in the title to this section, suggests a physicality to memory that the word "remembering" does not, and it thereby smudges the boundary between internal and external.

Coming from a variety of disciplinary perspectives and drawing on a variety of disciplinary methods, the following set of essays addresses many of the issues raised above. They explore how, why, and what we remember and how, why, and what we forget. Furthermore, they all foreground the integral role that time plays in these processes—whether the milliseconds it takes for temporal-order information to be coded and accessed in semantic memory or the decades it takes for a vivid experience to fade. The first three essays focus on the bodily inscription of memory, the second three on its loss.

The section begins with Christian Steineck's consideration of whether memory can survive brain death. Arguing for the interrelationship of mental and physical processes, Steineck claims that the past is inscribed in the body, and he draws on the examples of both a well-known brain-death case in Japan and the philosophy of the Neo-Confucianist Ogyû Sorai in order to make his case. As he suggests, a past may be inscribed in the body—a bodily memory if you will—without entailing "the mental performance of memories." When we read Steineck's argument, we may be struck by its relevance to the controversy generated by the notorious Terry Schiavo case that riveted United States citizens in the summer of 2005—a case that pitted Schiavo's husband against her parents in an effort to determine whether bodily reaction could be synonymous with brain death and that ultimately played out in the federal legislature and courts. As respondent Rémy Lestienne provocatively inquires, "Could a decerebrate person ride a bike?" Steineck's essay compels us to envision the possibility.

Karen Davies's contribution also addresses the issue of bodily memory. Examining narratives by professionals working in the heath care and

---

7 Derrida, 109.

information-technology sectors, Davies argues for the vital role that narrative plays in tapping into individual and collective memory. As she shows, these narratives involve gendered bodies, and, as such, they can have emancipatory potential. In essence, memories imprinted on the body in the past are narrated in the present in order to have an impact on the future. Davies's piece in itself constitutes "doing gender," its own narrative suffused with the emancipatory potential it describes. As respondent Linda McKie points out, "The research of Davies illustrates how individual and collective memories are dimensions of social change."

In the paper that follows, cognitive psychologists Elke van der Meer, Frank Krüger, Dirk Strauch, and Lars Kuchinike detail the results they obtained from experiments designed "to analyze whether and how temporal order information could be coded and accessed in semantic memory." Proceeding from the assumption that chronological order was stored in semantic memory, they combined several complementary research methods: behavioral studies, pupillary responses, and functional neuroimaging methods. As their experiments demonstrated, participants processed chronologically ordered events more rapidly than events in reverse order. These results may encourage us to consider the increased cognitive effort required as we deal with, say, a chronologically skewed narrative.

With Marlene Soulsby's contribution, we shift from inscribing to forgetting—specifically here to an examination of a situation in which what had once been inscribed in memory is no longer accessible. Soulsby reviews both fictional and nonfictional narratives of dementia and concludes with a discussion of the Time Slips project, whereby facilitators guide groups of people with memory loss to tell a story based on a picture (a project in which Soulsby herself has participated). Soulsby notes that people with memory loss are "trapped in a temporal world of simultaneities," and the narratives that they produce are fractured—indicative, it would seem, of their inability to make a bridge between past, present, and future. Soulsby's paper has particular poignancy in a time wherein we are increasingly aware of the depredations wrought by Alzheimer's Disease—upon both its victims and those who care for them. Yet as respondent Alison Phinney points out, "Positioning narrative as an embodied activity will enrich our ability to understand the experience of temporality and memory loss in dementia."

Marie-Pascale Huglo focuses on what we might call a purposeful effort of memory loss—Georges Perec's unfinished project *Lieux*. As Huglo describes, Perec planned to produce two texts a year on each of twelve Parisian places related to his past, each of which would then be sealed and not be opened for a particular span of time. Huglo examines the way in which *Lieux* highlights the relation between time, memory, and place and explains that, for Perec, memories needed to be inscribed in order to exist. In describing Perec's fascinating and tantalizingly incomplete project, Huglo prompts us to consider how its success depended on Perec's forgetting what he had described in the past so as to discover it anew in the future; as Huglo points out, however, the project was beset by tension between Perec's active memories and those he had preserved, a tension that Plato might have described as between the external inscriptions of memory and the internal ones. Stemming the tide of memory, it appears, may be as difficult as stemming the tide of memory loss.

We conclude with a narrative, one that evocatively illustrates the way in which time erodes memory. Brian Aldiss tells the story of two different sojourns in Sumatra—one he spent in 1945–46 as a young soldier and one he spent returning to his former haunts in 1978. The term "haunts" may be taken in its most literal sense, for what Aldiss discovers is that he is a revenant. Attempting to locate the quarters in which he passed a romantic interlude, he finds that the certainties of a remembered past are undermined. Yet the narrative he tells links the past of the young soldier with the present of the revenant, who will become the storyteller of the future—an elegant balancing of three different time periods that highlights the layered quality of memory.

Although Austen's Fanny Price surmised that "our powers of recollecting and forgetting do seem peculiarly past finding out together," the six papers that follow incisively and at times provocatively explore the mechanisms of those powers, helping us toward an understand of the "wonderful" faculty of memory.

CHAPTER ONE

THE BODY AS A MEDIUM OF MEMORY

Christian Steineck

Summary

When we attribute human actions and abilities to either the body or the mind, memory seems to fall neatly within the range of the mind. Still, the act of memorizing often has more to do with the body than with the mind, a fact strongly emphasized by, and reflected upon, in East Asian methodologies of cultivation. Furthermore, memories are more often than not evoked by sensual, corporeal experiences. And finally, the case of a brain-dead patient shows how the body serves as a subject and object of memories in the absence of all intellectual activities.

An explanation of such phenomena is possible when viewing the mind-body problem in the light of E. Cassirer's theory of expression. As an expression, the body is the external reality that signifies a (non-spatial) 'internal' state. If we analyze the correlation of body and mind accordingly, we may understand that, and how, the body functions as a primary medium of memory.

Introduction

Almost by definition, memory concerns something that is not physically present. Remember Shakespeare's Hamlet. As we see Hamlet contemplating a skull, we hear him remember Yorick, whose skull it is he is looking at. A lifeless skull is what is there, but to Hamlet, it conjures up a vivid image of the living person: "I knew him, Horatio; a fellow of infinite jest, of most excellent fancy; he hath borne me on his back a thousand times. [...] Here hung those lips that I have kissed I know not how oft. Where be your gibes now, your gambols, your songs, your flashes of merriment that were wont to set the table on a roar? [...]" (act 5, scene 1).

Because memory presents what is physically absent, or has even ceased to exist, it is small wonder that it counts as one of the mental functions par excellence. It may even be one of the critical points from whence our distinction between the mental and the physical starts.

However, my following argument will be mostly concerned with the body and its function as a medium of memory. In this, I do not mean to deny the intimate relationship of memory and mind. My point will rather be that our usual identification of the body with the physical and its separation from

Jo Alyson Parker, Michael Crawford, Paul Harris (Eds), Time and Memory, pp. 41–52

the mental is not fully adequate. Analyzing the body and its connection to memory may help us to clarify the meaning of this proposition and to correct some misconceptions of the human body.

In the following, I will first illustrate what I mean by talking about the body as a 'medium of memory' and then present my analysis before I come back to the more general picture. I shall give two examples from divergent fields of human experience, which may help to illuminate the scope of meaning involved in the formula 'the body as a medium of memory'.

EXAMPLES

Interpersonal Memories

My first example is taken from the field of modern biomedicine. It concerns the case of a young man in Japan, who fell into a state of persistent coma after an attempt at suicide. His condition deteriorated, and he was diagnosed as clinically 'brain-dead' after five days. According to the prevalent interpretation of that diagnosis, this meant that all parts of his brain had permanently ceased to function. At that point in time (1994), brain death criteria had not yet been accepted into Japanese legislation.[1] He was therefore treated as a terminally ill but living patient.

The father of this young man, Yanagida Kunio, is a renowned non-fiction writer of books about science. He published his observations and reflections in an article for the widely read monthly magazine *Bungei shunjû* (Yanagida 1994), and later in a book (Yanagida 1999). Two aspects of his report concern us here.

The first one is that even after the diagnosis of brain death, cardiovascular functions in Yanagida's son invariably stabilize whenever the father or brother is present. As the attending physician confirms, the patient's blood pressure and pulse rate regularly rise to a level closer to normal whenever relatives come to visit him. This fact, which remains without a medical explanation in the report, is interpreted by Yanagida and the attending nurses to mean that the patient perceives the presence of his kin and enjoys being with them (Yanagida 1994, 144; Yanagida 1999, 184). Such an interpretation runs

---

[1] This situation occurred in 1997. See Lock 2002 for an extensive description of the process of incorporating brain death criteria into the legal system of Japan, and the US. A printed version of the Japanese transplant law is available in Kuramochi/Nagashima 2003, 321–335.

contrary to the theory of brain death used in contemporary legislation around the globe. The diagnosis of brain death employed in Yanagida's case asserted that the patient, who did not suffer from intoxication or hypothermia, was unconscious, had lost reflexes mediated by the brain stem, and was incapable of spontaneous breathing. The irreversibility of that condition was established by repeating the diagnostic procedure after more than 24 hours. Such a diagnosis is generally assumed to mean that all functions of cognition, memory, and emotion have permanently subsided in the patient (Harvard Medical School Ad Hoc Committee 1968; National Conference of Commissions on Uniform State Laws 1980; Bundesärztekammer 1998). However, this theory has been disputed by some scientists, who state that none of the various brain death criteria employed in the United States, Britain or Germany can serve to indicate 'death', in the sense of physical destruction or irreversible organic disintegration, of the brain (Byrne et al. 1979; Byrne et al. 2000; the Japanese criteria do not differ substantially from those employed in Germany or the United States). Some also deny that meeting these criteria would equal complete dysfunction of the brain (for example, Nau et al. 1992, Heckmann et al. 2003). Yanagida's report seems to point in that direction, but we should keep in mind that we do not possess an independent medical record to substantiate its claims.

Instead of speculating to what extent there may have been neuronal activity in parts of Yanagida's brain after the diagnosis of 'brain death', I will focus on a conceptual issue involved in the case. Regardless of the neuronal pathways and connections involved, the physiological reactions purportedly witnessed in Yanagida were quite obviously connected to the deep personal relationships he held with his brother and father. However, I do not think that one has to assume that some form of conscious memory, perception or recognition was operative in order to explain this. What is important to keep in mind here is that such relationships include a long history of close communicative interaction on a corporeal level, of being with each other. The body as a whole is both involved in the building of this history, and influenced by it. Feelings of 'emotional security', 'belonging' or 'trust' relate to states involving the body *in toto*, with details such as the specific 'touch' of a person, as felt in various parts of the body, or the muscular relaxation triggered by his presence feeding into the perception and memory of a given situation. One may well imagine that their repeated experience over a long time may lead to a kind of habituated response to the presence of trusted ones, which does not depend on conscious perception and recognition. (A model to explain the neurophysiological workings of such a 'top-down' effect has been presented by Gerald M. Edelman, see, for example, Edelman / Tononi

2002, 241–262.) Without being capable of cognitive functions or explicit memory (which seem highly unlikely, if not impossible, in the state described as 'whole brain death'), the 'brain-dead' patient's physiological state could thus continue to be related to his individual past and personal life history. In such a case, one may say that the 'Brain-dead patient' in his remaining modes of activity is still related to his peculiar past, the past of a human person, without presupposing cognition or even 'emotions' in the usual sense of the term. The body (including the extremely impaired or dysfunctional brain) may act as a living medium, in whose workings this past is 'inscribed', beyond the mental performance of memories.

Some remarks may be in place about what is meant and what is not meant by this interpretation in order to prevent misconceptions that might arise. Firstly, in general the distinction between the 'mental' and the 'physical' should be kept separate from that between 'body' and 'brain'. The distinction between the 'mental' and the 'physical' is an ontological one that identifies different (but mutually non-exclusive) kinds, or ways, of 'being'. The differentiation can be made without hypostasizing separate kinds of substances, as a metaphysical dualism would do. The 'physical' is that which can be found and described in relations of space and time. The 'mental' is constituted through an 'awareness' of 'meaning', relating the 'mental object' to a 'subject' that relates the object to something else (with which it may or may not be connected in space and time) and relates itself to that relation. Its description requires categories of a different kind than that of physical reality. In a non-metaphysical view, the 'mental' is embedded in the physical through the 'subject' (at least), which performs the action of relating, as actor and action can be located in space and time. The subject has, or is, a body, which is 'physical' and 'mental', as its spatiotemporal actions are continuously related to, and interpreted in, the realm of 'meaning'.

In contrast, in the dichotomy of 'body' and 'brain', both sides are entities of one ontological kind. As the "Zentralorgan," the brain may be seen in contraposition to the extracerebral parts of the body from which it receives information and whose activity it regulates (Plessner 1975, 256–257). This contraposition is a physiological one, which identifies brain and extracerebral body as constituents of one single integrated entity, the body as a whole.

Secondly, to say that the process of building personal memories 'involves the body as a whole' or that 'the body serves as a medium of the personal past' does not imply that the mental function of memory can be performed by parts of the body other than the brain. Nor does it say that memories (and here we are concerned with 'personal memories', memories of personal relationships) are stored, say, in the hands or in the skin. Rather, the point is about what is being remembered in 'personal memories' and what may follow from this. As has been remarked by the Japanese bioethicist M. Morioka (2001), relationships that can last beyond the point of brain death are typically of a kind that involves close physical contact or coexistence over an extended period of time. In other words, they involve corporeal as much as verbal communication, emotional as well as 'rational' exchanges. However, the meaning of these corporeal aspects will be informed by all other levels on which the persons concerned interact. As words trigger emotions (which are, in one perspective, physico-physiological states of the body as a whole) and emotions lead to words again, a history evolves in which it is obviously impossible to take the 'mental' and the 'physical' neatly apart. On the one hand, this may help to understand why phenomena such as Yanagida's purported reaction to the presence of his father and brother could be possible without a mental memory being produced on his part. On the other hand, it would be overly reductive to interpret it as a mere physical reaction because the reaction as such is related to, or in fact made possible by, the full range of meaning that imbues the personal relationship in question.

The second aspect of Yanagida's report important to our topic is the fact that in the face of his comatose and, later, brain-dead son Yanagida is incited to delve into memories of the past. Large parts of his article and even larger parts of his book consist of reflections that connect memories of his son's previous life to observations made in the presence of his comatose, and, later, brain-dead son. As once again M. Morioka has pointed out, this is a common reaction of relatives and friends confronted with a brain-dead body (Morioka 2001, 74–77). In Morioka's view, it is the combination of an apparently living body and the absence of those channels of communication to which we normally pay most attention that provokes such powerful waves of reminiscence (Morioka 2001, 76). While the warm, breathing, and pulsating body continues to present the patient as a living person, she lacks the ability to transmit distinctive thoughts and intentions. Precisely because of this absence of a limited, actual content, her personality is presented in totality, in a similar manner as a whole relationship may be summed up in one single gesture. In this situation, the patient's body is not a white screen on which memories are projected. As long as it is perceived as an individual, living body, it continues to be part of an intercorporeal exchange, to which we cannot but attach communicative meaning, and, thus, the flow of memories will be influenced and mediated by the living presence of the brain-dead body (Morioka 2001, 76).

To summarize what can be gained from this example of the brain-dead, we found the brain-dead human body to function as a medium of memory that is shaped by personal life history. And we found that contact with this medium provokes memories in others who have shared in a close relationship with the brain-dead person, and influences their course. The human body appears to be a medium to memorize other persons and the relations we had with them in both ways—for ourselves and for others.

Memorizing virtue

The second example takes us again to Japan, but this time we add some temporal to the spatial distance. At the beginning of the seventeenth century, after a period of prolonged internal wars and strife, the Tokugawa dynasty completed the unification of the country and established a feudal and bureaucratic system that was to last for 250 years. The Tokugawa Shogunate fostered a school of Neo-Confucianism that taught a theory of correspondence between the natural order of the universe and that of society (see, for example, Ooms 1985). Emphasis was placed on insight into the principles that governed the cosmic order, and scholars concentrated on studying the grand systems of philosophers like Zhu Xi or Hayashi Razan on meditation and speculation to achieve virtue. However, the emphasis on words also led to a new emphasis on philology, which motivated some scholars to criticize Neo-Confucianists for their neglect and misrepresentation of the older Confucian sources (Ogyu 1970, 8, 9). Although these scholars called themselves the school of "old learning" (*kogaku*), they established a new way of thinking. One of the most revolutionary thinkers in this school was Ogyû Sorai, who was born in 1666 and died in 1728 (see Lidin 1973 for a biography of this remarkable scholar). He did away with the time honoured idea of the "heaven" or nature as a source of moral virtue, and separated the cultural world of man from that of nature as is evident in his saying: "The Way of the Early Kings is that which the Early Kings created. It is not the natural Way of Heaven and Earth" (Ogyu 1970, 24).

The way as taught by Confucius, Sorai said, was not to contemplate the universe and study philosophical systems; it was a way that was meant to bring peace and harmony to human society by people studying ancient models of virtue and, even more importantly, by practicing the rituals and ways of life they established. Sorai saw personal cultivation as the root source of all virtue, and this, to him, meant cultivation of and with the whole body, as is evident from the following passages:

To take lightly that which is on the outside and end up attaching importance
to that which is within is contrary to the ancient Way of the Early Kings and
Confucius. (Ogyu 1970, 14)
   When bringing peace and contentment to the world, to cultivate one's
bodily self is the fundamental basis. (Ogyu 1970, 13; translation from Japanese
kanbun [= Sinojapanese style] by the author).

To Sorai, the human body does not possess any innate virtue that will express
itself naturally if one reaches proper insight into one's nature—a theory held
by Daoists and some East Asian Mahâyâna Buddhists. But it can, and has
to be, employed to realize what has been established as human virtue, most
eminently through the combination of rites and music. In other words, the
body (that is, that which is felt and perceived of oneself) is a medium that
can be utilized to cultivate virtue, like any natural material may be used to
realize the products of human ingenuity—but this does not mean that the
cultural ends are intrinsic parts of the means: "When cutting wood in order
to build a palace, one must follow the nature of the wood and only thus does
one accomplish it. But how could wood in its nature be a palace?" (Ogyu
1970, 26).[2]
   Applied to the human body, this means that one must take the body's own
nature, its needs and inbuilt rhythms, into account in order to employ it
for cultural ends, but one cannot expect any virtue to grow 'naturally' out of
the body.
   The aspect of rites and music most emphasized by Sorai is the embodiment
of harmony—social harmony in the adherence to ritual, and emotional
attunement in the performance of music. According to Sorai, rites and music
can create an environment that enables people to live together in peace and
contentment. The emphasis on music is an old Confucian topos, but it has
specific significance in the light of Sorai's above quoted saying: as music works
with the body's own tendency to attune itself to environmental rhythms, it
may work as a medium to communicate a certain ethos—a function of music
effectively employed by modern advertisement and propaganda, but virtually
obliterated in 'serious' musical and aesthetical theory with its emphasis on
aesthetic freedom.
   There is another aspect to Sorai's theory which remains largely implicit,
but connects it to our theme of time and memory. It is apparent in his saying
that "Benevolence is the Way of nourishing" (Ogyu 1970, 54). In other

---

   [2] The striking similarity (plus the differences) of this passage to Francis Bacon's famous
saying, "Natura enim non nisi parendo vincitur" (Nature is defeated only by obeying her,
Bacon 1620: 1.3) is worth noticing.

words, moral cultivation is a way of memorizing and habitualizing virtuous conduct. This, of course, is a line of thought that has been pursued by classical European authors, such as Aristotle (*Nicomachaean Ethics* 2.2) or Descartes (1649), as well; however, we rarely find an explicit concern, as with Ogyû Sorai, for bodily performance as a way of realizing virtue.

My emphasis is not on Confucian concepts of social harmony and proper conduct. There is much room for doubt whether these specific ideas can really bring 'peace and contentment' to the world.[3] What I want to bring to attention here is only Sorai's reasoning that the whole body can be used as a medium of moral cultivation. I use the term 'medium of moral cultivation' in order to distinguish it from an 'object of discipline'.[4]

The body can become such a medium that in itself helps to generate a virtuous personality, however virtue may be defined. My hypothesis in this regard is that one of the central functions at play here is precisely the capacity to employ reiterated bodily performance to accumulate and store habits of perception and action on a level below conscious mental activity. Virtues have to be memorized in this way in order that we can intuitively, without much reflection, know 'what it was to be good' in a given situation. This is a necessity in human life, where in many instances there is no time for reflection when virtuous conduct is required. This is not to denigrate moral reflection and discussion. I simply want to point towards an exigency caused by the structure of human existence. In short, we should reflect on and argue about moral principles while there is still time to do so—and doing so is a realization of the virtues of moral sincerity, prudence, and perspicacity. We should then strive to cultivate ourselves through bodily performance, so that we will be able to act according to our principles. Cultivating the body is

---

[3] Speaking from the point of view of Japanese history, the Tokugawa dynasty certainly managed to end a period of internal war and strife and to impose a stable social order that lasted for 250 years. However, this order was based on fierce exploitation of the rural population, which was regularly living in duress and repeatedly ravaged by famines. Confucianists, with their emphasis on the virtues of frugality and their deep misgivings about trade (as an 'unproductive' profession), advocated anti-commercialist policies which did nothing to alleviate these conditions. As E. Honjo says in his survey of Tokugawa economics (1965: 32): "The keynote of the agricultural policy followed by the Tokugawa Shogunate during the 300 years of its régime was 'to keep farmers alive but not to allow them to live in comfort.'" In reality, this meant that bad harvests had a devastating effect on the rural population Honjo (1965: 105) quotes a saying from Yamagata Bantô as a typical example of Confucian misgivings about commerce: "Agriculture should be encouraged and commerce discouraged. Farmers are part and parcel of the state, while artisans and mechants can be dispensed with. [...] Efforts must be made to increase the farming population, while, on the other hand, the number of merchants should be reduced as far as possible."

[4] Consider Sorai's argument against discipline (Ogyu/Lidin 1970: 56).

a primary means to habitualize virtuous conduct because the body's state, structure and functions are shaped by what we do, and they feed back—via the paths of propriosensation—into the conscious perception of ourself and the world. This is why we can find 'in our body' what we want to remember once we need it.

A THEORETICAL FRAMEWORK

What we found in our examples was that, within various fields of human existence, the human body serves as a medium that communicates past experiences, actions, and emotions into the present. It does so in both a passive and an active manner. In other words, the body serves as a symbol, but as a living symbol, one that contributes to the meaning communicated through it. For this very reason, I have referred to it as a *medium*, and in our case, more specifically, as a *medium of memory*.

Some words on a theory of symbolic functions may be of help in further clarifying how the human body can fulfil this part. Such a theory has been developed by Ernst Cassirer in his famous *Philosophy of Symbolic Forms* (Cassirer 1964, English translation Cassirer 1953–1996). As Cassirer explains, there are three fundamental levels of symbolic functioning: Expression, representation, and pure meaning. Like the different forms of temporality in J. T. Fraser's theory of time, they are ordered in a 'nested hierarchy': the higher ones unfold within the lower ones at once remaining dependent on them and transcending their constrictions. Science, which is one of the symbolic forms that employ the function of pure meaning, is coined in language, the classical locus of representation, which itself makes use of expression. Expression and representation are indispensable for the functioning of science, but scientific theories are structured according to laws that transcend the specific constrictures limiting representation and language.

The symbolic function that is first and foremost at play in our examples is that of expression. According to Cassirer, expression is the first and most fundamental of all symbolic functions, in comparison to the more abstract ones of representation and purely notional meaning. On the level of expression, we can distinguish between the symbol and its meaning—this distinction being the *arche* of symbolicity and, therefore, meaning—but the perceptual form of the symbol is still closely correlated to what it denotes. In symbolic representation, there are the possibilities of translation, of using different signs to refer to the same meaning, of distinguishing denotation and connotations. In expression, a slight difference of shape, timing, or tone can mark a significant variance in what is communicated. Every aspect of an expression's *gestalt*, be it static or dynamic, contributes to the content it communicates.

Another feature of expression, which again is a result of its relative immediacy, is that it is incorrigible. One may say, "I didn't mean to frown," having listened somewhat incredulously to someone's expectorations, or "I didn't mean to shy away," after being stunned by an unwanted offer, but people will still impute that one's expression conveyed what one really felt. That is why learning to restrict facial expressions and corporeal gestures is a seminal part of education all around the globe: politeness and civility would be virtually impossible without such control.

Thirdly, the meaning of expressions is less determined (less specific and, therefore, more open) than that of other symbolic forms. This is a restriction as well as a potential of expressional communication, which has to rely on empathy and the power of mutual understanding through shared preconceptions much more than language or science.[5] Still, by virtue of this openness, the nod or frown of a scientist may sum up his theoretical outlook although you would have to share a lot of his knowledge to know exactly what it means.

This feature of expression, its openness for multiple layers of meaning, which may include those dependent on the more abstract forms of language, or science, is most important in explaining our examples. In this world we live in, even the most abstract forms of meaning have to be produced in some 'here and now' in spatio-temporal existence. Every representation needs a presentation, through some form of expression. The body is what provides for this 'here and now'. It determines where the present is. Conversely, as a medium of expression it is shaped by the meaning it has to convey. This is not magic but simply the way expression works: You have to smile to communicate your kind feelings, to talk or write to convince somebody of your theories. When you do so, nervous impulses are put to use, muscles have to work, and so on. All this happens on the physical-physiological level, according to that which we have come to call 'the laws of nature'. Still, if the

---

[5] It should be noted, however, that the communication of a new scientific theory depends on the fact that even abstract symbols do function as an expression—in this case, the expression of an attempt to say something scientifically meaningful. If we had to rely on representation and pure meaning alone, we could never *come to* understand something we didn't know in the first place because we wouldn't have understood *anything* before having grasped the theory as a whole.

movements do not follow the rules of articulation, there will be no smile or no words. Thus, like Sorai's timber becoming a palace, the body partakes of the 'world of meaning', which has its own rules—be they those of grammar, manners, mathematics, or morality.

And as it is continuously used and shaped by our communicative or deliberative actions, it becomes a medium in which our past emotions, predilections, and activities are 'inscribed', and communicated—to ourselves and to others.

References

Bacon, Francis. [1620] 1855. *Novum Organon.* Oxford University Press. English Translation: 2000. *The New Organon.* Ed. Lisa Jardine and Michael Silverthorne. Cambridge University Press. Cambridge Texts in the History of Philosophy.

Bundesärztekammer. 1998. Beirat der Bundesärztekammer (1998): Richtlinien zur Feststellung des Hirntodes, 3. Fortschreibung 1997 mit Ergänzungen gemäß Transplantationsgesetz (TPG). *Deutsches Ärzteblatt* 95 (30): 1861–1868.

Byrne, Paul A., Sean O'Reilly, Paul M. Quay. 1979. Brain Death. An Opposing Viewpoint. *Journal of the American Medical Association* 242 (18): 1985–1990.

———. 2000. Brain death: the patient, the physician, and society. Pp. 21–89 in *Beyond brain death : the case against brain based criteria for human death,* edited by Michael Potts, Paul A. Byrne and Richard G. Nilges. Philosophy and medicine 66. Dordrecht et al.: Kluwer .

Cassirer, Ernst. [1953] 1996. *The philosophy of symbolic forms.* Translated by Ralf Manheim. New Haven: Yale University Press.

———. 1964. *Philosophie der symbolischen Formen,* 3 vols. Reprint of the 2nd edition. Darmstadt: Wissenschaftliche Buchgesellschaft.

Conference of Medical Royal Colleges and their Faculties in the UK. 1995. Criteria for the diagnosis of brain stem death. *Journal of the Royal College of Physicians* 29: 381–382.

Descartes, René. 1649. *Les Passions de L'Ame.* Paris: Le Gras. Quoted from *Die Leidenschaften der Seele. Herausgegeben und übersetzt von Klaus Hammacher. Französisch-deutsch.* Hamburg: Meiner. 1986.

Edelman, Gerald M. and Giulio Tononi. 2002. *Gehirn und Geist. Wie aus Materie Bewusstsein entsteht.* München: Beck.

Harvard Medical School Ad Hoc Committee. 1968. A definition of irreversible coma. Report of the Ad Hoc Committee of the Harvard Medical School to Examine the Definition of Brain Death. Pp. 287–91 in *Bioethics: An Anthology,* edited by Helga Kuhse and Peter Singer. Oxford: Blackwell. 2001. [First published in *Journal of the American Medical Association* 205 (6): 85–88].

Heckmann, Josef G., C. J. Lang, M. Pfau, B. Neundorfer. 2003. Electrocerebral silence with preserved but reduced cortical brain perfusion. *European Journal of Emergency Medicine* 10 (3): 241–243.

Honjo Eijiro. 1965. *Economic Theory and History of Japan in the Tokugawa Period.* New York: Russell and Russell.

Kuramochi, Takeshi and Nagashima Akira. 2003. *Zôki ishoku to seimei rinri.* Tokyo: Taiyô shuppan.

Lidin, Olof G. 1973. *The Life of Ogyu Sorai, a Tokugawa Confucian Philosopher.* Lund: Studentlitteratur.

Lock, Margaret. 2002. *Twice Dead: Organ Transplants and the Reinvention of Death (Public Anthropology)*. Berkeley: University of California Press.

Nau, Roland, H.W. Prange, J. Klingelhöfer, B. Kukowski, D. Sander, R. Tchorsch, K. Rittmeyer. 1992. Results of four technical investigations in fifty clinically brain dead patients. *Intensive Care Medicine* 18 (2): 82–88.

Morioka Masahiro. 2001. *Seimeigaku ni nani ga dekiru ka*. Tokyo: Keisô shobô.

National Conference of Commissions of Uniform State Laws. 1980. *Uniform Determination of Death Act: Annual Conference Meeting*. [Annual Conference Meeting in its Eighty-Ninth Year on Kauai, Hawaii, July 26–August 1, 1980]. Chicago (Printed Matter).

Ogyu Sorai. 1970. *Distinguishing the Way* [*Bendô*]. Translated by Olof G. Lidin. Tokyo: Sophia University.

Ooms, Herman. 1985. *Tokugawa Ideology: Early Constructs, 1570–1680*. Princeton University Press.

Plessner, Helmuth. 1975. *Die Stufen des Organischen und der Mensch*. Berlin: De Gruyter.

Yanagida Kunio. 1994. Sakurifaisu: waga musuko nôshi no 11nichi. *Bungei shunjû* 72: 126–151.

———. 1999. *Sakurifaisu: waga musuko nôshi no 11 nichi*. Tokyo: Bungei shunjû.

RESPONSE

Remy Lestienne

Could a decerebrate person ride a bike? However cautiously written Christian Steineck's article is, a few comments might be in order so that no one be misled by its title. Steineck interestingly uses his knowledge of the Asian philosophies and civilization to support the idea that the body could be "a medium of moral cultivation", as opposed to "an object of discipline." "Virtues have to be memorized [. . .] in order that we can intuitively, without much reflection, know 'what it was to be good' in a given situation." However, from the examples given (particularly those of patients declared "brain dead" who seem to react to the presence of beloved ones), we understand that he refers to the periphery of the body as opposed to the central nervous system (CNS). Might a peripheral part of the body be a repository of memory?

From the point of view of neuroscience, this question is reminiscent of the question of formation of reflexes.

Let us take an example: that of learning to ride a bike. In such a learning experience, at the beginning all the conscious information—the feeling of being short, of falling on the road, etc.—is accumulated in the brain as an episodic memory (probably mainly in the cortex). But progressively the learned procedure becomes a habit, then a pure reflex, and the intervention of the complex system of episodic memory recall becomes no longer necessary. The activity of the brain tends to internalize itself, to engage the cerebellum or even the brain stem rather than the cortex.

In recent years, we have discovered that long term memory is built up in nerve cells by the activation of a cascade of genes. The ultimate effect of the genetic machinery is to modify the properties of the cell membranes, in particular changing (tuning) the efficacy of their synaptic contacts. It is quite possible that the storage of long term memories could be performed not only by CNS nerve cells (in the cortex, the hippocampus, the cerebellum, etc.) but also by the peripheral nervous system (but obviously not in non-nerve cells—the skin for example).

Steineck's examples also touch upon the role of affectivity in memory. As for the importance of communication and induction of the feeling of "being secure," it is quite likely that this very important effect of affectivity on memory be, at least in part, the consequence of the activation of the

Jo Alyson Parker, Michael Crawford, Paul Harris (Eds), Time and Memory, pp. 53–54.
© *2006 Koninklijke Brill N.V. Printed in the Netherlands.*

dopaminergic system. (See the Presidential Welcoming Remarks in this book.) However, the dopaminergic system characterizes the CNS, even more specifically essentially the limbic system and its associated prefrontal cortex, not the peripheral system.

On the other hand, when Steineck writes "as words trigger emotions (which are, in one perspective, physico-physiological states of the body as a whole)," he is certainly right, from the point of view of expression rather than from that of memorization, even though "memories are more often than not evoked by sensual, corporeal experiences." Cassirer, quoted by Steineck, insisted on the force of the symbols, on their power of creativity. If, following C.S. Peirce, we accept that symbols are the necessary intermediaries between the perception of facts, the experience, and the creation of the concept, it remains certain that the latter is the reserved domain of the central nervous system.

To summarize, one could say with Steineck that the role of the most peripheral sensitive systems in expression is clear. That the peripheral nervous system participate in some memorization tasks, as in the formation of reflexes, is quite likely. Of course, the article does not imply that that the body in itself, or even the peripheral nervous system, might be able to construct a symbol without the help of the CNS.

CHAPTER TWO

BODY MEMORIES AND DOING GENDER: REMEMBERING THE PAST AND INTERPRETING THE PRESENT IN ORDER TO CHANGE THE FUTURE

KAREN DAVIES

SUMMARY

Narrative is one method of tapping individual and collective memory where human experience is organized into temporally meaningful episodes and where a "point" is often made. In research projects that I have carried out about working life, "body stories" are often spontaneously related in making a "point." These stories exemplify "doing gender"—that is how gendered relations (that revolve around equality/inequality, domination/subordination, advantage/disadvantage) are *actually* created, recreated, contested and changed in everyday life. "Doing gender" is also linked to "body-reflexive practices"—the idea that the body is not only the object of symbolic practice and power but also participant. In other words, body-reflexive practices generate new courses of social conduct. In the second part of this paper, empirical examples taken from research projects on gender and professionals working in the health care and information technology (IT) sectors are provided to illustrate these concerns. The last example shows how a contestation of "doing gender" brings about change. More generally, one can say that without individual and collective memories, there would be no social change. Thus memory and the response that it invokes are pivotal in the unfolding of time.

The body as site of interaction of material and symbolic forces is the threshold of subjectivity; it is not a biological notion, but marks, rather, the non-coincidence of the subject with his/her consciousness and the non-coincidence of anatomy with sexuality. Seen in this light, the body, far from being an essentialist notion, is situated at the intersection of the biological and the symbolic; as such it marks a metaphysical surface of integrated material and symbolic elements that defy separation.

—Braidotti, 1991: 282

MEMORY AND NARRATIVE

Memory—understood as the individual's perception, processing, storage and retrieval of personally experienced events and knowledge of the world—has been the subject of extensive research and theorizing by psychologists and neuroscientists alike, and, according to Alan Baddeley (1989), the physical and biochemical basis of memory has elicited considerable interest in recent

Jo Alyson Parker, Michael Crawford, Paul Harris (Eds), Time and Memory, pp. 55–69

years. Memory in its more sociological connotations goes, however, beyond
this more narrow definition of memory as the retention of past experiences
and the ability to retrieve them (Jedlowski, 2001). Not all things are
remembered; indeed there is a constant sifting and sieving of information
leading to a specific distillation of memories. We remember selectively; what
is considered most salient to us is given priority over that which is not. Our
stories are not static either; they can change over Time, being influenced by
present deliberations. There is, then, not an objective account of how events
occurred but a personal interpretation of these events that is very much
linked to the historical context, to prevailing societal discourses and to the
way others also interpret these events. As Jedlowski (2001: 31), building on
the work of Halbwachs, points out, "[...] the memories of each individual
are inscribed within 'social frameworks' which support them and give them
meaning." These social frameworks are thus not divorced from the culture(s)
within which they are constructed.

Closely linked to memory as a store is also a *response* aspect—the response to
subsequent events is affected by previous acquisitions, which are individually
and collectively filtered and interpreted (Jedlowski, 2001). Memory thus
links the past via the present with the future, *but* within a social context and
within particular power structures. Memory allows us to use our knowledge
of the past to deal most effectively with present and future contingencies,
allowing us to utilize our knowledge base in a competent manner. It also aids
us in confronting the future. Without individual and collective memories,
there would be no social change. Thus memory and the response that it
invokes are pivotal in the unfolding of time.

Social scientists and historians have attempted to tap individual and
collective memory with the help of *narrative*, primarily in the form of interviews
or life-histories (see for example, Bertaux, 1981). As Polkinghorne (1988:
1) points out, narrative is "The primary form by which human experience
is made meaningful [...] it organises human experiences into temporally
meaningful episodes." In interviews then, stories are told; however they are
not told in "any old fashion." There are apparently narrative conventions. A
narrative is required to establish coherent connections between life events as
well as the fact that an acceptable story must first establish a goal, an event to
be explained, a state to be reached or avoided, an outcome of significance or,
more informally, a "point" (Gergen, 2001: 250).

In much of the research that I have carried out on working life from a
gender perspective—using qualitative methodology—unsolicited narratives
about the *body or bodily experience* have emerged. That is, research participants
often *spontaneously* recall incidents that revolve around the gendered body

in order to make a point—the point being that they feel they are unfairly treated and by extension would like this state of affairs to be changed, or that they have in fact accomplished change with regard to gender relations at their workplace. Narrative, then, is not just recalling a (filtered) past; it can also "form" the world with its emancipatory potential. Thus the "body stories" told in my research projects were not related simply to reveal the victim but told in the spirit of paving the way for dialogue and change. It is interesting to note that according to Hinchman and Hinchman (1997) the word narrative comes from the Indo-European root "gna," which means both "to tell" and "to know." In feminist terms, narrative provides the opportunity to tell "another story"—a story that questions the "master narrative." While the stories may be examples of "atrocity tales" (see for example, Goode and Ben-Yehuda, 1994; Jacobsson, 2000) or see women as victim, narrative may importantly be used as contestation or protest, as examples of different ways of seeing and knowing.

Doing Gender

According to Thomas Butler (1989) *emotion* is a good bonding agent for personal memories. In other words, incidents that are painful, energizing or ecstatic can be easily brought to mind (if the incident is severely painful rather than marginally so, then there is a risk of course that we will repress the memory). Gender struggles, sexual harassment, "doing domination" and "doing subordination" are of course all examples of painful and deeply unfair social practices. It is perhaps not surprising that these incidents are recalled and expanded upon in research interviews even if they are not specifically asked for.[1] And since the individual body and collective body have been the site of struggle in gendered power relations, it is even less surprising that "body stories" emerge.

These body stories (what is meant by a body story will be explained later in the paper) are closely linked to a central theoretical and analytical concept in feminist theory, namely "doing gender." Doing gender (see Fenstermaker, 1985; West and Zimmerman, 1991; Butler, 1993; Fenstermaker and West,

---

[1] Research participants in my studies were asked to give accounts of their work situation and how they experienced work. They were not asked to describe situations that they experienced as being gender struggles. Their accounts around such issues, however, emerged spontaneously and were not always phrased by themselves in terms of being a "gender problem."

2002; Davies, 2001, 2003) has emerged as an analytical tool in showing how gendered relations are *actually* created, recreated, contested and changed in daily life. The central question is: how are these relations actually done and accomplished? What are the deeds and who does the doing? Implicit is that in all interaction "doing gender" can and does happen (even between people of the same sex). The feminist concern is thus to unveil these doings—doings that are often so commonplace or taken for granted that the doers do not recognise the structures or consequences of their doing. Yet the doings reinforce gendered relations of power and often leave women in a disadvantaged position. In order to tap the actual doings, research methodologies linked to ethnomethodology and symbolic interactionism are required. As Dorothy Smith writes (in the foreword of Fenstermaker and West, 2002): "In contrast to the abstract nominalization of much sociology, ethnomethodology, in all its variants, is oriented to discovering how people produce what we can recognize as just this or just that everyday event, occasion, setting, act, or person." While observation of social conduct is one way of accessing these doings, they are also unveiled in narratives derived from interviews discussing everyday life. Such methods are often given the label of micro sociology, implying that we are only capturing interactions on an individual level. But the interactions on the micro level mirror the wider social structures that hinder or enable agency. Individual choice and actions are formed by the wider social structures, but at the same time individual choice and action form these social structures. As Fenstermaker and West (2002: 219) say: "We have taken the position repeatedly that the bifurcation of 'micro'/'macro' reduces to one of 'little' versus 'big,' or 'trivial' versus 'consequential,' or 'unstructured' versus 'structured' and precludes a vision of the bridges across the divide."

Doing gender then lets us understand that gender and gender relations are not a static pre-given but moulded in ongoing actions where at least two partners need to be present. As an analytical tool, the concept lets us examine how others (willingly or unwillingly, consciously or unconsciously) do gender when interacting with us, but equally how we are accomplices (willingly or unwillingly, consciously or unconsciously) in the doing. However, the active and continuous form implied in the "ing" of the doing equally signals its transformative status, indicating that relations can be changed. The doing can and does change form (for better or worse). Normative gender relations can be destabilised.

BODY-REFLEXIVE PRACTICES

A way to understanding how "body stories," "doing gender" and individual/ social change are linked can be achieved, I believe, by addressing more closely Connell's (1995) term *body-reflexive practices*. The strength of Connell's approach is that he wishes to see the body as both the object and *agent* of social practices. Connell (1995: 60) argues that while theories of discourse have succeeded in bringing the body back into social theory, the body is understood as "objects of symbolic practice and power but not as participants." He explores then, through his interviews with men about their lives in his attempt to understand masculinities, the idea of body as participant. He provides examples of how the knowing body—how bodily interaction and bodily experience—instigates new desires and new actions. Connell (1995: 231) asserts:

> [...] bodily difference becomes social reality through body-reflexive practices, in which the social relations of gender are experienced in the body (as sexual arousals and turn-offs, as muscular tensions and posture, as comfort and discomfort) and are themselves constituted in bodily action (in sexuality, in sport, in labour, etc.). The social organization of these practices in a patriarchal gender order constitutes difference as dominance, as unavoidably hierarchical.

In other words, an actual experience, which results in *both* a mental and bodily memory, triggers the desire for other or new kinds of actions that have not been experienced earlier; it importantly influences a *response*. Bodies then are implicated in agency in social processes, in generating and shaping new courses of social conduct. As Connell (1995: 58) notes: "Some bodies are more than recalcitrant, they disrupt and subvert the social arrangements into which they are invited."

Connell also strongly argues for the fact that it is not only a question of individual experience and agency; social relations and symbolism are also involved, and large-scale institutions may even be implicated (Connell, 1995: 64).

In order to understand more clearly what is meant by a body-reflexive practice (at least in my understanding) I will use an example from Johanna Esseveld's (1999) research, which is based on 'collective memory work' (see Haug, 1987). Participants were asked to write down a memory around the theme: when I felt good and when I felt bad. One woman wrote the following memory about getting an exam paper back:

> After a class [at university] we got called in to K's room one by one to receive feedback on an exam we'd done. I remember that I felt uncomfortable about

going there because he was usually so arrogant and superior. He started by asking me in an aggressive tone of voice: "*How* did you revise for this exam?" I felt very unsure and didn't know how to answer. "Tell me now *how* you revised!"

It wasn't very nice at all. I told him roughly how I'd gone through the books and then was quiet. He then said: "It was an excellent exam paper. I wish everybody used your study technique." Then he said that it was the best paper that he'd seen in this subject. "I think you should continue your studies and go in for this subject. You know we have post-graduate studies."

I left his room overjoyed, smiling. Felt that maybe the university had something to offer even me. This was something I couldn't tell my student buddies or friends since it would be showing off. But I felt really good, felt joy and strength.

Later the same day I was walking along the corridor outside the department and behind me was K. He came up behind me, grabbed hold of my bra at the back, pulled it out and let it snap back. He then laughed in a supercilious manner and rushed off. I was so shocked, insulted, that I was silent. (Esseveld, 1999: 107, my translation)

The narrative clearly highlights gendered power relations, and the body as the site of gender struggle is highlighted in the sexual harassment that occurs in grabbing the student's bra. While the student's bodily reactions are not spelled out as such in words in this narrative, one automatically imagines (and even feels oneself) how these undoubtedly change in this female student's day, leading up to what probably were muscular tensions, a possible headache coming on, the body diminishing in size, et cetera. And what were the consequences afterwards? We can only speculate (as the narrative does not give us more information): she never set foot in this department again; she lost interest in further studies; or she contacted an ombudsman and the professor was accused of sexual harassment and subsequently lost his position; the incident was so deep-rooted and painful that she devoted her work career later to working for sexual equality, et cetera. What is certain, however, is that the bodily knowing (as both a cognitive memory *and* a bodily reaction/memory, for example, muscle tension) led to some form of social action, whether emancipative or not. The body was not only then the object of symbolic practice and power but also participant.

From this example, it is clear that (an) individual memory is linked to wider societal structures, that bodily knowing and body-reflexive practices are involved in the incident, and that agency as well as objectification are part and parcel of the encounter. Doing gender is accomplished.

EMPIRICAL ILLUSTRATIONS

I will now turn to empirical material to illustrate certain body stories and doing gender. While the body has gained considerable interest in sociological theory in recent years, discussions with regard the body, gender and *working life* seem less prevalent. (See, however, Brewis, 2000.) Examples to be presented here are derived from two recent research projects about working life that I have conducted. One project examined the (somewhat problematic) working relation between doctors and nurses in Sweden seen from a gender perspective and problematised more generally the situation of women doctors in the medical world.[2] The other project examined the reasons for gender equality in the information and computer technology (ICT) sector in Sweden and Ireland.[3] A large number of body stories or body memories emerged spontaneously in the interviews. What I mean by a body story is a narrative that related how female bodies (or male bodies) are constructed or understood in work situations. What are the bodies thought to be capable of physically, are the bodies thought to be spatially mobile, or where are gendered bodies placed or where do they place themselves in time and space at work (what can be called *situatedness*)? The first examples given below will show how interviewees recalled situations that they felt were grossly unfair. I will then present an example of an account that led to change and empowered the subservient group. "Doing gender" and resulting body-reflexive practices are an integral part of the stories.

DOING GENDER—CATEGORISING WOMEN

Women in the Medical Profession

We have to take our starting-point in men as the norm—we have to adapt ourselves to their rules. We're only allowed in as guests. (Woman orthopaedic surgeon)

They [the male doctors] want tough boys for this job because they were tough boys themselves. (Woman surgeon)

---

[2] The research was financed by The Vardal Foundation during 1997–2000 with the author as the sole investigator. Results are presented in Davies (2001) and Davies (2003).

[3] The research was financed by the Swedish Agency for Innovation Systems (VINNOVA) during 2001–2003 with Chris Mathieu and myself as co-researchers. Results are published in Davies and Mathieu (2005) and Mathieu (forthcoming).

While women in Sweden have significantly increased their numbers in the medical profession from around ten per cent in the fifties (Nordgren, 2000) to well over fifty per cent in current medical education (http://nu.hsv.se/nu/index1.html)[4] the women doctors in my study related in a variety of ways how they were made to feel like outsiders in the medical academy, as suggested by the quotations above. The construction of the woman's body and what it was capable of was one of the strategies used to bring this about.[5]

Within certain specialities in particular, being accepted as a woman is a hard-fought battle, and the female body is used at times as a reason for a woman's exclusion, or at least for questioning her suitability in spaces that historically have been constructed as masculine, such as orthopaedic surgery and general surgery. It is the size and strength of the female body that is brought into question. It is not a body, it is argued, that can meet the demands of certain operations. Stamina may be lacking. It is not a body that commands authority and respect from staff assisting operations. Admittedly these were not opinions openly raised by the majority of respondents in my study—I would imagine that equality debate in recent years has inhibited what can be seen as open discriminatory talk. But the fact that it was mentioned spontaneously by a number of women doctors is evidence of its persistence notwithstanding.

A female orthopaedic surgeon showed the inconsistencies in the arguments put forward. Her example to be outlined below clearly highlights how the female body is socially constructed rather than simply biologically determined. Orthopaedics is a speciality where the number of women doctors is especially low—in fact, this is where, along with thorax surgery, we find the least number of women in Sweden; the figure for 2001–01–01 was 8%.[6] It could be argued that the reason for this lies in not enough women choosing this speciality. However, interestingly, many of the women preregistration house officers told me that they found the speciality exciting and said that they could seriously consider making this their choice of speciality. The woman orthopaedic surgeon I spoke to, however, described how women seldom stayed long—they disappeared into other specialities. Subtle forms

---

[4] Of course the higher positions are more likely to be held by men. Consultants (Sw: överläkare) consisted of 75% men and 25% women in 1995 (Nordgren, 2000: 70). Thirty-nine % of women doctors were at the junior doctor level in 1998, while the equivalent figure for men was 25% (Nordgren, 2000: 70).

[5] There are of course a variety of other strategies not linked to the body that are not discussed here.

[6] Physician statistics. Swedish Medical Association. (2001).

of exclusion appeared to be at work. An example was given to show how the body is used to question women's suitability.

Recently a woman house officer had started and was required to carry out an emergency operation on a hip fracture. The second on-call was contacted. The operation does require some strength, it was argued by my interviewee, but this is not what is most important; it is rather a question of technique. The senior doctor called in, though, commented afterwards upon the woman doctor's purported frailness. This caused my interviewee to react in the following way:

> They start straight away by measuring her physical strength. And yet we've had guys here that are much more 'delicate'. So I said to him, "Have you thought about the fact that when we competed in the Vasa race[7] the other day—here all the male doctors take part in this competition—that she beat them by two hours, she was much faster than any of the male doctors here—and you go around here and call her weak!"

Women in I.T. Professions

This section starts of with the narrative of a woman relating a discussion she'd had with her boss.

> At my performance appraisal I tried to say that I like working on the help-desk, but that I'd like to do more than this. "But you do such a good job. You can't finish there!" "OK, but I feel I could do an even better job if I was given more responsibility, I'd like to develop myself more." "But think about the fact that you have a family and children, it's very practical to have a job from 8 to 5." "I don't agree" is what I say. Why do they say this? Just because one has young children and are [sic] a mother? [...] So I say to him, "There are all sorts of things I would be prepared to work with, tell me where you need someone." They think this is very positive that you talk in these terms, "Fantastic... you're so flexible... but what about travelling?" "Well, I wouldn't want to be away the whole week."—But then there is *nobody* who wants to these days.—So I say, "It's no problem with the travelling, I'll solve it." And yet after talking for a couple of hours, they still say "yes, but don't forget that you have a family!" They would never have said that to a man. If you've chosen this line of work, to be an IT consultant, this doesn't mean that you should be a slave or do everything they demand. You *can* solve the travelling thing. If you've got a customer in Gothenburg, it doesn't mean that you've physically got to be there every day, every week. One or two days a week would be enough. You can do

---

[7] The Vasa race (Vasaloppet) is an annual, international, long-distance skiing competition where the distance to be covered is 90 km.

a lot of the work here. Here we are, working in an industry where we use the latest technology, and yet they can't see the possibilities! They're so bound by tradition! (Female IT consultant, early thirties, married, 3 children under 10 years)

This and similar narratives were expressed by women working at one of the Swedish companies included in our study. In many ways their opinions and experiences differ from what was expressed by women (and men) at other Swedish companies who argued that they assumed combining work and family would not be problematic. "Assumed" is central here. The IT sector is a young area, and in the majority of the companies we visited, the median age of the employees was low. Employees, especially women, had not started families yet. This is not surprising given that well-educated women in Sweden are postponing the arrival of their first-born until their thirties. Many of the Swedish companies in our study also displayed a "young, trendy image"— advocating buzz words such as diversity, opportunity, flexibility, "no old boy chauvinistic atmosphere here,"[8] they obviously wanted to disassociate themselves from what they saw as the burgeoning flotsam of traditional organizational life. They communicated a view of the company to the outside world and to their employees that "anything was possible." The company where the women cited above worked did not exactly fall into this category. It had existed for some time, and, while having always specialized in technical areas, it was only in recent years that an IT and telecommunication profile had been mantled. The median age of employees was also higher here and having families was a common occurrence. It could thus be argued then that the particular form of "doing gender" that emerges in the above narrative will apply to women in other companies later when the reality of combining work/career and children emerges.

The IT consultant's tale shows us that the women are constructed significantly in terms of careers. In no interview, with managers (men—or women for that matter), is it stated, or assumed, that the women lack technical or other competency. Technical knowledge or competence is not what is at issue here. What is at issue is that the work requires, it is argued, *mobility and accessibility both in space and time.* Having responsibility for children—at least for women—is *assumed* to jeopardize this mobility and accessibility. But the women strongly argue in the interviews that the arguments used against

---

[8] Willim (2002) found that in the young, fast-moving Swedish company Framfab, traditional and older IT companies, such as Ericsson, were disparagingly called "respirators."

them are somewhat a "sham." On the one hand, they're not asked *if* they can solve travelling /being-away-from-home "problems" (it is assumed that they can't);[9] on the other hand they argue that new technology can surely mitigate logistic headaches (that is, time and space are in a variety of ways "released" with the new technology). They emphasize again and again in the interviews that solutions can be found—but that they aren't given the chance to show their ingenuity. The result is that they get stuck in what they see as rather boring and dead-end jobs or work duties. They also point out that the male bosses don't realize that they are being discriminatory or narrow in their thinking. On the contrary, they say that the men argue that they are trying to take the women's situation into account in a fair way.

The body enters this particular narrative in two ways. On the one hand, the women's bodies are physically and geographically—in the managers' eyes— locked in space due to the women's caring function, and this affects what jobs it is assumed they can take (even if they are seen as being technically competent). On the other hand, it is their female bodies as such that signals diminished capacities. They are reproductive bodies. The men do not have reproductive bodies and thereby are not constructed as a certain type of worker, and thus they avoid certain assumptions and questions. Lack of competency may hinder the men in their careers, but the male body, as such, does not.

CHANGING THE AGENDA—THE EXAMPLE OF THE MORNING ROUND

> The appropriation and use of space are political acts. The kinds of spaces we have, don't have, or are denied access to can empower us or render us powerless. Spaces can enhance or restrict, nurture or impoverish. (Weisman, 2000: 4)

The example to be presented below is taken from the study examining the relations between doctors and nurses. Historically medicine and nursing have been constructed on certain notions of masculinity and femininity (see, for example, Davies, 1995; Davies, 2001, 2003; Gamarnikow, 1978) and embraced "doing dominance" and "doing deference." Doing gender

---

[9] Interestingly the husbands or partners of the women we interviewed in this company did not work in the IT sector but often had less qualified jobs (for example, carpenter, pre-school teacher), and it was argued by the women that it would be fairly easy for their husbands to take a larger share of child-care. Their husbands were not hounded by 'top-pressure jobs."

is done, one could say, collectively. Today, nursing staff—as a collective body—oppose doing deference, and the boundaries between nursing and medicine are both shifting and are unclear at times. There is, it would seem, an equalizing tendency where the medical profession unavoidably is losing some of its traditional power and authority, at least in Sweden, and I would imagine in other Western countries as well. But to what extent the nursing body can truly achieve an equal position remains to be seen. Boundaries are crossed but then reset. What can be agreed, however, is that the relations between the two professions are in flux and that change with regard the professions themselves is firmly on the agenda. An example of a changed morning hospital round will be taken to illustrate this.

Traditionally, space and place have unequivocally spelled out status and position. The long train of individuals trooping into the patient's room is not a haphazard formation. The consultant takes the lead; the lowest person in the hierarchy comes in last. Indeed, one problem I experienced in my fieldwork was to know where in this formation my own presence would be normatively correct so that I did not overstep a boundary. I will discuss below another type of morning round to show how the situatedness of bodies in space can provide a certain sense of empowerment for the traditionally subordinate group. The round, described above, is still in full force, at least on some of the wards I studied. But it should perhaps be pointed out that it is a far cry from an earlier regime where patients "stood (sat) to attention" when the consultant came in, where doctors did not talk directly to the patient but only communicated via the staff nurse and where even the paper baskets were emptied in advance so as not to offend the consultant!

On one of the surgical sections I studied, a different form of morning round had been introduced. Doctors (consultants, the ward surgeon, house officers and preregistration house officers) congregated in a room on the ward in the morning and seated themselves around a *round* table. Primary nurses— who were responsible for three or four patients—came in separately but in succession and also seated themselves at the table. Patients—their condition, their problems, their follow-up treatment—were then discussed. The nurse initiated discussion by first presenting a latest summary and appraisal of the patient. After the "conference," doctors, unaccompanied by nurses, visited each patient.

Nurses were overwhelmingly positive about the new routine; doctors were more wary in their comments: "It takes too much time, we need to get off to the operating theatre." "You need an old-fashioned round, so that the nurses can learn." In particular, the house officers felt they had difficulty in grasping the situation of all the patients. Some of the consultants drew out its advantages though: "We spend more time discussing the patient from various angles now, including social aspects and what will happen to the patient when he/she leaves the hospital—in addition to talking about various investigations and lab results. We can sit down and discuss more in peace and quiet."

The nurses felt that they had won time with this form of round, especially as they did not accompany the doctors later into the patient (unless it was felt that there was a special reason to do so). Earlier, time would be wasted waiting in the corridor for the doctor to get to *their* patient. Now the doctors might have to wait if the nurse could not drop everything on the spur of the moment when it was time to discuss her patients.

As I understood it, it had been the nurses that had pushed for the new order of things. The nurse manager gave the following reasons for its implementation:

> It's an attempt to minimise the number of people in the patient's room. What's the point of a whole flock? In part it's a question of confidentiality—when you're standing there in the corridor... And then the doctors, they get irritated when the cleaning trolley arrives or when the food trolley is on its way out. They, themselves, don't see that they stand there and take up a lot of space [in the corridor] during quite a long period of time. When the new electronic case records system was introduced, it became obvious that the old routine wasn't feasible any longer. I tried to argue that sooner or later they need to sit down somewhere when all the medication and case records are on the computer. We can't do a traditional round with a lap-top and everyone trying to peer into it. You have to be able to sit down at a larger screen.

Nurses frequently emphasized their patient advocacy role in their work as well as their concern for the patient, and this type of round, one could say, facilitates this. Improving patient care was an argument used by both doctors and nurses to justify a new form of round. Surprisingly, establishing more egalitarian ways of working between the two professions was not taken up by my respondents. And yet it seemed to me that this was one of the outcomes and an important one at that. Space was utilized in the traditional round to ensure medical authority. The culturally determined spatial rules, admittedly unwritten but none the less widely understood, saw to it that each person knew his or her place in the hierarchy (including doctors within the medical hierarchy). Sitting around a table, by contrast, places bodies on an equal footing and provides the opportunity for real discussion.[10] Sitting

---

[10] I am of course aware that spatial arrangements may not be sufficient to engender equal talk.

down diminishes the importance of size (and dominance)—which may be of relevance in relation to male doctors.

Concluding Remarks

Narrative, based on individuals' cognitive and bodily memories, showed in this paper how doing gender is actually accomplished in working life. Following Connell (1985), I would argue that the stories were not only about the body as object but also about the body as participant. The events were also experienced *in* the body—as muscular tensions, feelings of discomfort, anger, insomnia, et cetera—resulting in body-reflexive practices, that (could) le(a)d to changed praxis. In the last case we see an example of full-blown agency being accomplished. But even in the other examples, hoping for change was on the agenda in their relating their narrative to me, in their taking it up in the unions, in discussing it at their workplace at coffee-breaks. Cognitive and bodily memory were thus implicated in body-reflexive practices leading to a contesting of gendered social relations and working for social change—a small yet important part of the grander scheme of things without which time would stand still.

References

Baddeley, Alan. "The psychology of remembering and forgetting." In *Memory: History, Culture and the Mind*, ed. Thomas Butler, 33–60. Oxford: Basil Blackwell, 1989.
Bertaux, Daniel, ed. *Biography and Society.* London: Sage, 1981.
Braidotti, Rosi. *Patterns of Dissonance: A Study of Women in Contemporary Philosophy.* Oxford: Polity, 1991.
Brewis, Joanna. "When a body meet a body...': Experiencing the female body at work." In *Organizing Bodies. Policy, Institutions and Work*, ed. Linda McKie and Nick Watson, 166–84. Basingstoke: Macmillan, 2000.
Butler, Judith. *Bodies that Matter. On the Discursive Limits of "Sex."* London: Routledge, 1993.
Butler, Thomas. "Memory: a mixed blessing." In *Memory: History, Culture and the Mind*, ed. Thomas Butler, 1–31. Oxford: Basil Blackwell, 1989.
Connell, Robert W. *Masculinities.* Cambridge: Polity Press, 1995.
Davies, Celia. *Gender and the Professional Predicament in Nursing.* Buckingham: Open University Press, 1995.
Davies, Karen. *Disturbing Gender. On the Doctor–Nurse Relationship.* Lund: Department of Sociology, 2001.
———. "The body and doing gender: the relations between doctors and nurses in hospital work." *Sociology of Health and Illness* 25, no. 7 (2003): 720–742.
Davies, Karen and Chris Mathieu. *Gender Inequality in the IT Sector in Sweden and Ireland.* Work Life in Transition, 2005:3. Stockholm: National Institute for Working Life.

Esseveld, Johanna. "Minnesarbete" (Memory work). In *Mer än kalla fakta. Kvalitativ forskning i praktiken (More than Cold Facts. Qualitative Research in Practice)*, ed. Katarina Sjöberg, 107–127. Lund: Studentlitteratur, 1999.

Fenstermaker, Sarah. *The Gender Factory*. New York: Plenum, 1985.

Fenstermaker, Sarah and Candace West, eds. *Doing Gender, Doing Difference. Inequality, Power and Institutional Change*. New York: Routledge, 2002.

Gamarnikow, Eva. "Sexual division of labour: the case of nursing." In *Feminism and Materialism. Women and Modes of Production*, ed. Annette Kuhn and AnnMarie Wolpe. London: Routledge and Kegan Paul, 1978.

Gergen, Kenneth. "Self-narration in social life." In *Discourse Theory and Practice*, ed. Margaret Wetherell, Stephanie Taylor and Simeon J. Yates, 247–260. London: Sage Publications, 2001.

Goode, Erich and Nachman Ben-Yehuda. *Moral Panics: The Social Construction of Deviance*. Oxford: Blackwell, 1994.

Haug, Frigga. *Female Sexualization: A Collective Work of Memory*. London: Verso, 1987.

Hinchman, Lewis P. and Sandra K. Hinchman. *Memory, Identity, Community. The Idea of Narrative in the Human Sciences*. Albany: State University of New York Press, 1997.

Jacobsson, Katarina. *Retoriska strider. Konkurrerande sanningar i dövvärlden. (Rhetorical Struggles. Competing Truths in the Deaf Community)* Lund: Palmkrons förlag, 2000.

Jedlowski, Paolo. "Memory and sociology. Themes and issues." *Time & Society*, 10, no. 1 (2001): 29–44.

Mathieu, Chris. "Pushing and Pulling Female Computer Professionals into *Technology-plus* Positions." In *Encyclopedia of Gender and Information Technology*, ed. Eileen M. Trauth. Information Science Publishing, 2006.

Nordgren, Margreth. *Läkarprofessionens feminisering. Ett köns- och maktperspektiv (The Feminization of the Medical Profession. A Gender and Power Perspective)*. Stockholm Studies in Politics 69, Stockholm University: Department of Political Science, 2000.

Polkinghorne, Donald E. *Narrative Knowing and the Human Sciences*. Albany: SUNY Press, 1988.

Weisman, Leslie Kanes. "Women's environmental rights: A manifesto." In *Gender Space Architecture*, ed. Jane Rendell, Barbara Penner and Iain Borden, 1–12. London: Routledge, 2000.

West, Candace and Don H. Zimmerman. "Doing gender." In *The Social Construction of Gender*, ed. Judith Lorber and Susan A. Farrell, 13–37. Newsbury Park: SAGE publications, 1991. First published in *Gender and Society* 1, no. 1 (1987): 125–51.

Willim, Robert. *Framtid.nu Flyt och friktion i ett snabbt företag. (Future.now Flow and friction in a fast company)*. Lund: Symposion, 2002.

RESPONSE

Linda McKie

In this contribution Karen Davies draws upon theory and narratives from research projects to illuminate and explore "body stories" as these exemplify "doing gender." "Doing gender" refers to the varied and interweaving ways in which gendered relations are created and reinforced, and infuse the everyday. To that I would add that "doing gender" is reflected in structural and organizational entities; for example gender segregation in public and private sector bodies and policies ranging from government to education and employment. Body-reflexive practices are evident as the body is both an object of symbolic practice and power as well as a participant in these very processes. Davies asserts that memories can both chart these experiences and also prove "pivotal in the unfolding of time."

The construction of women's bodies, and bodily experiences, as infused with their reproductive potential and history, leads to assumptions about competency. Narratives on working in information technology highlighted the ways in which "doing gender" is evident in women's perceptions of their positioning in dead-end or boring jobs. Ironically, it would seem that women workers perceive male managers as trying to take women's caring roles into account but in ways that result in discriminatory practices.

The potential to shift gendered social relations and challenge discriminatory practices is developed through Davies's study of the morning rounds in several hospital wards. In one, nurses have negotiated across time and space to ensure that all medical and nursing staff met in a room at a round table. Nurses initiated discussion of each patient's case also presenting an appraisal. While doctors felt it took too much time and created an imbalance in the usual ward relations (in which doctors led a walk around beds) proposing that nurses might learn from medical staff, the nurses persisted with the "round table" arrangement. Nurses secured a more egalitarian way of working. Professional and gendered hierarchies illuminate the relationship between professions, organizations and patriarchy. In this example, the challenge by nurses achieved change and they worked to maintain these changes through the controlling the interplay of space—the room and the round table—and time—the appraisal of each patient's case.

Jo Alyson Parker, Michael Crawford, Paul Harris (Eds), Time and Memory, pp. 70–71

Through our collective memory we know there to be changes in body-reflexive practices. In most post industrial societies there are on-going debates on gender, caring and employment. Social scientists and economists have documented the increased participation of women in labour markets. Presumptions have been offered that as a consequence a more egalitarian society would evolve. Yet the reflexive practices of "doing gender" continue to trammel men and women into ideas about time and space that reinforce gender divisions. Notions of "doing gender" are all around us and, while potentially subtle, remain identifiable. The body becomes a centrifugal point for these practices. The research of Davies illustrates how individual and collective memories are dimensions of social change. Further research might consider how memory does and can lead to contestation of gendered social relations.

CHAPTER THREE

CODING OF TEMPORAL ORDER INFORMATION IN SEMANTIC MEMORY

Elke van der Meer, Frank Krüger, Dirk Strauch, Lars Kuchinke

Summary

The fundamental aspects of psychological time (duration, temporal order, perspective) are mediated by memory structures and processes. The present research examined the coding of temporal order information in semantic memory and its retrieval. To explore this question a recognition paradigm and a combination of several different research methods (behavioral, pupillometric, and neuroimaging studies) were used. The temporal orientation and the distance between events were manipulated. Reaction times, error rates and pupillary responses demonstrated that the temporal dimension in mental event representations showed directional and distance properties. In general, these findings supported the theoretical framework proposed by Barsalou (1999). Functional Magnetic Resonance Imaging (fMRI) data suggested that the processing of temporal-order relations depended on prefrontal brain regions. These results are discussed with regards to executive functions of the prefrontal cortex.

Theoretical Background

Time is one of the most central aspects of human life. However, no sense or sense organ is known by which time can be perceived directly. More than a century ago, the physicist Mach started with experimental studies on the sense of time. Based on the results he argued that "time is an abstraction at which we arrive by means of the changes of things" (Mach, 1886, p. 209). Ever since, researchers have been trying to disclose the secret of information processing responsible for the sense of time (cf., Gruber, Wagner, & Block, 2004; Ivry & Spencer, 2004; Kesner, 1998; Lewis & Miall, 2003). However, psychological time appeared to be extremely complex. No simple theory could adequately describe the multifaceted pattern of temporal experiences and behaviors.

At present, cognitive psychologists typically agree that the fundamental aspects of psychological time—namely, duration, temporal order, and perspective (cf., Block, 1990)—are mediated by memory structures and

Jo Alyson Parker, Michael Crawford, Paul Harris (Eds), Time and Memory, pp. 73–86
© *2006 Koninklijke Brill N.V. Printed in the Netherlands.*

processes. This chapter will focus on the aspect of temporal order relations in semantic memory. Following Tulving (1972, p. 386), semantic memory "is a mental thesaurus, organized knowledge a person possesses about words and other verbal symbols, their meanings and referents, about relations among them, and about rules, formulas, and algorithms for the manipulation of these symbols, concepts, and relations." The aim of our experiments was to analyze whether and how temporal order information could be coded and accessed in semantic memory. According to Michon (1993), temporal information cannot be completely separated from the content and the structure of events. That is, events and sequences of events are assumed to hold an intrinsic temporal structure. The theoretical framework adopted for representing highly familiar sequences of real-life events in semantic memory postulates conceptual entities in the form of *scripts* (Schank & Abelson, 1977). Scripts (for example, *going to a doctor*) have a relatively simple syntax that represents a succession of events defined by an initial and a final state. Within a script, individual events are thus primarily defined by their temporal position in the sequence. Other dimensions can be important as well. As we have argued elsewhere (van der Meer, Beyer, Heinze, & Badel, 2002) event sequences, about which we have background knowledge, typically have a causal relation of some sort. Trabasso, van den Broek, and Suh (1989) differentiated motivational, physical, psychological, and enabling relations. Nevertheless, most physicists and philosophers agree that all types of causal relations have one basic presupposition in common: temporal orientability (Earman, 1995, p. 274). The present chapter will focus on this property of *temporal orientability*.

Familiar sequences of real-life events are organized unidirectionally. Thus, temporal-order relations of our experiences are asymmetric relations. The order in which two or more real-life events occur obeys multiple constraints (Grafman, 1995). Some constraints are physical. You cannot drink coffee from a cup unless it is first poured into the cup. Some constraints are cultural. In Germany, people generally look left before passing the street. In Great Britain, however, one has to look right before passing the street without risk. Some constraints are purely individual. Some people drink hot milk in the evening before going to bed. Obviously, representing the chronological order of real-life events in memory and recognizing it later on contributes to human intelligence by supporting correct anticipations and adaptive behavior. Therefore, the chronological order of familiar sequences of real-life events is assumed to be stored in semantic memory. Several findings support this claim. For example, sentence and text comprehension is impaired if there is a mismatch between the chronological order and the reported order of events (cf., Ohtsuka & Brewer, 1992; van der Meer et al., 2002). Münte, Schiltz, and Kutas (1998) measured event-related brain potentials (ERPs) as participants read "before" and "after" sentences. "Before" sentences elicited, within 300 ms, greater negativity in the left-anterior part of the brain than "after" sentences. That is, deviations from chronological order increase cognitive effort.

On the theoretical front, it has been proposed that our mental representations of real-life events reflect the dynamic character of our environment. Dealing with very short-term memory effects, Freyd (1987, 1992) has hypothesized the internal temporal dimension to be like external time, that is, directional and continuous. Barsalou (1999) made the more general assumption that representations of events in semantic memory are not arbitrary and amodal, but are based upon physical experiences. According to Barsalou (1999), comprehension is the mental simulation of events, using perceptual symbols from all information modalities available. Because temporal order relations of our experiences are asymmetric relations, mental event representations should mirror the preferred temporal orientation of events in favor of future time.

The aim of this chapter is to discuss how temporal order is coded and accessed in semantic memory. In particular, we focus on whether the mental representation of temporal order relations between real-life events has directional and distance properties. With this in mind, a combination of research methods—behavioral studies, pupillary responses, and functional neuroimaging methods—was employed to study the coding of temporal order information in event knowledge.

DIRECTIONAL EFFECTS IN MENTAL EVENT REPRESENTATIONS: BEHAVIORAL STUDIES

To examine whether the temporal dimension in mental event representations has directional properties, a recognition paradigm was used (Krüger, Nuthmann, & van der Meer, 2001). Participants were presented probe-target pairs (for example, *bleeding—bandaging*). They had to decide as quickly and accurately as possible whether probe event and target event were related. Relatedness of probe and target was given when they both belonged to the same sequence of events. Probe and target were either related (50%) or unrelated (50%; for example, *bleeding—lubricating*). For related items, the temporal orientation between the probe event and the target event was varied, either being in the chronological order (chronological items; for example,

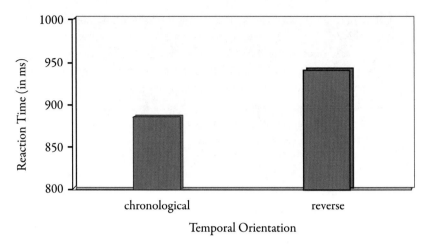

Figure 1: Reaction time in milliseconds required to accept related items depending on temporal orientation.

bleeding—bandaging) or being in reverse order (reverse items; for example, *bleeding—cutting*). The chronological and reverse items did not differ with regard to (a) typicality; (b) duration of the individual events in reality; (c) temporal interval between the two events in reality; (d) number of syllables; and (e) word frequency. We hypothesized: If event knowledge contains temporal information, it should be accessed in the recognition task described above. If this knowledge is directed toward future time, chronological items should be processed faster and more accurately compared to reverse items. The results are displayed in Figure 1 (cf., Krüger, Nuthmann, & van der Meer, 2001).

As hypothesized, chronological items were accessed significantly faster compared to reverse items. The error percentages indicated that there was no speed-accuracy trade-off.

In general, familiar sequences of real-life events show a stronger relatedness for chronologically oriented events compared to temporally reversely oriented events. Taking into account that the strength of relatedness is assumed to determine the speed of activation (Zeelenberg, Shiffrin, & Raijmaker, 1999), the findings provide empirical support for an associative account of access to mental event representations. Similarly, Grafman (1995) assumed that the preferred temporal orientation toward future events is mentally coded by a higher association strength between an event and its successor as compared to an event and its predecessor. Higher association strengths between events are assumed to cause lower activation thresholds, and in consequence, a faster access (cf., Friedman, 1990; Sirigu et al., 1995). In general, our findings correspond with this associative view. They support the hypothesis that the temporal dimension in mental event representations has directional properties (cf., Barsalou, 1999).

DISTANCE EFFECTS IN MENTAL EVENT REPRESENTATIONS: BEHAVIORAL STUDIES

A second question was whether the temporal dimension in mental event representations also has distance properties. The notion of "distance" can either refer to an ordinal measure, namely, the distance between the positions of events on the mental temporal dimension, or to a quantitative measure, namely, the metric distance between events on the dimension. The metric type of representation is highly detailed or fine-grained, while the ordinal type of representation consists of less detailed, coarsely grained categorical representations. We were interested whether people base their judgements of relatedness on the ordinal or metric distance of events. Thereby, we considered two hypotheses: According to Freyd (1987, 1992), who assumed the internal temporal dimension to be continuous, the temporal dimension in mental event representations should be metrically structured. On the other hand, retrieving the rough position of two events on the internal temporal dimension (that is, the region) can already provide critical information on relatedness. Detailed fine-grained metric information might therefore not crucially improve performance in this situation.

Experimental materials were retrieved in controlled association experiments and were tested in detail. Participants were presented individual events (verbs) and had to generate succeeding events. Event pairs with a minimal association frequency of 25% were selected. A second group of participants had to rate each event pair on a 7-point scale with regard to (a) typicality, (b) duration in reality, (c) temporal distance between individual events in reality, (d) abstraction level. Then, participants were asked to group these events to sequences of events and to order the events of each sequence chronologically. To explore, whether the temporal dimension in mental event representations has distance properties, a recognition paradigm like the one described above was used (Strauch, 2002). Fifty percent of the items were related, the remaining 50 % of items were unrelated (for example, *painting—catching*). For related items, the rated metric distance (small versus large) was manipulated as well as the ordinal distance (1 or 2) between probe event and target event within the two metric distance groups. For example, the rating

Table 1: Mean Reaction Times (M, in Milliseconds), Standard Deviations (SD), and Percentages of Errors (E%) Dependent on SOA (Upper Half: 200 ms; Lower Half: 1,000 ms), Metric Distance, and Ordinal Distance

| | SOA = 200 ms | | | | | |
| | Ordinal Distance 1 | | | Ordinal Distance 2 | | |
	M	SD	E%	M	SD	E%
Metric Distance Small	893	217	12.3	917	219	10.0
Metric Distance Large	890	211	6.9	931	236	9.2

| | SOA = 1,000 ms | | | | | |
| | Ordinal Distance 1 | | | Ordinal Distance 2 | | |
	M	SD	E%	M	SD	E%
Metric Distance Small	851	197	3.8	928	223	9.2
Metric Distance Large	852	190	7.7	929	298	11.5

revealed that both *ringing—opening* and *ringing—entering* have a small metric distance, while *painting—exhibiting* and *painting—selling* were rated as having large metric distances. However, *ringing—opening* and *ringing—entering* differ with respect to ordinal distance. *Ringing—opening* was rated as having an ordinal distance of 1 (as opening is a direct consequence of ringing); *ringing—entering* was rated as having an ordinal distance of 2 (as there is another event, namely opening, between *ringing—entering*). Similarly, *painting—exhibiting* has an ordinal distance of 1; *painting—selling* has an ordinal distance of 2.

The related items did not differ with regard to (a) typicality; (b) association frequency; (c) level of abstraction; (d) number of syllables; and (e) word frequency. Apart from the manipulation of relatedness and metric as well as ordinal distance, the time interval (stimulus onset asynchrony; SOA) between the probe and the target was manipulated (200 ms or 1000 ms). The characteristic time constants for automatic activation mechanisms are a mere of 200 ms (Fischler & Goodman, 1987). If, however, the time interval between probe and target is considerably longer, strategic processes that reflect the expectations of the participants may also be involved and modify results of automatic activation (Neely, 1991).

Table 1 illustrates the main results. For reaction times (RTs), the analysis revealed no significant main effect of SOA [F (1,17) = 3.71, $\underline{MSE}$ =

117,557, p = 0.71] and metric distance [F < 1], but a significant main effect of the ordinal distance between probe event and target event [F (1,17) = 10.16, MSE = 4,470, p = .005], with longer RTs for more distant events. All two-way interactions and the three-way interaction were not significant. In addition, the analysis of error rates revealed no significant effects.

These data closely match our second hypothesis. Retrieving the rough position of the two events on the internal temporal dimension provides critical information on relatedness. Participants primarily accessed knowledge of the distance of the position of events on the mental temporal dimension. An SOA of 1,000 ms did not help in speeding up and improving the recognition of temporally related items compared with an SOA of 200 ms. In accordance with Fischler and Goodman (1987) and Neely (1991), it is argued that the SOA interval of 200 ms stands for automatic activation, whereas the longer SOA interval of 1,000 ms stands for controlled, elaborative processing. Therefore, the access of ordinal distance information in event knowledge is assumed to be due to automatic spreading activation mechanisms. In the recognition task, rough accuracy in relatedness judgement could be derived from coarse-grained categorical (position) knowledge. However, taking into account other tasks, for example distance estimation tasks, knowledge of fine-grained metric type might crucially improve performance. Thus, future research will have to determine whether the temporal dimension in mental event representations might contain some event related metric distance information, too.

DIRECTIONAL EFFECTS IN MENTAL EVENT REPRESENTATIONS: PUPILLARY RESPONSES

Whereas recordings of behavioral measures have been useful for mapping out the temporal dimension in general event knowledge, additional psycho-physiological measures could provide valuable supportive evidence. Pupillary responses have been shown to be a reliable and sensitive psychophysiological index of the processing load induced by a task (cf., Beatty & Lucero-Wagoner, 2000; Just & Carpenter, 1993; van der Meer, Friedrich, Nuthmann, Stelzel, & Kuchinke, 2003). Beatty (1982) argued that pupillometric measures reflect within-task, between-task, and between-individual variations in the demands imposed by a variety of tasks (for example, mental arithmetic problems, language processing, memory tasks). He suggested the peak dilation of the pupil to be the most relevant characteristic of the pupillary response: The higher the mental resource consumption, the more the pupil dilates.

Table 2: Means (M), and Standard Deviations (SD) for Pupil Parameters Peak Dilation (PD, z-score) and Latency to Peak (LP, in Milliseconds) Dependent on Temporal Orientation of Items

	Chronological Order		Reverse Order	
	PD	LP	PD	LP
M	**0.19**	438	**0.43**	482
SD	0.29	170	0.31	206

Therefore, we asked ourselves whether the preferred temporal orientation of event-sequences toward future time is also reflected in pupillary responses. To explore this question the paradigm and the experimental materials of the first experiment described above were applied (Krüger, Nuthmann, & van der Meer, 2001). Taking into account the finding that chronological items led to shorter reaction times and smaller error rates compared to reverse items (cf., Figure 1), reverse items should be associated with higher pupillary dilations compared to chronological items. Pupillometry was done with an iView system (Sensomotoric Instruments). Table 2 illustrates the results. Chronological items showed significantly smaller pupillary dilations and shorter latencies to peak compared to reverse items (cf., Krüger, Nuthmann, & van der Meer, 2001).

Thus, the present study demonstrated that pupillometric indices provided a converging measure that confirmed the pattern of results found at the behavioral level. Pupillary response showed to be sensitive to the influences of temporal orientation of events (chronological or reverse) in a recognition task. In summary, both the behavioral measures and the pupillary responses supported the assumption that the temporal dimension in mental event representations has directional properties. These findings closely match Barsalou's (1999) view.

TEMPORAL INFORMATION IN GENERAL EVENT KNOWLEDGE: NEUROIMAGING STUDIES

Human and animal learning studies have provided evidence that the prefrontal cortex is important for temporal sequencing (Kesner, 1998). Moreover, frontal patients have been reported to experience impairments in event and action processing at both the schema and the script level. For example, patients with damaged frontal lobes often experience difficulties in remembering, managing, and planning everyday situations (Shallice & Burgess, 1991) due to difficulties in correctly retrieving chronological information from memory (Allain, Le Gall, Etcharry-Bouyx, Aubin, & Emile, 1999; Humphreys, Forde, & Riddoch, 2001; Sirigu et al., 1995). Some recent neuroimaging studies assessed brain activation during the processing of script events and found strong prefrontal cortical activation (Partiot, Grafman, Sadato, Flitman, & Wild, 1996; Crozier et al., 1999).

Wood and Grafman (2003), on the one hand, argued that highly typical sequences of real-life events are processed and stored in prefrontal cortex. On the other hand, researchers proposed that the prefrontal cortex has a more general executive function. It is assumed to bias, filter, or modulate posterior representations of information in certain contexts or situations (cf., Miller & Cohen, 2001). The aim of the present investigation, therefore, was to examine whether the prefrontal cortex is specifically involved in processing temporal-order relations in general event knowledge in healthy individuals. Using functional magnetic resonance imaging (fMRI), the hypothesis was tested that recognizing the temporal orientation of pairs of events would be associated with higher activation of prefrontal cortex compared to recognizing non-temporal semantic relations (Kuchinke, 2002).

The experiment consisted of three tasks (block-design). The first task (the baseline control condition) was a discrimination task. Participants were presented single words. They were required to decide whether the presented word was a verb or a noun. The second task consisted of items, each containing two script events (for example, *cooking—eating*). The temporal order between these events could either correspond to the chronological order of events (chronological items), or to the reverse order (reverse items). Participants had to decide whether an item described the chronological orientation of events or the reverse temporal orientation of events. The third task consisted of items representing non-temporal semantic relations. Participants had to decide whether an item described a coordination (e.g., *rose—tulip*) or a sub-superconcept relation (for example, *lily—flower*). Given the block design, a relatedness judgment task (cf., Experiments 1–3) would not allow for differentiation in processing temporally related and unrelated items. For that reason, an explicit judgement of temporal order (and of type of non-temporal semantic relation, too) was required in the current experiment. The experimental materials were controlled for relation typicality, number of syllables and word frequency. Functional magnetic resonance images were acquired during task performance with a 1.5 Tesla magnet. Statistical

Table 3: Cortical regions (Brodmann's area) with significantly greater signal
intensity (Subtraction Method)

Cortical Region	Stereotaxic coordinates*			
	X	Y	Z	Voxel
I. Non-Temporal Semantic Relations Minus Baseline				
Left middle frontal gyrus (BA 6)	−34	−6	42	56
Left inferior frontal gyrus (BA 9)	−47	11	29	433
II. Temporal-Order Relations Minus Baseline				
Left precentral gyrus (BA 6)	−36	−11	60	41
Left inferior frontal gyrus (BA 9)	−44	8	27	17723
Right precentral gyrus (BA 6)	23	−14	48	598
Right middle frontal gyrus (BA 46)	48	18	26	44
Right inferior frontal gyrus (BA 47)	46	19	−2	791
Medial frontal gyrus (BA 32)	−2	7	44	2553
III. Temporal-Order Relations Minus Non-Temporal Semantic Relations				
Left superior frontal gyrus (BA 6)	−22	−5	63	33
Left superior frontal gyrus (BA 9)	−45	38	30	46
Left inferior frontal gyrus (BA 9)	−46	7	28	16261
Left precentral gyrus (BA 9)	−39	18	39	53
Left middle frontal gyrus (BA 10)	−45	44	20	47
Right inferior frontal gyrus (BA 45)	46	21	3	121
Right inferior frontal gyrus (BA 45)	52	22	14	267
Right inferior frontal gyrus (BA 47)	49	20	−11	38
Right precentral gyrus (BA 6)	26	−15	49	117
Cingulate gyrus (BA 24)	−3	4	47	3654

* Coordinates of voxels with maximum t-value within the significant region of activation
from Talairach & Tournoux (t = 7,6; p < 0,0001): x (right/left), y (anterior/posterior), and
z (superior/inferior) coordinates in mm

parametric mapping was used to contrast changes in signal intensity during
the temporal-order trials to those during the control trials and the non-
temporal semantic relations trials (subtraction method).

Table 3 illustrates the main results. In general, processing of temporal-
order and non-temporal semantic relations trials was associated with left
mid-dorsolateral prefrontal cortex activation (BA 9) (cf., Table 3, I and II).
Such activation has consistently been found in studies of semantic memory
retrieval for verbal material (Gabrieli, Poldrack, & Desmond, 1998). In
addition, for temporal-order trials prefrontal activation was confined to the

right ventrolateral cortex (BA 47), the posterior lateral cortex (BA 6), and the medial cortex (BA 32) (cf., Table 3, II). These patterns of activation have been observed under a variety of conditions, suggesting that they are related to generic semantic retrieval operations, too (Cabeza & Nyberg, 2000).

Contrasting the temporal-order and the non-temporal semantic relations conditions (cf., Table 3, III) resulted in enhanced activation in the anterior cingulate cortex (BA 24) for temporal-order relations. This finding fits the results of several imaging studies that have shown an involvement of the anterior cingulate cortex (ACC) in conflict monitoring (Botvinick, Braver, Barch, Carter, & Cohen, 2001). It has been suggested that activation in ACC during complex cognitive processing reflects, among others, the initiation of appropriate responses while suppressing inappropriate ones. The cognitive processes involved in the temporal-order discrimination task might be very similar to those seen in Stroop studies: For reverse items, the prepotent mental representation (that is, the chronological orientation of sequences of events) must be inhibited. Instead, the trial-appropriate mental representation (the reverse temporal orientation of sequences of events) and the corresponding response must be initiated. Note, that due to the block design it was not possible to compare the patterns of activation for chronological and reverse items directly in this study.

In addition, contrasting the temporal-order and the non-temporal semantic relations conditions resulted in enhanced activation in prefrontal cortex (BA 6, 9, 45, 47) for temporal-order relations. Because the temporal-order condition led to longer RTs and higher error rates than the non-temporal semantic relations condition, this result pattern was assumed to reflect the higher task difficulty of temporal-order trials compared to non-temporal semantic relations trials (Smith & Jonides, 1997). The higher the task difficulty, the more areas in the brain are activated. This would be consistent with evidence that temporal information-processing generally taxes more cognitive resources than processing of other types of information (Hälbig, von Cramon, Schmid, Gall, & Friederici, 2002).

In consequence, these findings provide evidence for a non-specific involvement of the prefrontal cortex in executive monitoring functions, which seems to be closely related to temporal-order memory performance, too (Shimamura, Janowsky, & Squire, 1990). Future research should examine this issue in more detail by using a variety of temporal and non-temporal tasks.

CONCLUSION

On the methodological front, there are currently methods available that allow us to study our cognitive system while it is performing cognitive tasks. To explore the temporal dimension in mental event representations in more detail we have used three of these methods that are complementary in strength: behavioral methods, pupillary responses, and functional magnetic resonance imaging. Behavioral studies provide global information on the duration and on the difficulty of access to temporal order information in semantic memory. Pupillary responses reflect the resource consumption induced by the task. Pupillary response has a high temporal resolution on the order of milliseconds. fMRI, however, has relatively low temporal resolution on the order of seconds, but provides a fine-grained insight into where processes occur in the brain. fMRI is used to record the amount of neural activity (intensity and areas in the brain) that is generated by the access of temporal order information in semantic memory. The neural activity is assumed to be dependent on the computational demand that the task imposes.

We have combined these methods to study coding and retrieval of temporal order information in semantic memory and have found consistent evidence. In summary, the results demonstrated that the temporal dimension in mental event representations has directional and distance properties. Mental representations of sequences of real-life events were accessed in chronological order. Further, the ordinal position of events on the internal temporal dimension rather than the metric distance appeared to be relevant for recognizing the relatedness of events. The fMRI data were consistent with ideas generally linking the prefrontal cortex with executive functions supporting the acquisition, maintenance, monitoring and organization of information in memory. Whether there are executive functions that are specific to the temporal domain remains an issue for future research.

REFERENCES

Allain, P., D. Le Gall, F. Etcharry-Bouyx, G. Aubin and J. Emile. "Mental representation of knowledge following frontal-lobe lesion: Dissociations on tasks using scripts." *Journal of Clinical and Experimental Neuropsychology* 21:5 (1999): 643–665.

Barsalou, L. W. "Perceptual symbol systems." *Behavioral & Brain Sciences* 22 (1999): 577–660.

Beatty, J. "Task-evoked pupillary responses, processing load, and the structure of processing resources." *Psychological Bulletin* 91 (1982): 276–292.

Beatty, J. and B. Lucero-Wagoner. "The pupillary system." In *Handbook of psychophysiology,*

2nd ed., ed. J. T. Cacioppo, L. G. Tassinary and G. G. Berntson, 142–162. New York: Cambridge University Press, 2000.

Block, R. A. "Models of Psychological Time." In R. Block (Ed.), *Cognitive Models of Psychological Time*, ed. R. Block, 1–35. Hillsdale, NJ: Erlbaum, 1990.

Botvinick, M. M., T. S. Braver, D. M. Barch, C. S. Carter and J. D. Cohen. "Conflict monitoring and cognitive control." *Psychological Review* 108 (2001): 624–652.

Cabeza, R. and L. Nyberg. "Imaging cognition II: An empirical review of 275 PET and fMRI studies." *Journal of Cognitive Neuroscience* 12:1 (2000): 1–47.

Crozier, S., et al. "Distinct prefrontal activations in processing sequence at the sentence and script level: An fMRI study." *Neuropsychologia* 37 (1999): 1469–1476.

Earman, J. "Recent work on time travel." In *Time's arrows today*, ed. S. F. Savitt, 268–310. Cambridge: Cambridge University Press, 1995.

Fischler, I. and G. O. Goodman. (1987). "Latency of associative activation in memory." *Journal of Experimental Psychology: Human Perception and Performance* 4 (1987): 455–470.

Freyd, J. J. "Dynamic mental representations." *Psychological Review* 94 (1987): 427–438.

———. "Dynamic representations guiding adaptive behavior." In *Time, action and cognition*, ed. F. Macar, V. Pouthas, and J. Friedman, 309–323. Dordrecht: Kluwer, 1992.

Friedman, W. J. *About Time*. Cambridge, MA: The MIT Press, 1990.

Gabrieli, J. D. E., R. A. Poldrack and J. E. Desmond. "The role of left prefrontal cortex in language and memory." *Proceedings of the National Academy of Sciences USA* 95 (1998): 906–913.

Grafman, J. (1995). "Similarities and distictions among current models of prefrontal cortical functions." In *Structure and Functions of the Human Prefrontal Cortex*, ed. J. Grafman, K. J. Holyoak, and F. Boller. *Annals of the New York Academy of sciences*, Vol. 769 (1995): 337–368.

Gruber, R. P., L. F. Wagner and R. A. Block. "Relationships between subjective time and information processed (reduction of uncertainty)." In *Time and Uncertainty*, ed. P. A. Harris and M. Crawford, 188–203. Leiden, Boston: Brill, 2004.

Hälbig, T. D., D. Y. von Cramon, U. D. Schmid, C. Gall and A. D. Friederici. "Processing of temporal duration information in working memory after frontodorsal tumour excisions." *Brain and Cognition* 50 (2002): 282–303.

Humphreys, G. W., E. M. E. Forde and M. J. Riddoch. "The planning and execution of everyday actions." In *The handbook of cognitive neuropsychology*, ed. B. Rapp, 565–589. Ann Arbor, MI: Taylor and Francis, 2001.

Ivry, R. B. and R. M. C. Spencer. "The neural representation of time." *Current Opinion in Neurobiology* 14 (2004): 225–232.

Just, M. J. and P. A. Carpenter. "The intensity of thought: Pupillometric indices of sentence processing." *Canadian Journal of Experimental Psychology. Special Issue: Reading and language processing* 47:2 (1993): 310–339.

Kesner, R. P. "Neural mediation of memory for time: Role of the hippocampus and medial prefrontal cortex." *Psychonomic Bulletin and Review* 5:4 (1998): 585–596.

Krüger, F., A. Nuthmann and E. van der Meer. "Pupillometric indices of temporal order representation in semantic memory." *Zeitschrift für Psychologie* 209 (2001): 402–415.

Kuchinke, L. *Zur Dissoziierbarkeit von Exekutiven Funktionen und Zeitfolgemanipulation mittels fMRI* [Dissociation between executive functions and processing of temporal order: An fMRI study]. Master thesis, Humboldt University at Berlin, 2002 (unpublished).

Lewis, P. A. and R. C. Miall. "Distinct systems for automatic and cognitively controlled time measurement: Evidence from neuroimaging." *Current Opinion in Neurobiology* 13 (2003): 250–255.

Mach, E. *Beiträge zur Analyse der Empfindungen (Contributions to the Analysis of Perception)*. Jena: Fischer, 1886.

Michon, J. A. "Concerning the time sense: The seven pillars of time psychology." *Psychologica Belgica* 33:2 (1993): 329–345.

Miller, E. K. and J. D. Cohen. "An integrative theory of prefrontal cortex function." *Annual Reviews of Neuroscience* 24 (2001): 167–202.

Münte, T. F., K. Schiltz and M. Kutas. "When temporal terms belie conceptual order." *Nature* 395 (1998): 71–73.

Neely, J. H. "Semantic priming effects in visual word recognition: A selective review of current findings and theories." In *Basic processes in reading: Visual word recognition*, ed. D. Besner and G. Humphreys, 264–336. Hillsdale, NJ: Erlbaum, 1991.

Ohtsuka, K. and W. F. Brewer. "Discourse organization in the comprehension of temporal order in narrative texts." *Discourse Processes* 15 (1992): 317–336.

Partiot, A., J. Grafman, N. Sadato, S. Flitman and K. Wild. (1996). "Brain activation during script event processing." *Neuroreport* 7 (1996): 761–766.

Schank, R. C. and R. P. Abelson. *Scripts, plans, goals, and understanding*. Hillsdale, NJ: Erlbaum, 1977.

Shallice, T. and P. W. Burgess. "Deficits in strategy application following frontal lobe damage in man." *Brain* 114 (1991): 727–741.

Shimamura, A. P., J. S. Janowsky and L. R. Squire. "Memory for the temporal order of events in patients with frontal lobe lesions and amnestic patients." *Neuropsychologia* 28 (1990): 803–813.

Sirigu, A. et al. "Planning and Script Analysis following Prefrontal Lobe Lesions." In *Structure and Functions of the Human Prefrontal Cortex*, ed. J. Grafman, K. J. Holyoak and F. Boller. *Annals of the New York Academy of Sciences*, Vol. 769 (1995): 277–288.

Smith, E. E. and J. Jonides. "Working memory: A view from neuroimaging." *Cognitive Psychology* 33 (1997): 5–42.

Strauch, D. *Repräsentation zeitlicher Distanzen in typischen Ereignissequenzen* [Representation of temporal distances in highly familiar sequences of events]. PhD Thesis, Humboldt University at Berlin, 2002 (unpublished).

Trabasso, T., P. W. van den Broek and S. Y. Suh. "Logical necessity and transitivity of causal relations in stories. *Discourse Processes* 12 (1989): 1–25.

Tulving, E. "Episodic and semantic memory." In *Organisation of memory*, ed. E. Tulving & W. Donaldson. London: Academic Press, 1972.

van der Meer, E., R. Beyer, B. Heinze and I. Badel. "Temporal Order in Language Comprehension." *Journal of Experimental Psychology: Learning, Memory, & Cognition* 28:4 (2002): 770–779.

van der Meer, E., M. Friedrich, A. Nuthmann, C. Stelzel and L. Kuchinke. "Picture-Word-Matching: Flexibility in Conceptual Memory and Pupillary Responses." *Psychophysiology* 40 (2003): 904–913.

Wood, J. N. and J. Grafman. "Human prefrontal cortex: Processing and representational perspectives." *Nature Reviews/Neuroscience* 4 (2003): 139–147.

Zeelenberg, R., R. M. Shiffrin and R. W. Raijmaker. "Priming in free association tasks as a function of association directionality." *Memory & Cognition* 27 (1999): 956–961.

CHAPTER FOUR

TELLING THE TIME OF MEMORY LOSS:
NARRATIVE AND DEMENTIA

MARLENE P. SOULSBY

SUMMARY

Narratives of dementia evoke a human experience typically associated with loss—the loss of self, the loss of a hold on reality, the loss of the past. This paper explores the more complex layers of self, reality and time that we encounter through the reading of several different narratives: biographies by Eleanor Cooney and Sue Miller, an autobiography by Thomas DeBaggio, the novel *Man in the Holocene* by Max Frisch, and group stories created through the Time Slips project. These narratives tell the time of realities as they are experienced by those afflicted with memory loss. They also tell about the frustrations and the possibilities that can emerge when the experiences of different temporal realities meet. These narratives not only give us insights into memory loss and the experience of time but also create a complexity of time experiences through the very process of reading.

With the use of memory, we construct narratives that lift the events and experiences of life into a meaningful and coherent order. Our narratives give shape to time and define a continuity of self. For those who are losing the ability to remember due to Alzheimer's Disease or other dementias, however, narrative becomes more difficult if not impossible to construct. The sense of a continuous self appears to unravel or dissolve, and attempts to communicate deteriorate. Yet the impulse to tell is a vital way to confirm, or affirm, that the self still exists in the midst of this experience. It is also a way to connect to the world outside the self.

Telling the story of memory loss involves telling the time of an experiential world. In this paper I examine several examples of narratives that attempt to convey the temporal experience of those whose memory is failing. They include biographical and autobiographical perspectives, a work of fiction, and samples of stories created in group settings by people with mild to moderate dementia. In each case we encounter realities shaped by temporalities that are not linear or continuous. We enter worlds that are difficult to traverse because of their temporal structure. The encounter with these realities and their temporalities takes us on a journey through many layers of the human story. The attempt to understand, to make connections, or to bridge gaps between realities is indicative of the human attempt to touch one another.

Jo Alyson Parker, Michael Crawford, Paul Harris (Eds), Time and Memory, pp. 87–97
© *2006 Koninklijke Brill N.V. Printed in the Netherlands.*

The ability to remember and to engage in the process of narration defines who we are in time and reflects on our very humanity. According to James Olney in *Memory and Narrative*, memory provides "the self an evolving stability in the face of the Heraclitean flux in all things." (343–344). Ricoeur has written extensively of the correlation between the narrative act and the temporality of human experience. "Time," he writes, "becomes human to the extent that it is articulated through a narrative mode" (52).

Time in a human sense is associated with linearity—we learn from the past, integrate the present, and plan for the future. In J. T. Fraser's hierarchical theory of time, the human mind brings a consciousness of direction and order to life, without which we would have no civilization, no morality, no values, and certainly, no pension plans. Time is an arrow with a clear head and tail, moving from past to future. Alzheimer's, Fraser writes, steals from us this "sense of human time" (179). In a very real sense, it also steals from us the world which relies on mind and memory to exist.

Narratives of memory loss tell the time of other worlds or other realities, where the arrow of time disintegrates and causality gives way to unconnected succession. These realities are nested in human experience, below the surface, a frightening possibility for each one of us. Many works, including Eleanor Cooney's *Death in Slow Motion* and Sue Miller's *The Story of my Father* narrate the experience of Alzheimer's Disease from the perspective and memory of a caregiver. These stories tell as much about forgetting and shared memories as they do of the tension resulting from confrontations of vastly different temporal worlds: on one hand the linear, arrow-of-time perspective of the caregiver and on the other, the less structured temporalities where the afflicted person dwells.

Indeed, these accounts are characterized by the attempt of the caregiver to remember and recreate a past life that preserves the person's identity. But they also speak of the great frustration, anger and guilt that arise from the inability to establish communicative links to the person in the present. In the following passage, Eleanor Cooney's frustration is clearly evident as she conveys the repetitive, immediate and purely successive temporality of her mother's world:

> The front door to the new house bursts open twenty, thirty, forty times a day, every day: my mother, searching for her basket (which she carries instead of a purse, always has), asking if she's having dinner with us, showing us a grocery list, asking if there's vodka in the house, telling us she's going to take a nap, asking the name of the doctor she saw last time she was here, showing us a grocery list, searching for her basket, telling us she wishes she knew when she was going to die, asking if there's vodka in the house, showing us a grocery list,

asking if there's a hospital in the next town, telling us she's going to take a nap, asking the name of the doctor she saw last time she was here, asking if she's having dinner with us, telling us she wishes she knew when she was going to die, asking if there's a hospital in the next town, asking if there's vodka in the house, searching for her basket. (55)

In *The Story of my Father*, Sue Miller recounts an episode with her father when he insisted that he needed to get in touch with "Sue Miller" even though she was sitting in front of him. Their short exchange reveals that Miller's sense of linear, ordered, noetic time confronts a temporality with no direction, an abiding present of repetition and pure succession that blurs with the past. Upon reflection, Miller speculates that her father's memory of her and the person he saw before him had somehow separated from one another. There were multiple Sues in his mind based actually on a "deeper reality"—that Alzheimer's had changed her too and made her into another person. For him the memories of his daughter at different times coexisted in the present (137).

Miller's biography, however, is clearly a narrative of her own memory—it is *her* story of associations and feelings, *her* attempt to preserve the identity of her father and in a broader way to make sense of life. The time that she tells is *her* time based on her memory. For example, the recollection of a little girl watching her father work merges with a memory in the more distant past when she herself was a girl going to her father's study (53). This blending of past memories is narrated in the present tense. They exist now, simultaneously, not linearly, in her mind. They *are* memory in the Augustinian sense of "the present time of things past" (11.20, 273). The difference between Sue and her father is that she is not trapped in a temporal world of simultaneities. She can pull back and order this experience through narrative.

Writers such as Diana Friel McGowin in her book *Living in the Labyrinth* and Morris Friedell in his internet journal record their own progression of memory loss through a constantly evolving story of past memories and present reality. These writers juxtapose an awareness of the "present time of things past" with an awareness of the present of things slowly slipping away.

Thomas DeBaggio in his autobiography, *Losing my Mind*, employs several narrative lines to reveal who he *is* and *was*. But throughout the text is the unavoidable realization that "Alzheimer's works to destroy the present and the past" (xi). One narrative line traces long-term memories and projects into the future. It reflects on and analyzes experience. It is characterized by a noetic arrow of time and the ability to move easily back and forth along this line, reconstructing a life that changed dramatically with the diagnosis of Alzheimer's. Thus in one paragraph he speculates, "I have been thinking of my obituary all day, wondering what will be written, if anything." The next paragraph turns back to the past, "Going to school was important, I was told, but school made me anxious" (62).

A second narrative line frequently interrupts this account with brief, often metaphorical statements such as, "*Once in a while I hear wind whistling through my brain*" (62), or "*I am suspended in time, hanging by a rotting thread of memory*" (59), or "*I am being gobbled up in time*" (20). The statements are not logically connected to one another and are set off in the text by italics. DeBaggio says that this line "intersects the first" with "a mind-clogged, uncertain present" (xi), a present charged with the immediacy of emotion.

A third line, identified by text indented five spaces from the rest, is not really narrative. It consists of selections and excerpts from other texts that provide information about Alzheimer's Disease, for example: "Due to the complex changes occurring in their brain, patients with Alzheimer's may see or hear things that have no basis in reality" (117). Many of these statements are taken verbatim from publications of agencies such as the National Institutes of Aging and Health and The Alzheimer's Disease Research Center. These objective and factual items suspend time altogether by objectifying the personal experience and stepping outside of the narrative.

Thus DeBaggio interweaves three time-lines in his story: one that depicts the continuum of his life from the past through the present and towards the future; one that taps into the emotion of the present; and one that is atemporal in both content and tone. The juxtaposition and interweaving of these times-lines generates the temporality of the work as a whole; they are simultaneous aspects of the writer's increasing memory loss while he is still able to reflect on the experience and articulate it through language. Eventually, however, he projects that his self-narrated story will come to an end. As he says, "Alzheimer's sends you back to an elemental world before time [...] I sense reality slipping away, and words become slippery sand" (89).

Arthur Frank in his book *The Wounded Storyteller* writes about the storyteller who can no longer narrate experience. As memory fails, and with it the ability to determine sequence and causality, experience becomes more chaotic. Frank says, "In telling the events of one's life, events are mediated by the telling. But in the lived chaos there is no mediation, only immediacy" (98). Thus, narratives that tell the story of memory loss must deal in some way with the time of an abiding present. Authors like DeBaggio attempt to give narrative order to a life that is progressing into greater disorder, like narrating the edge of an ever-widening hole or wound. Those who are living the chaos, in the hole so to speak, cannot gain the reflection, distance, or articulation of language necessary to form narrative; therefore their stories are told through the voice of someone else, as we have seen in the narratives of Cooney and Miller.

Works of fiction present another possibility for narrating the experience of memory loss and giving voice to "lived chaos." A fictionalized account can create a world, a self, and a sense of time without stigmatizing the author. I will give particular attention to one such work, *Man in the Holocene* by Max Frisch. This novel records a week in the life of a 74-year-old man with failing memory, a week of relentless rain and thunderstorms. The only access road to the remote Alpine village where he lives is blocked; telephones are not working, and the power fails repeatedly.

One day follows another. Not much seems to happen. Geiser, the protagonist, isolated in his home, "has time to spare" (3). Whatever he does, time passes. Whether he remembers what occupies his mind from one moment to the next, time passes.

The narrator sustains this sense of passage through a perspective situated in the moving *now* of the present moment. He observes and records whatever comes to his awareness without making causal or logical connections:

> The hot plate is turned off.

> Cats always fall on their feet, but in spite of that she is now yowling outside the front door; perhaps Geiser said: Get out—but after that not a word in the house.

> Outside it is raining.

> There is no ladder in the house.

> It is true the gray cobwebs on the ceiling have been there a long time [...]. (59)

At times it is difficult to distinguish narrator from protagonist, creating a uniquely dual perspective as though we were watching from the *inside*. Michael Butler points out that this technique enables the reader both to share in Geiser's "experience of progressive disorientation and simultaneously to observe it almost clinically" (574). It also allows for an intimate interweaving of objective and subjective time. (See also Donahue, Dahms, Probst.) In the following passage, the narrative tracks the objective, inexorable ticking of the clock, juxtaposing it to Geiser's perception that time is standing still:

To keep looking at one's wristwatch, just in order to convince oneself that time is passing is absurd. Time has never yet stood still just because a person is bored and stands at the window not knowing what he is thinking. The last time Geiser looked at his watch it was six o'clock—or more exactly, three minutes to six.

And now?

—one minute to six.

There is always something one can do.

So one would imagine.

The portrait of Elsbeth that he recently took from the wall and put in the hallway is out of place there. Where should it go? This indicated that, since last looking at his watch, Geiser has been in the hallway; otherwise the portrait would not be in his hands, and he is now standing in the bedroom.

Presumably his watch has stopped.

. . .

When Geiser goes back to the window to convince himself, by watching the slowly gliding raindrops, that time is not standing still—and in the whole of history it has never done that!—and when he cannot resist looking at this again, it reads seven minutes past six. (66–67)

The narrator has recorded ten minutes in a text which in its entirety numbers 279 words. It takes a reader much less time to read those words. During this span, Geiser's thought leaps from one impression or thought to another, leaving gaps in the causal chain. Without a memory of the past continuum, he looks for evidence of change in the external world (the watch as well as the rain) and uses clues in the present (his position in the hallway, the portrait in his hands) to reconstruct and fill in a bridge of linked changes between moments. The contrast between the passage of clock time and Geiser's inner experience, as well as between the time that is narrated (ten minutes) and the time that it takes a reader to read, generates the impression that time has stopped. We, as readers, know the minutes are ticking away but there is little or no content to fill them up—no description of action, no flow to the thoughts, no recollection of time lost—only short paragraphs, fragments of sentences and white space in between. Time feels empty.

In another passage, the narrator alternates a record of chronological clock-time with sensory perceptions of the rain: the one is a human construct of measurement like a digital jump from number to number. The time periods they delineate are unconnected and irregular in duration. The other is a record of human impressions linked, not causally, but rather through the common element of the rain. The one suggests the arrow of time as well as its fragmentation; the other evokes an abiding present without direction. The direct juxtaposition of these two temporalities generates the time experience of emptiness which Frisch associates with memory loss.

> Sunday:
> 10:00 A.M.
> Rain as cobwebs over the grounds.
> 10:40 A.M.
> Rain as pearls on the windowpane.
> 11:30 A.M.
> Rain as silence; not a bird twittering, not a dog barking in the village, the noiseless splashing in the puddles, raindrops gliding slowly down the wires.
> 11:50 A.M.
> No rain. (41)

Because of Geiser's inability to sustain temporal continuity, his focus narrows with greater frequency to the present. Yet he is also *aware* that the focus is narrowing and with it the store of knowledge that he has accumulated in a life-time. He, therefore, conceives of ways to compensate for the gaps. By writing notes or cutting out sections of encyclopedias and other informational texts and tacking them to the walls, he surrounds himself with a construction of important knowledge that he does not want to forget. But his attempt to build an enduring edifice of facts mirrors his attempt to build a pagoda out of crispbread—both efforts are doomed to failure. The pagoda typically begins to wobble with the fourth floor, and his spatial, timeless structure of a memory turned inside out is just as unstable. At the end when his daughter arrives to look in on him, she opens the shutters and a puff of air blows the slips of paper to the floor in a "confused heap that makes no sense" (106).

After returning home from a mountain hike, Geiser presumably suffers a stroke which impairs his memory even more. He wakes up lying on the floor with no recollection of how he got there. Also, he cannot remember the names of three grandchildren. But he experiences a memory from 50 years ago that is vivid and real *now*. The narrative style used to describe this memory is more continuous and linear than the shifting, fragmentary style characteristic of the rest of the narrative. At first the past merges in the telling with the flowing present so that it is not immediately apparent that this is a memory; then the past recollection alternates with present thoughts. The lived memory, immediate and powerful, is experienced simultaneously with the confusion of the *now*. Linear sequential time alternates with the associative randomness of an abiding present.

In the experience of this memory, Geiser recalls a climbing expedition up the Matterhorn with his brother Klaus: the triumphant ascent, the treacherous descent, standing alone against the face of the glacier, unable to move, waiting for a rope from his brother to save his life. "That remains in his memory. Klaus was a good brother" (104). The reality and emotional impact of this long-term memory and the sense of an enduring human bond between brothers combines with other experiences: the intensifying frustration of trying to remember the names of his grandchildren and the inability of figuring out the sequence of events leading to the appearance of his daughter in his house.

The daughter wants to know what events have led to this moment. She asks why—"Why the closed shutters, why all these papers on the wall, why his hat on his head?" (105) He wants to know about *now*: "Why does she talk to him as if he were a child?" The questions speak of the painful divide between two experiential worlds, two temporal realities and the frustrating inability to bridge the gap. She wants to reconstruct a linear causal timeline in order to understand what has happened in the past; he wants her to acknowledge who he is. His identity and his humanity are grounded not only in the ability of recollection but also in the fleeting experiences of life as they occur, in the unpredictable human encounters and relationships with others, and in the triumphs, large and small, that are part of his own unique existence. Yet without communication, he is essentially alone.

Narratives that tell the story of memory loss typically tell of the disintegration of time's arrow and the implications of that disintegration for the sense of self and community. If the "real self" is the one who *was* able to function in linear time but can no longer do so, then that self ceases to exist, dissolves, when that temporal world dissolves. Geiser's story, however, reminds us that human time encompasses multiple levels of experience, and the self does not forfeit its reality or its humanity because it is trapped in a lower temporal order. It just needs a lifeline to enable communication and connection to someone else.

Other types of stories and story-telling situations direct our attention to the communal needs of the self and the possibility of making connections between people living in different time realities. Anne Davis Basting has devised a story-telling project called Time Slips that reinforces the sense of self as more than linear time and memory. She explains that "one's personal control of memory is just one of several components of identity" and that persons are defined by "more than their ability to link past, present, and future" (135). Time is the great "divider." There are those who can master linear time and those who cannot; those who know the date and the year and who can draw numbers on a clock and those who cannot (136). The difference between the two experiences of time is indeed the difference between worlds. The Time Slips project avoids the futility of trying to impose the temporality of one world on another. Instead, it establishes a "shared present" where people from both temporal worlds can communicate and interact through creative give and take.

The story telling situation is as follows: groups of people with memory loss meet with facilitators for an hour or so and tell a story together based on a picture. The facilitators assist with the process and write down the story as it unfolds, but are trained not to interfere or to impose their own temporality, "sense of linear logic," or idea of what a story should be on the group (136). Stories, just as human experience itself, can take many shapes and tell of many worlds. These stories are the product of the combined efforts of all persons in the group; and time can flow as it will. Basting describes the time of the story-telling situation: "Staff, volunteers, and storytellers were under the spell of the present moment and our ability to communicate in it, occasionally in complete nonsense. For all of us, story-time was a place where fragments of memory could launch us into new places that we built together rather than lock us into labels of loss or badges of control" (142).

The stories are inspired by a picture—usually something funny or provocative that is not so realistic that people fear they have forgotten its "real" story. The members of the group "tell" the story as it occurs to them spontaneously or in response to questions posed by the facilitators: for example, "What should we call her?" or "Where are they?" The narrative is recorded as it evolves from the group without forcing it to make sense or to assume a certain direction with a beginning, middle and end. If a person in the group cannot think of the words but uses sound or gesture—that too becomes part of the story. The stories, therefore tell the time of the moment and, in doing so, reflect the temporality of the group's experience.

I have facilitated several story-telling sessions in a residential health care facility and in an adult day-care center. Participants were identified by facility administrators as having mild to moderate dementia. For comparison purposes, I also collected stories told by elderly without memory loss and by college students. During the sessions, I worked in the questions: "What is happening now? What happened before? What might happen after this?"

The stories told by groups with memory loss tell of present experience and time that is discontinuous. In fact, my questions about the past and the future were typically followed by silence or disinterest. If there was reference to the past it might be a trace of remembered experience from a person's own past: " We used to have a light like that." Or it might be a general impression of the future: "Somebody could get hurt doing that." The stories are driven for the most part by associative leaps, simultaneous threads, ruptures, repetition.

The stories told by elderly without memory loss incorporated more personal memories from the past and would often include narrative diversions tangential to the picture. College students were more concerned that the story had a beginning, middle and end reflective of linear thinking and an arrow of time. For example, the following sentences establish a definitive and complex time line: "They're returning home from a class reunion and stopped along the way in Las Vegas to get married by an Elvis impersonator. Because he listened to her, he turned down the wrong road and then lost his toupee about ten miles ago." When telling this story, students paid particular attention to the plot, and they frequently worked together through several revisions to get it "just right"—that is, causally logical. A story told by a group with memory loss about the same picture reveals no such concern with establishing a thread. Instead it relies on the power of emotion felt *now*, as is evident in the repetitive language and the merging together of storyteller and story:

> The car must have broken down.
> The man is holding his eyes and saying, "Oh no!"
> He's got his hands full and he's the driver of the car.
> He's saying, "Oh no," because the car broke down.
> To no avail, he's saying, "Oh no."
> …
> He has a headache.
> Which way are *we* going, "Oh no."
> You can see the dirt and sand.

The narratives that were created as part of Time Slips reflect the time of human experience without the imposition of narrative order. Arthur Frank might refer to them as chaos stories told by "wounded storytellers." Indeed, they do tell of life as it is experienced, "without sequence or discernable causality" (97). They often tell of repetition, unrelated succession, and even silence. But unlike chaos stories, these are not threatening or anxiety-provoking. They do not tell "how easily any of us could be sucked under" (97). Instead, they tell that a story may emerge from many "worlds" and that there are ways to bridge the temporal rifts, thus acknowledging the person, the connection to community, and the story.

Many of the biographies, autobiographies, diaries, e-journals, fictional texts and group narratives of memory loss that are available today bring us in touch with realities that are frightening yet strangely familiar. We fear loss— the loss of threads that hold the world together and the loss of timelines that give order to experience; we fear the social and spiritual isolation that accompanies such loss. Yet the narratives of loss take us on journeys into depths and times of our own selves. As J. T. Fraser's hierarchical model of time illustrates, human beings encompass many levels of temporal experience; even though we identify the experience of noetic, linear time as the common measure of our identity and our humanity. For those who dwell in a less ordered temporality, it is difficult if not impossible to function effectively in the everyday world, but the self and the humanity of the person are not lost. These narratives attempt to give a voice to the experiencing self and to tell the time of that experience. In this way, they not only promote understanding of an affliction but also help us to recognize and make contact with the many dimensions of the human story.

References

Augustine of Hippo. *The Confessions of St. Augustine*. Translated by Rex Warner. New York: Signet, 1963.

Basting, Anne Davis. " 'It's 1924, and Somewhere in Texas, Two Nuns are Driving a Backwards Volkswagen': Storytelling with People with Dementia." In *Aging and the Meaning of Time*, eds. McFadden, Susan H. and Robert C. Atchley, 131–149. New York: Springer, 2001.

Butler, Michael. "Max Frisch's *Man in the Holocene*: An Interpretation." *World Literature Today: A Literary Quarterly of the University of Oklahoma* 60:4 (1986 Autumn): 574–580.

Cooney, Eleanor. *Death in Slow Motion: My Mother's Descent into Alzheimer's*. New York: HarperCollins, 2003.

Dahms, Erna. *Zeit und Zeiterlebnis in den Werken Max Frischs*. Berlin: de Gruyter, 1976.

DeBaggio, Thomas. *Losing My Mind: An Intimate Look at Life with Alzheimer's*. New York: The Free Press, 2003.

Donahue, Neil H.: "Age, Beauty and Apocalypse: Yasunari Kawabata's *The Sound of the Mountain* and Max Frisch's *Der Mensch erscheint im Holozän*." *Arcadia: Zeitschrift fur Vergleichende Literaturwissenschaft* 28:3 (1993): 291–306.

Frank, Arthur W. *The Wounded Storyteller: Body, Illness, and Ethics*. Chicago: University of Chicago Press, 1995.

Fraser, J. T. *Time the Familiar Stranger*. Amherst: University of Massachusetts Press, 1987.

Friedell, Morris. "Home Page." <http://members.aol.com/MorrisFF> (Accessed 12 May 2005).

Frisch, Max. *Man in the Holocene*. Translated by Geoffrey Skelton. Orlando: Harcourt, 1980.

McGowin, Diana Friel. *Living in the Labyrinth: A Personal Journey through the Maze of Alzheimer's*. New York: Delacorte, 1993.

Miller, Sue. *The Story of My Father*. New York: Knopf, 2003.

Olney, James. *Memory & Narrative: The Weave of Life-Writing*. Chicago: University of Chicago Press, 1998.

Probst, Gerhard. "The Old Man and the Rain: *Man in the Holocene*." In *Perspectives on Max Frisch*, ed. Gerhard F. Probst and Jay F. Bodine. Lexington: University Press of Kentucky, 1982.

Ricoeur, Paul. *Time and Narrative*, 3 vols. Translated by Kathleen McLaughlin and David Pellauer. Chicago: University of Chicago Press, 1985.

RESPONSE

Alison Phinney

Through examples drawn from auto/biography, works of fiction, and narrative performance, this essay offers a nuanced portrayal of how dementia shatters assumptions about narrative and the temporal world. As memory fails, time seems to crumble into a series of disconnected moments and narrative breaks down. One's story comes to be told by others, or becomes a story that cannot be told at all.[1]

As the author suggests, this is in large part because the person with dementia has no reflective distance from which to tell his story. By existing in the moment, he is dwelling in a liminal space where he cannot recall the past, nor project toward the future. Being trapped in the immediacy of lived time, his story has no sense of historicity.

And yet narratives of memory loss are becoming increasingly common, especially in the context of earlier diagnosis and treatment of dementia. Dementia Advocacy and Support Network International is an association of people with dementia who are "making their voices heard," using the internet to convey their stories to the world.[2] And there is an expanding literature in the health and social sciences exploring first person accounts of what it means to live with dementia.

This essay has shown how in these narratives we find new possibilities for understanding what it means to suffer memory loss and ultimately what it means to be human. I want to briefly consider one of these in particular. In the Time Slips project, we saw how, even as words fail, a shared narrative can emerge from "sounds and gestures." This idea that narratives can be conveyed through the body offers important possibilities not only for telling the "time of the moment," as the author suggests, but also for telling the time of the past. Our bodies are deeply historical. In my own research of persons with Alzheimer's disease, I have watched a retired music professor lead a choir

---

[1] Alison Phinney, "Fluctuating Awareness and the Breakdown of the Illness Narrative in Dementia." *Dementia: an International Journal of Social Research and Practice* 1, no. 4 (2002): 329–44.
[2] Dementia Advocacy and Support Network International www.dasninternational.org (accessed September 30, 2005).

Jo Alyson Parker, Michael Crawford, Paul Harris (Eds), Time and Memory, pp. 98–99
© 2006 Koninklijke Brill N.V. Printed in the Netherlands.

in a moving performance of the national anthem. "My arms to this day, can just conduct," he said. I have listened to a man explain how he is "in the old habit of being a teacher," watching over the person he has hired to build a fence in his backyard, offering words of encouragement and guidance for the work he can no longer do himself.[3] These "old habits" of the lived body may constitute something of a bridge across the "temporal rifts" created out of memory loss, thus contributing to a sustained sense of self.

To say that we understand our lives through narrative is to say that we are each in the process of living out a story that runs from our birth to our death, and we are accountable for providing an intelligible account of the events and experiences of that life to ourselves and to others.[4] Dementia forces us to recognize that intelligibility resides not only in language but in the habits and practices of the lived body. Positioning narrative as an embodied activity will enrich our ability to understand the experience of temporality and memory loss in dementia.

---

[3] Alison Phinney and Penny Brown, "Embodiment and the dialogical self in dementia". Symposium paper, Canadian Association of Gerontology Annual Scientific and Educational Meeting, Victoria, BC, Canada, (October 2004).

[4] Alistair MacIntyre, After Virtue (Notre Dame, IN: University of Notre Dame Press 1981).

CHAPTER FIVE

GEORGES PEREC'S "TIME BOMBS": ABOUT *LIEUX*

MARIE-PASCALE HUGLO

SUMMARY

In considering Georges Perec's unfinished project *Lieux*, I reveal the relationship between archives, memory and memory realms. First, I expose the frictions between time and memory in a project which aimed to collect and to preserve present traces for future use. From there, I show that even if *Lieux* is not homogeneous, it clearly refers to a lack of memory related to Perec's early childhood. I try to see how such a lack *inscribes* itself into the project. But *Lieux* is not solely autobiographical. It highlights the instability of urban places and can thus be connected to the disappearance of a memory milieu rooted in tradition and space. Taking Perec's *Lieux* as memory realms, I re-examine the strong opposition between records and memory that governs the well-known reflections of Pierre Nora on *Lieux de mémoire*.

Georges Perec's unfinished project called "Lieux" is an experiment that was intended to last for 12 years.[1] Perec described this project many times between 1969 and 1981,[2] and he saw it as a device enabling him to measure the passing of time: it incorporated the time of writing—which usually does not count—into his descriptions of 12 given Parisian places related to his past experience. Perec planned to produce two texts a year on each of the 12 places, one labelled "Real" written *in situ* and another labelled "Memory" written anywhere, at another time of the year, according to a schedule established beforehand. Each text was enclosed in an envelope, which was then sealed. Perec was not supposed to unseal the envelopes before 1981. He compared *Lieux* to Time Bombs. My purpose here is to examine more precisely what his project consisted of. I would like to see if, rather than just "incorporating time," Perec's bombs *framed* time in order to produce a sense of ageing, which was apparently not self-evident. Of course, only time would have told what would have come out of this experiment, but many

---

[1] This paper is part of a research project on the Poetics of archives in contemporary literature, funded by FQRSC (*Fonds québécois de la recherche sur la société et la culture*).
[2] "Lettre au mathématicien Chakravarti"; "lettre à Maurice Nadeau"; "Lieux, un projet (confié à Gérard Macé)"; "Les Lieux (notes sur un travail en cours)"; "Entretien avec Georges Perec et Bernard Queysanne". Perec mentions *Lieux* many times in his *Entretiens et conférences*.

Jo Alyson Parker, Michael Crawford, Paul Harris (Eds), Time and Memory, pp. 101–114
© *2006 Koninklijke Brill N.V. Printed in the Netherlands.*

publications directly[3] or indirectly[4] related to *Lieux* came out before and after 1982 (the year of Perec's death), and they enable us to better understand what was at stake. I used those publications as well as Perec's descriptions of his project to study *Lieux*.

Even though Perec programmed himself to "capture" and to record time as objectively as possible, his project was haunted by deeply buried memories. It leads us to see differently this rigorous—even maniac—program. Moreover, one can see that Perec's project implies not only autobiographical but also sociological concerns. In fact, individual and collective memories go hand in hand. Finally, the fact that Perec rooted his experiment in carefully chosen Parisian places enables us to relate *Lieux* to Pierre Nora's memory realms and to put into question Nora's clear-cut division between records and memory. Indeed, in *Lieux* as well as in Perec's *œuvre*, memory and records as so closely intertwined that they invite us to ponder on their supposed opposition. Records may "freeze" time as well as memories, but they also shape our memories, as it is the case with photographic memory in *Lieux*. They even may transform themselves into memory realms. Eventually, *Lieux* enables us to gain an insight into the relationships between time, memory and place, but also between memory and recording media which, whether we like it or not, are part of our contemporary landscape.

I shall isolate three features of Perec's Time Bombs—planning, recording, differing—in order to highlight the complex relationship between time and memory in *Lieux*.

---

[3] Perec published five Real places : "Guettées"; "Vues d'Italie"; "La rue Vilin"; "Allées et venues rue de l'Assomption"; "Stations Mabillon". Philippe Lejeune posthumously published a Memory place: "Vilin souvenir". Andrew Leak translated four "Reals": "Scene in Italy"; "Glances at Gaîté"; "Comings and Goings rue de l'Assomption"; "Stances on Mabillon" into English (*AA files*, nos 45–46).

[4] Perec insisted on the fact that, after he had given up *Lieux* as a whole, he wanted to develop each *Lieu* and to produce different types of publications: films, an Album (photographs and poems), a radio program("Georges Perec, Le Paris d'un joueur", *Entretiens et Conférences II*, 130). Perec refers to (respectively): *Les Lieux d'une fugue* (1976); *La Clôture* (1976); *Tentative de description de choses vues au carrefour Mabillon le 19 mai 1978*. He also relates the film *Un homme qui dort* (1974) to *Lieux* ("Entretien avec Georges Perec et Bernard Queysanne", *Entretiens et conférences I*, 161). Of course, *Tentative d'épuisement d'un lieu parisien* extends from *Lieux* as does *W ou le souvenir d'enfance* (see Philippe Lejeune, "La lettre hébraïque. Un premier souvenir en sept versions", *La Mémoire et l'oblique*, pp. 210–231). In fact, many publications have to do with *Lieux: Je me souviens* brings forth collective memories and transient manifestations. The same goes for *La Vie mode d'emploi* (Perec began to write it after he had given up *Lieux*) and *Récits d'Ellis Island*. As we can see, *Lieux* is not a failure. It is a matrix and figures among Perec numerous unfinished projects, which are part, *per se*, of his *œuvre*.

1. Planning

Lieux is a program which involves the future. It is a time projection which plans the next twelve years in advance and somehow shapes them. The future needs to be scheduled, it is *a plan*. The program is not conceived according to an aim that one should try to achieve by following a "line"; rather, it displays itself, in the long run, according to a formal constraint that physically bound Perec, who had to visit one Parisian place each month of the year. "Reals" and "Memories" are tabulated; they are distributed according to a permutation rule. Perec asked the mathematician Chakravarti to send him a model of a "carré latin d'ordre 12" (orthogonal latin square), and he drew two tables, which took the form of a square:

> [It] distributed the numbers 1 to 12 around a 12 × 12 grid in such a way that no single pair of numbers recurred in any of the 144 squares. The "real" places were numbered 1 to 12 , as were the "memory" places, so that in each square there would be a pair of numbers, one referring to a place to be described *in situ*, and one referring to a place to be evoked from memory. All that remained was to distribute the months of the year across the absciss, and the years themselves down the ordinate: the result was a *mode d'emploi* of the next twelve years— from January 1969 in the top left corner, to December 1980 in the bottom right.[5]

This grid resembles a calendar planning-table that encompasses twelve years at a glance. It is interesting here that the planning goes together with tabulation, regulation and overall view.

Tabulation and regulation enabled Perec (1) to permute his "Reals" in order to visit the 12 given places at different times of the year; (2) to ensure that the "Reals" and the "Memories" did not overlap.

The Parisian Places were chosen for personal reasons: Philippe Lejeune gives a list of the 12 Parisian places and exposes Perec's autobiographical relationships to them in detail (*La Mémoire et l'oblique*, 164). Interestingly, one of the places (*Passage Choiseul*) has nothing to do with Perec: only 11 places are autobiographical. This number could well be a cryptic allusion to his mother: the declared date of her death is the 11 February 1943. Bernard Magné showed that the number 11, related to Perec's mother, structures many books or exercises and is mentioned in many of Perec's works (*Georges Perec*, "Le 11", 58–65). *Lieux*, with its "empty place" (in fact a *passage*), would be no exception.

---

[5] Andrew Leak, "Paris created and destroyed", 26. The second table, reproduced in Philippe Lejeune, identified the places by name (*La Mémoire et l'oblique*, 156).

As for the timetable, it is clearly mathematical. It imposes a rule in order to exhaust the combinations of places and months, which necessarily takes 12 years. However, the orthogonal Latin square (which Perec used to plan *La Vie mode d'emploi*) *also* has to do with memory or, at least, with a desire to root memories into a place. In Vilin *Memories* written in 1970 in Annecy, Perec stresses the importance of the tiled floor of the WC in the house where he stayed, a tiling that "à lui seul, suffirait à enraciner une existence, à justifier une mémoire, à fonder une tradition [...]."[6] In order to understand why Perec gave such extreme importance to mere tiling, one must keep in mind that *Rue Vilin* is the Street where he was born in 1936. He lived there with his family until 1942. At that time he fled from Paris by train and did not come back to rue Vilin after the war: as an orphan, he grew up with his Uncle's family, rue de l'Assomption (another Parisian place included in *Lieux*).[7] In the first "Vilin Memory" he mentions that he does not remember rue Vilin at all and even completely forgot its location for many years. Moreover, he does not trust his early childhood memories: they are unreliable and unsettled. The grid, then, takes the place of an original place whose memory has vanished; it traces a memory pattern on paper (the tiling) that is out of place.[8] The tiling-like grid functions as a metonymy: it stands for the house, and the house symbolizes a period and a continuity that Perec wants.[9] We can see the grid both as a mathematical organisation of time and as a recurrent fantastical pattern which gives a contour to memories which did not take place.

It leads me to another aspect of the planning table, namely that of visibility. By providing an overall view of 12 years, the orthogonal latin square *visibly* encloses the future. Indeed closure is at work here: the grid makes completion possible; it delimits time and space; it frames them into a big square and distributes them into small squares. Nothing goes beyond; nothing remains shapeless. Thus, time can be captured, it is projected into space. Moreover, by connecting time and places, the future is clearly dedicated to the past since each Parisian place metonymically represents a span of time in Perec's life: early childhood (rue Vilin), childhood (rue de l'Assomption), et cetera. Time stops

---

[6] "[It] would be enough to root an existence, to legitimate a memory, to establish a tradition" ("Vilin souvenirs", 136 [my translation]).

[7] His father died in 1940. A Jew, his mother was sent to Auschwitz in 1943 and never returned.

[8] See Bernard Magné's comments on the grid pattern: "Quelques pièces pour un blason, ou les sept gestes de Perec", 209–211.

[9] Rue Vilin is also the metonymical site of Perec's early childhood which, according to him, also took place elsewhere in Paris (see "Vilin souvenirs").

in 1969 on *l'Île Saint-Louis*: the plan does not allow room for a new "place"; there is no future opening for it. As Philippe Lejeune extensively shows, *l'Île Saint-Louis* is related to S., with whom Perec was involved. The emergence of *Lieux* goes along with the end of his relationship with S. Lejeune considers that *Lieux* was initiated as a love memorial (*La Mémoire et l'oblique*, 146). As far as the plan is concerned, the importance of 1969 and *l'Île Saint-Louis* is stressed by the fact that it is there that time stops. Perec's publications, however, never put forward that place (and time). They focused instead on Rue Vilin and early childhood, as if the immediate loss gave way to a more remote loss that was extremely difficult to deal with. This move "backwards" goes along with the psychoanalysis that Perec undertook in 1971. As we can see, it is difficult to make a clear-cut division between Perec's life and his project. The cryptic autobiography involved in *Lieux* is partly responsible for that, but the fact that Perec left his project uncompleted inevitably confronts us with a work in progress where writing and "life" go together.[10]

The scopic *impetus* that governs Perec's description of "real" places also governs his planning. The square encompasses future and past projections in its grid, which represents a dis-located memory as well as a suspended future.

2. RECORDING

Whether Perec remembered a place or described it, he labelled it with great precision: he gave it a number and a name, he mentioned the date and the time of writing. When he was behind schedule (which became increasingly frequent after 1969), he mentioned both the planned date and the actual date of the writing. This means that the rule he gave himself was never lost from sight: even when he did not comply with it, it was part of the ritual of writing. As for the mention of the actual place and time, it assimilated writing with instant recording. "Memories" and "Reals" were invariably projected into time and space, as if Perec attempted to photograph his memories in the same way that he took snapshots of the streets of Paris. He produced time-traces of "real places" and "memory places." As Jacques Derrida reminds

---

[10] I chose to analyse Perec's concrete "apparatus" (the grid, the records) and images (the time bombs) in order to avoid immediate biographical reduction, and I connected Perec's work and life *via* a textual nexus: the recent publication *of Entretiens et conférence* enabled me to establish a number of connections between the *Œuvre* itself (including films and photographs), the project (*Lieux*) and what Perec said about it.

us in *Mal d'archive*, to remember and to record are two different operations. Perec, however, attempted to record what he remembered, his memories *had to* be inscribed, tested and attested in order to exist. Perec recognized his tremendous lack of spontaneity when he watched a scene: for him, a scene was real if it was remembered and remembering was real if it had been recorded, if there remained a trace (*Entretiens et conférences II*, 236). He mentioned more than once that he wished to "get three kinds of [...] ageing: the ageing of the places, the ageing of my writing and the ageing of my memories" ("The Doing of Fiction", *Entretiens et conférences II*, 249). But why should one try *to get* such an ageing when ageing is a constant process? As far as memories are concerned, it means at least that such a process was not self-evident. We already saw that duration and continuity were not *given* to Perec. Moreover, the wish to get not one but three kinds of ageing obviously fights the wish to "freeze time." The conflicting "wishes" that appear to fuel *Lieux* (1. to produce a sense of ageing, 2. to freeze time and history) are, however, cleverly combined into Perec's use of records. The ageing process had to be exteriorized and registered; there had to be visible traces of the same place at different times in order to ensure that the past really *took place*, time after time. *Lieux* resists the "mnesic" ability to forget and to alter the past by regularly producing kinds of time punctures which would eventually give evidence, after 12 years, of an ageing process.

From 1974 onwards, Perec insisted on the sociological dimension of his project which was initially presented as autobiographical ("Lettre à Maurice Nadeau," 58). This dimension directly concerns the Real "punctures." Perec continued to claim his interest in everyday life after he had abandoned his project : "The interest was to try a very close view of what can be seen in a street or in a street scene. Most of the time we don't pay attention to what I call quotidienneté…everyday. [...] So when I do those texts, I try to be most precise and…flat [...] like if I was a Martian going through a city…: I don't try to interfere and I don't try to put myself in a position other than an eye looking [...]" ("The Doing of Fiction", *Entretiens et conférences II*, 249–250). I shall return to this reference to the Martian since it relates to the Time Bomb, but first I would like to emphasize the fact that Perec tried to be a camera eye, he tried to see but not to recognize, that is to say not to *remember*. He wanted to "estrange" himself from the street milieu, to neutralize his memories in order to record what we usually do not notice because it is ordinary. Perec framed his glance in order to dissociate vision and memory related to habits or emotions, but in fact, the everyday environment that he attempted to record had to do with another kind of memory: the memory of what has vanished. Perec gave the example of Parisian underground tickets, which ticket punchers used to... punch (!): nobody paid attention to *that*, but after ticket punchers had disappeared, it became, said Perec, a memory.[11] Therefore, the unnoticed present is a potential memory on which Perec capitalized. As Philippe Lejeune puts it, *Lieux* was a Time-saving Bank.

The relationship between records and memory goes even further than that. Perec attempted to neutralize his gaze and to produce records instead of *impressions*, but since he systematically wrote down written traces scattered in the streets (announcements, names of shops and films, slogans), the intimate relationship between (external) traces and (internal) memory comes to the forefront. To remember and to record are indeed two different operations, but a strict separation between memory and records does not make sense. Their intertwining is quite clear in *Lieux*: Perec remembered what he recorded (or wrote down), and he recorded what already was a trace (letters or numbers) or what would become a trace precisely because he had "recorded" it. His "Reals" have to do both with the written and the visual, they are *impressed* (or framed) by various forms of recording. The numerous descriptions and enumerations that Perec made can be seen as a way to combine visibility and scriptibility into records that have to do with the "memorizable."

3. Differing

Perec enclosed his descriptions in envelopes which he ritually sealed. Just as the square grid is a fantasmagorical memory pattern, the gesture of sealing with wax clearly refers to the past: it used to be an ordinary practice and became a symbol of "the" past. These envelopes also contained snapshots that friends took for him here and there as well as scrap papers such as transport-tickets, bills, et cetera. They enabled him to isolate "Reals" and "Memories" into kinds of time capsules: he was not supposed to open the envelopes until 1981—hence the seal—and did not even look at the photographs that had been taken *in situ*. Such a device can be seen as a way to ensure blindness in order to avoid interferences. Sealed envelopes hermetically isolated "Real" and "Memory" places. The "samples" would not have been fully representative

---

[11] Georges Perec, "Le travail de la mémoire", *Entretiens et Conférences*, vol. II, 48. In order to become a memory that can be shared with others, the punched ticket has to be *recorded*. Perec did it in his book *Je me souviens* (memory no 185), but before that, he did it in his film, *Un homme qui dort* (1973) (see note 5). From *Un homme qui dort* to *Je me souviens*, a *network* is in progress: it discreetly connects patterns which, however (apparently) insignificant, are related to Perec's life and places.

if they had been contaminated by what had already been recorded. Perec *had to forget* what he had already written in order to be in accordance with each "portion of time."

The running of the experiment therefore depended on a simple device which can be compared, according to Lejeune, to a banking operation. "Reals" and "Memories" are like currency taken out of circulation, accumulated into a savings account, and yielding profit over time. Such a comparison, however, misses something crucial: in order to gain value, current time must change into memory. An ordinary underground ticket is not of much value, and yet an obsolete and useless underground ticket is valued as a memory of the past. Memory of times (things, places) gone by is the value on which Perec bets, a value that had to be secured because things (as well as memories) change or disappear *with time*.

Perec expressed what he had in mind when he spoke of Time Bombs in 1981 at a conference "About description": "I want to build my own time Bombs," he said.[12] He explained that time bombs are ordinary objects, such as a Coca-Cola bottle, that are "very very deeply buried" so that in a million years, when Martians may discover them, they will be able to know that we drank Coca-Cola and so on (*Entretiens et conférences* II, 236). It means that what is over *now*, what is enclosed, sealed and buried, shall be disclosed later as a *testimony*. To make your *own* time bombs also means that you produce *memories*. Time traces had to be very deeply buried in order to be discovered afterwards as good as new: the found past would be like a found treasure. Of course, to plan to undertake this yourself, not in a million years but in twelve years, also implies an acceleration of history which is, as Nora puts it, a sign of our times (*Les Lieux de mémoire*, vol. I, XVII). According to this, the time Bomb device is a deferred "explosion" of memory, a tentative redemption of the past: Perec probably expected it to provoke an anamnesis of what he already anticipated as his previous existence. Anamnesis is, in fact, related to places themselves. In "La Ville mode d'emploi," Perec explains that he had almost completely forgotten that he ran away as a child until he found himself *in situ*. Concrete details played the same role as Proust's madeleine (*Entetiens et conférences II*, 28).[13] In *Lieux*, the fact that he noted

---

[12] We can see here than the image of the time bomb was still active and productive after Perec had given up *Lieux*.

[13] The same goes for a journey he used to do twice a week by train as a child: it came back later as he was once more on that train ("Entretien Georges Perec/Bernard Pous", *Entretiens et conférences II*, 191).

down the price of the coffee he drank or recorded the slogans of the day into his "Reals" indicates that he was well aware of the potential memory value of these ephemeral manifestations. The time bombs enabled him to pin down *both* the swift passage of time and its fixation in a lasting trace. *Lieux* is a matrix of Perec's poetics of transience, which is at the core of his sociological approach of everyday life.

However, there is a dark side to this: what is "very deeply buried" reminds us that grief is at stake. And what Perec's scenario does not tell is the catastrophe, the destruction of our world. It is in this way that the time bomb device makes sense. We understand, then, that what is "very deeply buried" is destruction itself, which *already* took place. Perec's previous existence is not ahead of him but *behind* him, buried on Rue Vilin, which he did not quite remember, where he did not belong. We can say that Perec was deeply divided: on the one hand, he was the Parisian astronaut who built his own time bombs and dreamt of ageing; on the other hand, he was the Martian who collected the remnants of an estranged world. It produced unforeseen interferences between active memory (which Perec did not *plan*) and preserved memory which had to be captured and frozen into time capsules. After the first year, Perec found that the sealed envelopes did not prevent him from remembering, and this greatly annoyed him ("Lieux, un projet (confié à Gérard Macé)"). In the same way, whereas the astronaut's eye managed to catch transient manifestations, the supposedly neutral Martian's eye was insecure: Perec was obsessed by what he called his incapacity to look correctly ("Lieux, un projet (confié à Gérard Macé)"). He was not certain to observe the right things, he felt deprived of vision, disturbed by the fact that, after the first record, he noted what had changed rather that what was there.[14] The "monstrous" device that he planned was an attempt to bring together past, present and future, but it somehow suffered from the lack of memory it intended to replace and from the "interferences" and frictions inherent to

---

[14] In fact, when Perec notices that it is difficult to "look again" at a place that he has already seen, he is blind to the fact that his attention to changes is at work from the start, from the very first records. The same goes for the interferences between descriptions and memories. One good example of such "interferences" is the first record of Rue de la Gaîté (1 December 1969): "Coming from the new Montparnasse Station, I arrive in rue de la Gaîté by avenue du Maine. I was intending to stop in a café I've known for many years, 'Aux Armes de Bretagne' (fried sausage, pinball machines) but it is closed, not, apparently, for refurbishment; it looks more as if it has gone bankrupt or been sold (in anticipation of a radical transformation of the quartier over the next few years: the new expressway will wipe out the whole of rue Vercingétorix).

I fall back instead on 'Les Mousquetaires' café, almost opposite, on avenue du Maine (I came here one day in '55 or '56 to look for Jacques and his father who were playing billards). The billiard tables are still here." ("Glances at Gaîté", 44).

the project. In that sense, the intimate destruction that Perec kept "deeply buried" probably contributed to the weariness he developed over time, but it also took part in another framing that came out of *Lieux*.

In order to understand this, it is important to realize that short-lived manifestations have different implications when it comes to places. Unlike prices, slogans, pedestrians or cars, places and buildings are supposed to remain the same, they are supposed to *last*. When places become evanescent, memory looses its landmarks; it is indeed dislocated. In this way, the image of the time bombs reminds us that if such a dislocation goes unnoticed here and now, it is destructive in the long term because it affects a stability which memory needs. It is the very faculty of memory that is at stake here. What Perec accumulated into his envelopes were not only ephemeral signs but also signs of destruction: his Parisian places did not age; they changed; they were (and still are) renovated: "out with the old in with the new!" (Andrew Leak, "Paris created and destroyed," 28). What he recorded was the demolition of old buildings. In "Glances at Gaîté," for example, Perec *anticipates* the "radical transformation" of the neighborhood (see note 14). Rue Vilin is particularly exemplary of that: it began to interest him once he learned that it was doomed for destruction, as if the planned demolition of that particular street and the emergence of *Lieux* went hand-in-hand. Remarkably enough, Perec writes that the place *will blow up* [sautera] like a bomb ("Vilin souvenirs," 135). Along with the fixation of mobile and transient scenes, *Lieux* attempted to preserve a trace of vanishing places and to record the visible signs of a double disappearance: that of Parisian places and buildings and that of memory rooted in space and transmission.

CONCLUSION

A close reading of the grid, of the time-bomb and of the records that Perec imagined and produced enables us to expose the complexity of the relationship between time and memory in *Lieux*. If indeed Perec aimed to "secure" future memories, the various temporal layers of each of the twelve Parisian places that he intended to archive one by one did not fully meet the archaeological model which apparently inspired him. He wanted to preserve the (future) memories of a few Parisian places from alteration and destruction, but the very organization of the project was haunted and disturbed by "very deeply buried" memories. The urge to freeze time, to plan, to treasure, and to protect has to do with a grievous event which occurred in early childhood but never *took place*: the loss of Perec's mother was impossible to *localize*.

Even the fact that Perec was behind schedule can be read as the repetition of an endless delay surrounding his mother's death, a delay that the time-bomb device reproduced but also attempted to *unbury*. To archive and to record the signs of destruction of old Parisian places can thus be understood as an attempt to give a place to a destruction that was out of place, but it would be a simplification to see *Lieux* as a mere product of a traumatic past. The growing importance of rue Vilin, during the writing of *Lieux* and after Perec had abandoned it,[15] enables us to see that it was indeed crucial, but the conflicting forces at work that I pointed out also make it clear that there was more than one time bomb involved in the project. One of them is closely related to time and memory: *Lieux* managed to produce what "exploded" later as *lieux de mémoires* (or memory realms) and are related to the loss of collective memory.

Pierre Nora shows that the multiplication and diversification of archives is related to the decline of a memory *milieu*. Along with the disappearance of a rural society rooted in space and tradition, along with the accelerated mobility of people and the instability of places, memory loses ground to more and more sophisticated recording devices. Memory as a *milieu* used to give rhythm and continuity to everyday life, and Perec's project coincides with the decline of a collective and spontaneous memory. In fact, *Lieux* failed to replace that *milieu*, but it produced *Lieux de mémoire*, which are kinds of sacred Islands isolated from the rest, where every single little thing, however trivial, is supposed to convey meaning (Pierre Nora, *Les Lieux de mémoire*, vol. I, XLI). Perec's records produce *a sense of the past* and of the *common place*. As Nora puts it, the production of *lieux de mémoire* is an answer to the acceleration of history (which complicates the disappearance of a past which is still part of our personal experience) and to a sense of continuity rooted in space (which, however "mobile" we might be, we basically need). To that extent, Perec's attention to Parisian places and to their fast transformation (destruction) can be seen as part of our post-war cultural and historical landscape. However, his *lieux de mémoire* are not produced in spite of archives and records: they *are* archives and records. As such, rue Vilin is again exemplary: it has been almost erased from (the map of) Paris, it has almost completely disappeared, but Perec's photographs and texts transformed it

---

[15] Cf. "La rue Vilin"; *La Clôture*; *W. ou le souvenir d'enfance*; *Un homme qui dort* (the film) (the last image of this film is an image of Paris and of rue Vilin). In *Récits d'Ellis Island* (the film) the photograph of his mother's house on rue Vilin can be seen among others in the album that Perec glances through…

into a memory realm which can now be part of our memory even if we never lived or went there.

Perec is sensitive both to the loss of traditional memory (mediated by the family) and profoundly curious about the creative powers of recording media. After he had given up *Lieux*, he chose to concentrate on each place through different recording media. It enabled him to relate each memory realm to a form of *description*: "The experiment stopped in 1975, and has been taken up and continued by other sorts of descriptions: poetic and photographic (*La Clôture*, about rue Vilin), cinematographic ("Les Lieux d'une fugue," about Franklin Roosevelt), radiophonic (about Mabillon, in progress; Georges Perec, "Scene in Italie." 34). In *La Clôture*, rue Vilin is photographed in black and white, silent, deserted and still ("as if," writes Pierre Getzler, "the space they show is already dead, as if *a neutron bomb* had fallen on the city [...]"),[16] whereas carrefour and Mabillon is noisy, lively and colourful. There is no separation, in his memory realms, between media and memory. The different framings of parisian places involve different types of memory. The materiality of the various time traces (or records) is not simply taking the place of a more or less idealized spontaneous memory; it gives new contour to the "memorizable," which *can* be, in turn, internalized and transformed (that is, memorized). Contrasting with Nora's longing for a lost collective form of life, Perec (however nostalgic) shows that different arts of memories go along with various recording techniques. In fact, techniques are part of the so called "spontaneous" memory: songs, for example, display repetitive patterns that enable us to remember them. Furthermore, recorded images, slogans *and* old songs that we know by heart are, more often than not, part of our individual *and* collective memory. So if records tend indeed to supplant memory, they also transform and diversify it. The question of whether we can digest our countless records is left open, but Perec's "failure" and what came out of it gives us a hint of the *potentiality* of records as far as memory is concerned.

This is perhaps what emerged after *Lieux*. Perec's persistent experimentation with different "sorts of description" brings the relationship between memory and image to the forefront. Even if the "visual proof" is most

---

[16] Jean-Charles Depaule & Pierre Getzler, "A City in Words and Numbers", 124 (my emphasis). Getzler adds that Georges Perec asked Chistine Lipinska—who photographed rue Vilin for *La Clôture*, "to photograph front on" and also asked him to take the façades "where the openings had been bricked up". Getzler did not want to to do it this way, so that "he could imagine the movement of the city" (*Ibid.*, 124 and 119).

important to him, it is not a *given* which he naïvely believes. On the contrary, Perec's photographs and films, as well as his written and oral descriptions of places, always pay attention to the *framing* which both constructs, "feeds" (or *impresses*) memory. Records and imagination fuse in the image; images and written traces go together; each memory realm becomes a network. In an era sensitive to "the colonization of the visual by the word" (James A. Knapp, "'Ocular Proof' : Archival Revelation and Aesthetic Response," 699). Perec's experimentation with the visual *and* the written is most interesting, but it also invites us to reconsider the clear-cut division that Pierre Nora makes between records and memory. In the works that Perec presented as a "continuation" of *Lieux*, the records do not fight against fantasy and imagination but compose with them. Memory, with its power to alter the past (through fantasy and imagination), and archives, with their power to preserve and to produce different recordings of "the" past, manage to merge into art and to produce memory realms that involve plural memories and that allow change to take place.

REFERENCES

Derrida, Jacques. *Mal d'Archive*. Paris: Galilée, 1995.

Depaul, Jean-Charles et Pierre Getzler. "A City in Words and Numbers", translated from French by Clara Barrett, *AAfiles*, no 45–46, 2001, 117–128.

Knapp, James A. "'Ocular Proof': Archival Revelations and Aesthetic Response", *Poetics Today*, 24, 4, Winter 2003, 695–727.

Leak, Andrew. "Paris created and destroyed", *AAfiles*, no 45–46, 2001, 25–31.

Lejeune, Philippe. *La Mémoire et l'oblique. Georges Perec autobiographe*. Paris: P.O.L., 1991.

Magné, Bernard. "Quelques pièces pour un blason, ou les sept gestes de Perec", in Paulette Perec, *Portrait(s) de Georges Perec*, 199–233.

———. "Carrefour Mabillon: 'ce qui passe, passe'", *Georges Perec*. Paris: André dimanche éditeur/INA, 1997, 55–67.

———. *Georges Perec*. Paris: Nathan Université, 1999.

Nora, Pierre. *Les Lieux de mémoire*, vol. I. Paris: Gallimard, 1984.

Perec, Georges. "Lettre du 10 juin adressée au Mathématicien Chakravarti" in Philippe Lejeune, *La Mémoire et l'oblique, Georges Perec autobiographe*, 205–206.

———. "Lettre à Maurice Nadeau", *Je suis né*. Paris: Seuil, 1990, 51–66.

———. "Lieux, un projet (confié à Gérard Macé), *Georges Perec*. Paris: André dimanche éditeur/INA, 1997, CD 3.

———. "Guettées", *Les Lettres nouvelles*, no 1, 1977, 61–71.

———. "Glances at Gaîté", Andrew Leak (trans.), *AAfiles*, no 45–46, 2001, 44–53.

———. "Vues d'Italie", *Nouvelle Revue de psychanalyse*, no 16, 1977, 239–246.

———. "Scenes in Italie", Andrew Leak (trans.), *AAfiles*, no 45–46, 2001, 34–41.

———. "La rue Vilin" in *L'Humanité*, 11 novembre 1977 (also in *L'Infraordinaire*, Paris: Seuil, 1989).

———. "Allées et venues rue de l'Assomption", *L'Arc*, no 76, 28–34.

———. "Comings and Goings in rue de l'Assomption", Andrew Leak (trans.), *AAfiles*, no 45–46, 2001, 56–67.

————. "Stations Mabillon", *Action poétique*, no 81, 1981, 30–37.

————. "Stances on Mabillon", Andrew Leak (trans.), *AAfiles*, no 45–46, 2001, 70–71.

————. «Vilin souvenirs (presented by Philippe Lejeune)», *Genesis*, no1, 1992, 127–151.

————. *Espèces d'espaces*. Paris: Galilée, 1974.

————. *Tentative d'épuisement d'un lieu parisien*, Paris: Christian Bourgois, 1982 (1975).

————. *W ou le souvenir d'enfance*, Paris: Denoël, 1975.

————. *La Clôture*, dix sept poèmes hétérograpmmatiques accompagnés de dix-sept photo-graphies de Christine Lipinska. Paris: imprimerie Caniel, 1976.

————. *Je me souviens. Les choses communes* I, Paris: Hachette/POL, 1978.

————. *La Vie mode d'emploi*, Paris: Hachette/POL, 1978.

————. "Tentative de description de choses vues au carrefour Mabillon le 19 mai 1978 (A.C.R.), *Georges Perec*. Paris: André dimanche éditeur/INA, 1997, CD 3.

————. "Inventaire", *Georges Perec*. Paris: André dimanche éditeur/INA, 1997, 71–78.

————. *Les Lieux d'une fugue* (couleur, 38'), 1976 (film).

———— avec Bernard Queysanne, *Un Homme qui dort* (nb, 82'), 1974 (film).

———— avec Robert Bober. *Récits d'Ellis Island : histoires d'errance et d'espoir*. Paris: P.O.L., 1994 [1980].

————. "Le travail de la mémoire (entretien avec Frank Venaille)", *Je suis né*. Paris: Seuil, 1990, 81–93.

————. *Entretretiens et conférences I*. Paris: Joseph K, 2003.

————. *Entretretiens et conférences II*. Paris: Joseph K, 2003.

Perec, Paulette. *Portraits de Georges Perec*. Paris: Bibliothèque nationale de France, 2001.

Virilio, Paul. "Paul Virilio on Georges Perec (Interview with Enrique Walker)", *AAfiles*, no 45–46, 2001, 15–18.

CHAPTER SIX

SEEKING IN SUMATRA

Brian Aldiss O.B.E.

Some late news about time. Well, I'm late with it.

The University of Oxford awarded an Honorary Degree of Doctor of Science to Professor Ahmed Zewail in 2004. Professor Zewail has divided up Time—and I quote—"into units so tiny that while we are pronouncing a single letter, a million billion units have elapsed." These are femto seconds. I must try and write a short story about that—as short as possible.

In the 2004 tennis tournaments in Wimbledon, the BBC were using a new TV camera, the Typhoon. The Typhoon takes 5000 frames per minute. So one can study exactly the techniques of players.

Both these achievements are awesome. However, I am going to talk about a larger slice of time—thirty years, no less.

English poetry has always been involved with time. The most famous example is probably Andrew Marvell's line "Had we but world enough and time," the opening of a poem that goes on to extol whole epochs. Thomas Hardy uses time for its pathetic qualities, for example in "She to Him": "When you shall see me in the toils of time, / My lauded beauties carried off from me [. . .]," C. Day Lewis, once poet laureate, celebrates the pleasures of time:

> Now the full-throated daffodils,
> Those trumpeters of spring,
> Call Resurrection from the ground,
> And bid the year be king!

What were once the nation's two favorite poems, in the days when people enjoyed poetry that rhymed, are time-bound: "The curfew tolls the knell of parting day," which begins Thomas Gray's "Elegy in a Country Churchyard," and, at the other end of the day

> Awake! For morning in the bowl of night
> Hath flung the stone that put the stars to flight.
> And lo! the Hunter of the East hath caught
> The Sultan's turret in a noose of light.

Jo Alyson Parker, Michael Crawford, Paul Harris (Eds), Time and Memory, pp. 115–121
© *2006 Koninklijke Brill N.V. Printed in the Netherlands.*

Time must be expressed because it's in our bones. And our intellects. I have always expressed it in novel form.

I am no philosopher. I am a story-teller. It happens that I have a true story embedded in historic time and geography. It takes place in what some would call "my time," although I make no proprietorial claims for it.

By the seventies, I had become moderately prosperous, certainly in comparison with the naive youth from whom I had developed, that youth in World War II who took the King's shilling and was posted to the East to serve his country. By the seventies, my books were published, first in hardcover, and then, when published in paperback, could be bought in Sydney, Samoa, Singapore and Sarajevo, as well as San Francisco (if nowhere else . . .).

In 1978, I took a week's package tour from Singapore to Sumatra. Sumatra!—Once I had lived there for a year, in my army days. I had never expected to be able to visit it again; it lay like a dream at the back of my mind. Off I went. Sumatra is speared by the equator like a sausage on a skewer. My novel set in Sumatra, *A Rude Awakening*, had just been published. I was curious to see how the country had changed in thirty years. And, although I was happily married, I hoped to revisit the scene of an old love affair.

I was to discover that history is fossilized time.

In August 1945, the atomic bombs had been dropped on Hiroshima and Nagasaki, Japan had surrendered, peace was in the offing. If it was end of an epoch for Japan, so it was—although few of us realized it—for both the British Empire and the Dutch Empire in the East. Those two great galleons were sinking fast in the seas of time—and, as it happened, were inextricably locked together. They and we were victims of history. History is what happens when you are not looking.

In November of that momentous year, 1945, the 26th Indian Division arrived by ship in Sumatra. Our mission was three-fold: to return the occupying Japanese army to their homeland, to release from prison camps the Dutch and other nationalities imprisoned there, and to reinstate the Dutch in authority. When we landed, how green the country seemed, how well-dressed the people, how friendly—all this in contrast to the India we had left behind.

But time was against us. Sumatra was no longer part of the N.E.I.—the Netherlands East Indies: it had become, under Soekarno, part of Indonesia, a sovereign power, and the people longed for freedom—the freedom to go to hell in a bucket in their own way.

As long as the British Army went about disarming and shipping the Japanese home, all was well. As soon as we welcomed in the first Dutch contingent, the natives started firing at us. The situation resulted in deadlock. We were too few in number to be effective. In the end, we gave up and went home.

In 1978—thirty years later—I flew in to Medan airport. When I had flown out in 1946, the airport had been in the wilds, outside Medan, the capital. Now it was surrounded by squalid suburbs.

Uncomfortable as the political situation was in 1945 and 46, I was able to better my position. Medan was enjoyable and semi-functional. There were restaurants, shops, clubs and two operating cinemas. One cinema was the Deli, named after the local river, forbidden to troops, one, the Rex, open to troops and showing old films which had been in storage for a long time. The Hollywood films of the thirties carried Dutch subtitles and Chinese side-titles, so that half the screen was covered by writing. The films belonged in another time-frame, with stars like Wallace Beery and Jean Harlow still young. They spoke of a distant epoch, but they were something to watch, fairy tales of the past.

One evening, as we were all marching out after the show, a group of civilians standing by spoke up and asked, "How was the film this evening?" It happened that of all those troops, the question was addressed to me as I passed. I stopped to talk.

So I became acquainted with some friendly Chinese families. The ladies of one group had been educated at Hong Kong University and spoke excellent English. I saw some of them most days, I ate with them in the evening, I took the ladies shopping in my Jeep, and I escorted them to the cinema. I fell in love with one of those ladies and she with me. She had exquisite manners and looked exquisite. Was exquisite.

We would sit outside her house at night—those warm starry nights! That humble, tumbledown town! Just down the street, guitars were playing where the Ambonese lived. The *sateh* man would come along with his mobile wooden kitchen and serve us soup and *sateh* on sticks. I wished for nothing better. The agreeable lady would smile at me and dream of distant London.

By 1978, Singapore, like Indonesia, had become an independent country. But Singapore flourished as an entrepreneurial port, bolstered by the fact that English was its official language for the diverse nationalities living there. In Singapore, the *sateh* were big and fat and sizzling with peanut sauce.

Because I had some artistic talent, I got the job in 1946 of looking after a theatre—a thatched barn, in effect. I decorated it with a series of screens depicting local life. Sunsets and dancing girls. We got a band together, held dances there (local lady friends invited), musical evenings and film shows. The barn became a social centre, open every evening. I had private quarters there, with a bed, and two sepoys to do all the hard work for me.

I became virtually a man of leisure, and drove about in a company Jeep with my Chinese lady friend. I had already served three hard military years overseas—which included warfare in Burma—and had at last found a niche in time that suited me. For many a year afterwards, I regarded this equatorial year as the happiest period of my life.

But political difficulties piled up. In those immediate post-war years, there were few airborne troop carriers. Getting the Japanese forces back home by steamer was a slow business, and became slower. As time wore on, the local opposition—the Merdeka army—grew stronger. The British force, small in number, relied on the Japanese for policing duties. Japanese soldiers were allowed to keep their carbines. Everyone became armed to the teeth.

On one occasion, I was sitting at a table in a restaurant with army pals. We had our sten guns. At a neighboring table were Indian troops armed with rifles. Present at other tables were Japanese with their carbines and, nearby, two Dutch soldiers armed with American carbines; while at a fifth table sat Indonesians with miscellaneous varieties of fire-power.

The Chinese were serving us all, making a living! Nervously neutral, they accepted our Japanese guilders which acted still as currency. The guilders were in fact worthless, but we insisted on their having value. You always need a currency for any semblance of order or civilization. The Japan Government "promised to pay the bearer"—but there was no Japanese government any more. The time indeed was out of joint.

By 1978, thirty years later, many of the Chinese had disappeared from Indonesia. Bertrand Russell claimed that Indonesia had murdered fifty thousand of them. Was that not genocide—for which Indonesia has never been called to account? Those Chinese surviving to trade were forced to adopt Muslim names on their shop fronts. "Mohammed Wong Fu" and so on. What happened to my dear fragile lady is a matter for sad speculation. It seems all too likely that the bad times gobbled her up.

In 1946, time had turned to jelly. No progress. No seasons either. All quiet on the Eastern front—or almost. 1946 went on for ever. It suited me well enough. The Indonesian forces grew stronger; we referred to them by the mild English appelation of "Extremists." We became increasingly reluctant to rid the island of the well-disciplined Japanese army. When a pitched battle was fought at Palembang, Japanese troops joined with the British and Indians against the Indonesian Extremists. This, less than a year after the end of the war against the Japanese. How mind-sets had changed!

Revisiting Sumatra in 1978, these old scenes returned to mind. In the end, of course, we had been forced to pull out. It was not practical to reinstate the Dutch in Indonesia when Britain was in the process of handing over India to an Indian government. We left in a hurry. The Dutch fought on. Their empire, like the British Empire, had begun in an earlier century as a modest East India Company, in quest of trade and spices.

In the Sumatra of 1978, no one remembered that the British had ever been there. The record had been wiped. Memories must be shorter in the tropics. It's the heat, you know! We had not been belligerent enough; otherwise, they would have remembered us ...

I walked the streets of Indonesian Medan, the capital, once more—this time unarmed. It was no longer the quiet shady place I had known, its avenues disturbed only by the creak of a bullock cart, the odd rifle shot. Over-population had overtaken the town. The streets teemed with people and the roads with mopeds. This was a different place from the one I had known and delighted in: now it was filthy, noisy, broken, crowded, inhabited by a different generation. What a contrast to the spick and span city of Singapore, only one hour's flight away!

I hoped to find where my lady love had lived and where I had had my billet. But Dutch names for the streets had long been obliterated: now the street names were all in Bahasa Malay, named after revolutions or massacres or the more brutal generals. New roads had been built, new buildings were erected. No street maps could be found.

I discovered a taxi driver who was Christian, not Muslim, and who spoke English. We did a slow tour of parts of the city. All seemed to me confusing. My billet had stood at the end of a neat row of houses, overlooking a wide open space. There were now no open spaces, only a clutter of houses. I remembered what Marcel Proust says regarding the quest for such destinations, that one is in search of one's youth rather than one's house. Time is a great stealer.

We drove about. The taxi driver, a good man, became as eager as I was. Eventually, we came on a street that seemed vaguely familiar. The house in which I thought I had lived now had a garden—if it was the house, a garden was only to be expected. Mainly because I did not wish to disappoint the driver, I said I thought that it was the very house I had known thirty years previously. He came with me, to act as interpreter. A family lived there now. They welcomed me in.

It was a puzzle. I was a revenant. Could this indeed be the place? If so, it was now richly over-furnished and aglitter with mirrors. I assured the family that, yes, it was indeed the house where I had lived. All the same, hadn't there been a downstairs shower room where now their kitchen was ...? The longer I stayed, the less certain I became.

I found that I was not as clear about the layout as I had imagined. What had once been vivid was dimmed by thirty years of eventful life. The charming family and I sat and talked about freedom—Merdeka!—and the way times changed. I said I had lived in an upper room. They were excited. They showed me upstairs. But hadn't the stairs run the other way? However, I thought I recognized the landing, with a window at the end of it. So *that* had been my room, there! Still misgivings... We entered. Had it indeed been here that, thirty years earlier, a delectable Chinese lady had crossed the floor, put her arms about my neck, and kissed me?—The opening shot of an intense love affair?

Ah, but this room had no balcony. Mine had had one. But balconies in the tropics tend to fall off like fruit from a twig . They tend to rot like over-ripe fruit. All the same... "Was this your very room?" they asked, eagerly in chorus, these pleasant people. I had not the heart to disappoint them.

"Yes," I said. "My very room."

Never be a time-traveller. To this day I cannot tell if I had managed to work my way back through the interstices of time to where that saving idyll had blossomed.

On the day following, I tried to find her house, nearby, where Dinah had once lived. I walked and walked in the heat, seeking my way through a maze of streets. Here was a concrete analogy of our life, as we wend our allotted path through unknown circumstance.

I came at last on a familiar curved stucco wall. Behind the wall, a cinema, the Deli Cinema!—Unmistakable! Much as it had been, thirty years back. Despite all other changes, the Deli was still operating. Its posters announced the same kind of fantasy horror it had screened in the old days: *The Night of the Rivers of Blood, Shark Raiders of the Pacific, Queen of the Python Men*.

Some things at least had not changed in thirty years: in particular, a human desire to be amazed, to fantasize—as I myself was pursuing a fantasy.

Indeed, it seems that human brains are built better equipped for fantasy than intellect, for Tolkien rather than Aristotle, for imagination rather than logic. How painful are the years of education for most children, as we school them from dreamtime into real time, from the world of play to fish-cold reality.

As far as I know, official histories make no reference to that old British campaign in Sumatra. One hopes our occupation of Iraq will not be as long or as fruitless. Certainly I find no mention of the battle where British, Indian and Japanese troops were united in a fight against the Extremists, as indeed they did also in neighboring Java.

No trace is left of our brief regime in that tropical island. No memorials on land, nothing in the history books. The soldiers who were involved have grown old, and memories fade. One solitary account of our occupation of

Sumatra is recounted in my novel *A Rude Awakening*. And we all know how unreliable are novels as evidence. Although—truly—just recently a scholar by name Richard McMillan rang me to ask about Sumatra and my time there because he is writing a book about the campaign.

Time is busy at its work, like a batch of invisible earthworms, burying organic matter and aerating and draining the soil, preparing for the next crop of events.

In closing, I want to revert to the curious case of the Deli Cinema, the longevity of fantasy, and evolution. Evolution: the surname of Time on Earth. How is it that human skulls are so solidly built they survive thousands of years buried in the earth? Was it not an evolutionary blunder to devote so much energy to bone and so little to brain? Were not the contents of the skull more precious than the skull itself? No wonder thinking is so hard; but bone—even harder...

Had I had but 10% more brain I would never have gone to Sumatra in the first place!

SECTION II

INVENTING

PREFACE TO SECTION II

INVENTING

Paul Harris

Approaching the theme of time and memory from the standpoint of "Inventing" might elicit knowing smiles. Personal memory tends to be selective in nature, often to the point that recollection involves embellishment, if not outright invention. Humans have long experienced the visceral reality or realism of memories, though, in many quite serious ways. Once a memory is felt or recalled as if it were a record of past experience, the fluidity of time gets frozen; the reality of that experience becomes quite fixed for the subject. The implications of this simple fact are evident in the brief history of modern psychology, from Freud's own shifting speculations concerning female patients' memories of sexual abuse to the furor surrounding many cases of "recalled memories" of abuse from the 1980's onwards. The elusive nature of what happens 'in' time is itself a product of the complexities arising from how temporality conditions the mental lives of human beings. The amazing plasticity of memory as such, the ways in which memory moves between a passive recall of events and an active formation of impressions of the past, underscore the indelible link between time and memory. Henri Bergson famously said that "time is invention, or it is nothing at all."[1] One might introduce the nuanced issues of this section of the volume by saying that 'memory is invention, or it is nothing at all.'

As the following essays attest, questions of memory and inventing are not confined to assessing the truthfulness of individual memories. Thanks to the human capacity to invent new tools, memories form and circulate in increasingly mediated, distributed ways. Over time, new techniques and technologies for storing memory have changed the very meaning of "memory," and memory has become a less predominantly cerebral property of human beings. The earliest examples of writing from many cultures consist

---

[1] Henri Bergson, *Creative Evolution* (London: MacMillan, 1911), p. 361.

of lists—inscription was often used to store records and data that exceeded the mnemonic capacities of individuals. A stirring evolutionary and historical account of memory along these lines is presented by Merlin Donald in *The Origins of the Modern Mind*. Donald argues that human brains function within a larger "cognitive architecture," that encompasses human minds, the representational systems they develop, and the tools through which such systems are deployed. He concludes that, "We act in cognitive collectivities, in symbiosis with external memory systems. As we develop new external symbolic configurations and modalities, we reconfigure our own mental architecture in nontrivial ways."[2] In other words, time, mind and memory are strongly intertwined: emerging mnemonic tools recursively impact the ways in which we learn and think, which in turn modifies brain development, and so on in entangled loops.

The essays in this section of *Time and Memory* investigate different ways in which inventing has entered into human memory. Mary Schmelzer revisits the history and practices of the memory palace, one of the most renowned and fascinating of mnemonic inventions. Rather than emphasizing their power to help a person store and retrieve information, however, Schmelzer demonstrates how memory palaces ultimately reveal the temporal contingencies that intervene in mnemonic practice, seen in "the constructedness, selectivity, and interpretive status of memory." She then goes on to explore how contemporary theoretical views of memory and language shed light on the dynamics of memory's selective and preferential predilections. Heike Klippel's essay examines how psychological writings on memory in the early 20th century yield productive terms in which to understand new mnemonic formations emerging in conjunction with the early cinema of the period. The cinema in fact substantially effected human perceptions and conceptions of time in many complex ways.[3] Klippel resuscitates the work of Bergson and Freud's contemporary August Gallinger, who described the quality of "self-evidence" inherent in the process of remembrance, in which the Ego experiences itself without differentiating between past and present. Klippel compares the experience of self-evidence, which entails "the presence of a past that as such has never been real and a sensual shimmer that owes its

---

[2] Merlin Donald, *The Origins of the Modern Mind: Three Stages in the Evolution of Culture and Cognition* (Cambridge, Mass.: Harvard University Press, 1991), p. 382.

[3] For a thorough treatment of this issue, see Mary Ann Doane, *The Emergence of Cinematic Time: Modernity, Contingency, the Archive* (Cambridge, Mass.: Harvard University Press, 2002).

existence to precisely this irreality," to the kind of self-recognition enabled in cinema, where the spectator sees others' experiences from within, as it were. The cinema thus stimulates a kind of psychological experience, in which it becomes a kind of external memory system that does not store but creates images with a deeply mnemonic quality.

Memory and inventing inevitably raises complex ethical and political issues as well. Ever since Theodor Adorno argued that "writing poetry after Auschwitz is barbaric,"[4] questions of memorializing history have been contested among artists and critics in a number of ways. Clearly, the Holocaust, and on a different level, the atomic bomb, demanded that new kinds of memory be invented, that could be adequate in the face of the potential annihilation both phenomena threaten. In dramatic terms, prospective apocalypse presses for preservative commemoration. Claude Lanzmann's *Shoah* and Marguerite Duras' and Alain Resnais's *Hiroshima Mon Amour* could be seen as representing polar models of memory in the wake of disappearance and destruction: the one an exhaustive set of testimonies and the other an evocative, impressionistic set of images and words. Michal Ben-Horin explores memory devices invented by writers clearly confronting questions of representation and history in the wake of World War II and the Shoah: Yoel Hoffman and W. G. Sebald. Ben-Horin shows how these writers use techniques borrowed from music to disrupt the strict linearity of "absolute time," and create unique "memory poetics" that play on the phonetic and patterning aspects of music as translated into narrative expression. Their techniques create new ways of commemorating that do not treat the past as set in stone but open different forms of dialogue with it.

Historians might be quick to respond that aesthetic inventions that open past events to reconsideration cannot, unfortunately, reverse the effects of these events. Memory in a historical sense often appears as irreversible as time itself. Once a certain impression of an event or series of events is formed, it is notoriously difficult to dislodge it from popular perception—especially if the formation comes in a context of specific urgency or importance. Polls taken in 2003 found that 70% of Americans believed Saddam Hussein was involved in the attacks of September 11, 2001, despite their being no proof or conclusive evidence of such a link; these are only a recent example of the stubbornness of impressions formed over contentious issues. Katherine Sibley examines a fascinating case of the institutionalization of memory, in carefully

[4] Theodor W. Adorno, "Cultural Criticism and Society," in *Prisms*, trans. Samuel and Shierry Weber (Camb. Mass.: MIT Press, 1967), p. 19.

tracing the U.S. government's investigations of J. Robert Oppenheimer and its decisions about his security clearance. Sibley uncovers several layers of memory construction, starting from the widespread general impression that Oppenheimer was treated unfairly because of McCarthyism. While fully acknowledging that Cold War politics played a large part in Oppenheimer's treatment by the government, she also argues that his own behaviour some eleven years before, when asked about Soviet intelligence interest in the Manhattan Project, was in fact more decisive in shaping conclusions reached about him. Sibley thus ultimately shows how inventions of historical memories may swing from one extreme to the other—Oppenheimer vilified for Communist ties, Oppenheimer as victim of McCarthyism—but take persistent hold thanks to institutional and social constructions of memory.

In this section's final essay, Erik Ringan Douglas attempts to identify the foundations of the forming and/or inventing of memories in the first place. In its disciplinary orientation and ambitious scope, Douglas's essay hearkens back to work within the nexus of philosophy and physics especially prevalent in the first several volumes of *The Study of Time* series. Douglas contrasts two definitions of memory that point to different, mutually exclusive concepts of time. Memory as a mode of preserving the past rests on an assumption of the existence of the past, a tenet that forms part of what Douglas terms the chronological conception of time. If memory is rather understood as a species of intentionality, then it may be integrated into what Douglas calls a rhealogical notion of time. Douglas's aim is, finally, speculative and synthetic: he seeks to show how debates between these conceptions of time and memory rests on a common ontological foundation.

The essays in this section of *Time and Memory* reflect an interest in memory and inventing that remains distinct from questions of the factuality or verifiability of individual, psychological memories. But while the thorny question of the extent to which specific individual memories are inventions is bracketed, the complexity, malleability, and constructedness of memories in the hands of various "external memory systems" becomes starkly clear. It is most likely inevitable, and by turns exhilarating and exasperating, to be caught in the loops of inventing new mnemonic tools and techniques, and struggling to separate invented memories from ostensibly real ones. The human experience of time stretches and folds back on itself, unfolding in complex textures, as past itineraries form the embedded context in which present events take shape.

CHAPTER SEVEN

FURNISHING A MEMORY PALACE: RENAISSANCE MNEMONIC PRACTICE AND THE TIME OF MEMORY

Mary Schmelzer

Summary

Renaissance memory devices, the treasure house that Erasmus adumbrates in *de Copia* the memory palace that Matteo Ricci constructed as a Jesuit missionary in China, the extensive iconography of Renaissance portraiture or even Ignatian Spiritual Exercises, mark the period's keen interest in holding on to memories over time.

Erasmus and Ricci both understand the instability and complexity of memory over time and their mnemonic devices seek to stabilize memory on one hand, but to free it from stasis and calcification over time. No strict constructionists they! The path of their desire shifts the objects in the memory palace or treasure house over time. Some once honored objects are sent off to the attic to be held onto, but to be used less often and to be sometimes forgotten. Desire and memory shift over time.

What interests me most about Renaissance construction of memory is the temporal space in which an item is prepared for the palace, the instant in which a decision is made about the mnemonic assignment of a place in a memory system. This moment signifies more than a curious practice born of a need of a specific time. I propose it as a model for the construction of all memory. The question I pose, "What happens every moment that memory is being constructed in a liminal temporal space over-determined by conscious and unconscious desire that is itself constucted by every moment that has preceded it?"

Time is as tricky, elusive, and variable as memory. It is difficult to take the measure of either, but we keep trying to fathom time and naturalize memory. In this essay I argue that the complicated mnemonic practices of Renaissance culture adumbrate the synchronic density of every moment of linear time, while, at the same time, figuring forth the structural fullness of unconscious time. These were never important issues for memory practitioners from Quintillian until Matteo Ricci, but their concerns with the connection between time and memory through the long history of rhetorical practice provides me with a fertile starting place for my work.

The frequency and variety of Renaissance memory devices mark a keen awareness of both the difficulty and necessity of holding on to memory over time, especially when writing and the dissemination of manuscripts was technologically limited. The treasure house that Erasmus adumbrates in *de Copia*, the memory palace that Matteo Ricci constructed as a Jesuit

Jo Alyson Parker, Michael Crawford, Paul Harris (Eds), Time and Memory, pp. 129–143

missionary in China, the extensive iconography of Renaissance portraiture, or even the Ignatian Spiritual Exercises, each in its discrete way and for different reasons, seeks to connect mnemonic practices with rhetorical performance.

Across the ages, much controversy about the efficacy of memory practice has arisen. Much of it has been seen as cumbersome, unwieldy, and non-productive by rhetoricians from Cicero to the present. To many it made more sense to remember facts and situations directly, without diverting one's energy to the construction of vast mnemonic systems. I enter the dialogue here by challenging what seems the assumed possibility of a direct relation between what is experienced and any subsequent record of that experience and suggest that those engaged in memory practice understood a more convoluted relationship between words and things. Mnemonicists insisted on a temporal space between the material reality of an event or idea and its insertion into memory as well as between the matter of memory and its deployment. Their practice foregrounded the interpretive status of memory, and, in so doing, offers me across the centuries a model for the temporal construction of all memory that embodies post-structural linguistic and psychoanalytic discursive theory. From this per-spective, the unwieldiness of these constructions disappears, as their synchronicity is emphasized. Put simply, it takes more linear time to speak or write about the endlessly complicated moment of memory construction or for the moment of the deployment of memory into speaking or writing than it does to think it. There is always interpretation going on in every memory act. I wish to look closely at the time of memory construction inherent in Renaissance mnemonic practice and argue that every moment of remembering is as constructed, complex, and interpreted as an Erasmian notebook or Matteo Ricci's memory palace, and that this offers invaluable insights into the contents and temporality of the unconscious.

In the early sixteenth century, c. 1510–12, Erasmus stressed the value of a vigorous reading and writing program that would encourage the students at St. Paul's School who would become the future political, religious, and literary luminaries of sixteenth century England to construct a treasure house of citations, figures, and exempla that they might "have at the ready" to use powerfully and extemporaneously in the rhetorical performance which was the goal of their education. Such competence comprehended the necessity of memory in an episteme delimited not by the availability of absolute truth or knowledge, but one derived from Aristotle's *Topica*, and inside the sub-lunar or post-Edenic limitations that shaped epistemological expectations allowing only the possibility of "knowing about" the "The Things of the World," to use Michel Foucault's chapter title from *The Order of Things*.[1]

Where Erasmus practiced the mnemonic of the notebook and encouraged students to construct a figurative treasure house of apt citation to be re-deployed performatively in discrete circumstances, Matteo Ricci chose the imaginative construction of a virtual memory palace as a way of domesticating and managing at least that aspect of individual interiority. These palaces, in their long history from the Greeks to the late Renaissance, relied on connecting especially striking objects, with a whole constellation of ideas that the artifact would embody and recall. For example, blind justice holds balance scales and figures forth the complexities of the virtue. Moreover, in a more carefully delineated object every fold of a robe, every ornament, and every physical attribute could be used as a mnemonic. That way, vast quantities of information might be stored in a single image.

This object was then placed in an architectural space that could be derived from something familiar like famous buildings in a city or a real street or neighborhood; or the area could be an imaginative amalgam of places that would produce a structure that facilitated the storage of memory images. Such places were to be quiet, uncrowded, and of appropriate amplitude, lighting and differentiation. The rooms of a memory palace should not be especially similar one to the other. Each object was situated relationally with all the others so that they formed a chain of signification. If a viewer recognized Justice in one corner of a room, she might more easily recognize Fortitude in the next chosen space. Often, the architectural spaces held ideas about a single subject. The rules of navigation might best be remembered by filling a sailing ship with countless nautical objects. A church might be furnished with statues, stained glass, and paintings that served as memory prompts. Elaborate memory constructions might include many buildings, each assigned to a discrete area of inquiry. Objects in the memory palace could be moved around as their significance increased or diminished. Some objects were sent to an attic for storage. Other rooms might be sealed off as their usefulness became marginalized.

---

[1] Michel Foucault, *The Order of Things: An Archeology of the Human Science* (New York: Vintage Books, 1973). Two other works helped me to understand the importance of Erasmiam rhetorical practice in a world that questioned the availability of truth in the episteme of the English Renaissance. I recommend them both: Terrance Cave's *The Cornucopian Text* (Oxford: Clarendon Press, 1979) as well as Marion Trousdale's *Shakespeare and the Rhetoricians* (London: Scholar's Press, 1982).

In choosing this method, Ricci placed himself in the tradition of mnemotechnicians who found value in storing memories in architectural places. Credit for developing the first memory palace mnemonic is given to the poet Simonides of Ceos, circa 556 to 468 BC. Legend has it that he used the memory of individual places in a space to identify the remains of fellow guests at a banquet who were killed when the roof fell in on the dining hall of the home of Scopas in Thessaly at a moment when he was called away. Scopas had reneged on a promise to pay Simonides for a lyric that he wrote praising Castor and Pollux. Myth has it that it was the celestial twins who called him out of the room before the catastrophe. Frances Yates tells the story this way:

> The corpses were so mangled that the relatives who came to take them away for burial were unable to identify them. But Simonides remembered the places at which they had been sitting at the table and was therefore able to indicate to the relatives which were their dead.... And this experience suggested to the poet the principles of memory of which he is said to have been the inventor.[2]

While this Art of Memory trails through early Western history with recommendations on, and warnings against, its use by both Plato and Aristotle, the sources for Renaissance as well as earlier Medieval practice were largely Roman and almost exclusively limited to three sources: Cicero's *De Oratore*, another textbook, *Ad Herennium* circa 86–82 BC which was incorrectly attributed to Cicero until the Renaissance, and Book XI of Quintillian's *Institutio Oratoria*. Each of these sources describes techniques that depend on the intensification of sense memory, especially the visual, in order to create a three dimensional, interior space that might be entered, furnished, and explored.

Frances Yates, whose 1966 *The Art of Memory* has become the reliable gold standard for anyone wishing to work with mnemonics, discourages her readers from thinking that particular practices of Classical memory are fully recoverable. She does, however, suggest that its general principles are described most clearly by Quintillian in the first century AD. She quotes this passage from Quintillian about the construction of memory places:

---

[2] Francis Yates has written the quintessential text on the history and use of memory practice in *The Art of Memory* (Chicago: University of Chicago Press, 1966), 2. More recent and important explorations of the subject have been done by Mary Carruthers in *The Book of Memory, A Study of Memory in Medieval Culture* (Cambridge: Cambridge University Press, 1991), and by William Engel in two valuable works: *Mapping Mortality: The Persistence of Memory and Melancholy in Early Modern England* (Amherst: University of Massachusetts Press, 1995), and *Death and Drama in Renaissance England* (Oxford: Oxford University Press, 2003).

Places are chosen and marked with the utmost possible variety, as a spacious house divided into a number of rooms. Everything of note therein is diligently imprinted on the mind, in order that thought may be able to run through all parts without let or hindrance. The first task is to secure that there shall be no difficulty in running through these, for that memory must be firmly fixed which helps another memory. Then what has been written down [Erasmian treasuries] or thought of [Riccian palaces], is noted by a sign to remind of it. The sign may be drawn from a whole 'thing', as navigation or warfare, or from some 'word'; for what is slipping from memory is recovered by the admonition of a single word. However let us suppose that the sign is drawn from navigation, as, for instance, and anchor; or from warfare as, for example, a weapon. These signs are then arranged as follows. The first notion is placed, as it were, in the forecourt; the second let us say in the atrium; the remainder are placed in order all around the impluvium, and committed not only to bedrooms and parlours, but even to statues and the like. This is done when it is required to revise the memory, one begins from the first place to run through all, demanding what has been entrusted to them, of which one will be reminded by the image. Thus, however numerous are the particulars which it is required to remember, all are linked one to another as in a chorus nor can what follows wander far from what has gone before to which it is joined, only the preliminary labor of learning being required.

What I have spoken of as being done in a house can also be done in public dwellings, or on a long journey or going through a city with or without pictures (22–23).

Here, it seems to me, Quintillian obscures as much as he enlightens. At the same time, he uncovers notions about the status of memory that help us with the project at hand, connecting the temporalized memory construction of mnemonic practice with contemporary discursive and psychoanalytic understanding of memory.

Whether one is reading Roman Quintillian or Renaissance Matteo Ricci, someone who shares my theoretical position will not fail to be struck with qualities that were, perhaps, of little interest to either of them. Most importantly, the method uncovers the constructedness, selectivity, and interpretive status of memory. An architectural space is entered by its creator who then places a mnemonic object in it, an anchor, in the bow of a sailing vessel, for example. When that object is returned to, it will serve as a reminder of what one needs to know to place an anchor properly precisely because it is properly placed and weighted in the memory image. In addition, a person could use the sailboat figuratively as a life journey or a ship of state and understand that anchor from an entirely different position. No two ships embody the same set of ideas. One ship can be read from many places.

Memory, furthermore, is as selective as it is constructed. Memory palaces serve functional utility. What is deemed *un* or less important may be discarded.

Every act of remembering becomes, at the same time, an act of forgetting in a decision made by an individual.[3] Memory is further relational, "linked one to another in a chorus," Quintillian says (Yates, 22). Context drives mnemonic technique as much as does the vivid imagery attached to an individual object. The place in the memory palace is like any place of signification, sensible only as it connects with other signs in a field.

Finally, memory is not necessarily the record of an event. Quintillian notes that even in the act of constructing memory what is actually occurring is the recovery of that which is already slipping away. Memory is never in the present and in this classic model marks absence and loss. Memory practice illuminates a desire to hold onto a moment in time that has never had any staying power to begin with. Memory has no grasp on the present of an experience; it rather only records the circuitous construction that can never be present to what it seeks to circumscribe. Mnemonic memory is the constructed absence of an unstable and unavailable present; as such, it seems to me, to model every act of memory.

Erasmus and Ricci both understood the instability and complexity of memory over time. In common with their Classical and Scholastic forbears, they focused on the utility of mnemonic practice and were concerned about the temporal problematic of memory. Erasmus's program for copious discourse called for a vast reading program and was understood as kinetic and dynamic. He saw memory and memory technique as essential for rhetorical power. Ricci, on the other hand, had few books in China and was forced to use the more traditional techniques of architectural memory as a tool for evangelization. His method recovers the Renaissance Memory Palace for me in ways that will help connect the conscious formation of memory over measurable time to the discrete, but constant, construction of memory in the synchronic time of unconscious.

In 1984, Jonathan Spence published *The Memory Palace of Matteo Ricci*, introducing his readers to both Ricci and his mnemonic technique. Spence recounts the facts of Ricci's life by organizing his story around the few mnemonic pieces he has recovered. After he introduces his reader to the

---

[3] John Willis wrote on the significance of forgetting in *Mnemonica; sive Ars Reminiscendi* (London, 1618), in English: *Mnemonicas or The Art of Memory* (London: Leonard Sowersby, 1661). Grant Williams use of Willis in "Introduction: Sites of Forgetting in Early Modern English Culture," coauthored With Christopher Ivec , in *Forgetting in Early Modern English Literature and Culture: Lethe's Legacies*, edited by Williams and Ivic (London: Routledge, 2004), provides an inventive model for thinking about the connection between remembering and forgetting.

memory image for war, for example, he proceeds to examine the stormy wars in Italy at the time of Ricci's birth in 1552. He performs mnemonic technique as he explains it and in so doing produces a rich history of Counter Reformation Europe and sixteenth century China.[4]

In 1582, during the Ming dynasty, the Jesuit Matteo Ricci arrived in China where he was to spend almost thirty years introducing the Christian faith and Western thought to scholars, rulers, and literati, a highly sophisticated, albeit alien, society, one in which he thrived. He learned Chinese, but was limited in the ways he could bring the Gospel message without a Bible in that language. To overcome that difficulty he conceived of two strategies: using Chinese ideographs to construct a memory palace in his 1596 draft of *The Treatise on Mnemonic Arts* and later working with printer Cheng Dayue (1606) publishing an Ink Garden with illustrations of Bible passages accompanied by brief explanations of their content.

Ricci's use of sense-memory devices is especially apposite for a Jesuit since Ignatius of Loyola the order's founder saw the vivid restructuring of memory as vital to spiritual training. Spence says, "In order that his followers might live the biblical narrative in all its force, Ignatius instructed them to apply their five senses to those scriptural passages that they were contemplating" (15). Here, Ignatius follows the lead of Medieval Dominicans who believed that if a memory is to be effective, it must be must be firmly imprinted . The firmest impressions were vividly sensual even to the point of being shocking; the soul must be moved for memory to be effective (Yates, 65).

Ricci's desire to move souls colored every decision he made about imbuing an object with meaning. To serve his purpose he engaged in conscious acts of restructuring the physical aspect of an object in order that he might redact its meaning. The first image in Spence's study, The Warriors (Fig. 1), shows the Chinese ideograph for war, pronounced *wu*.

To prepare it for placement in the reception hall of his memory palace, Ricci divided it into two separate ideographs one for spear, and another that carried with it a sense of "to stop" or "to prevent." This second sign is under the left arm of the overreaching image of the weapon. Ricci then anthropomorphizes each into warriors, the larger with a spear ready to strike, the smaller grasping at the spear trying to save himself from falling victim to it. Each is given facial expression, dress, and myriad qualities that distinguishes each from the other, as well as suggesting aspects of both warriors. In so doing Ricci encourages

---

[4] Jonathon Spence, *The Memory Palace of Matteo Ricci* (New York: Penguin Books, 1984).

Figure 1: Ideograph of *wu* (warrior).

his audience to internalize a concept of war that also holds out a possibility of peace. Spence notes that in this mnemonic, Ricci lights upon a dichotomy that resonates both Christian and Chinese intellectual and moral positions:

> In dividing the ideograph in this fashion Ricci follows—whether wittingly or not—a tradition among Chinese scholars reaching back almost two millenia, a tradition that allowed one to see buried inside the word for war, the possibilities, however frail, for peace (24).

We need to pay attention to the temporal complexity of the construction of this mnemonic. In linear time Ricci analyzes an image, decides how it can best serve his needs, divides it into two parts, adorns each part with mnemonically helpful particulars, carefully places it in his memory palace, decides how it will be lit and approached, and how it will connect with other objects placed in close proximity. While it is not possible to say how much time Ricci spent thinking about this image or the other images no longer available to us, that time, nonetheless, is more than instantaneous. This memory is constructed in a measurable time that negotiates the space between two moments: the time of the event and the time of its encoding into memory.

The second image from what remains of the Memory Treatise might shed more light on Ricci's work. This ideograph, yao, (Fig. 2), is elusive. It can mean something desired, needed or important.

Ricci in his 1605 *Fundamental Christian Teachings* uses the image to translate the word "fundamental." The journey of this ideograph points to the connection between the meaning of a memory and the desire and discursive circumstances of the rememberer.

Again Ricci divides the image, this time in half horizontally. The top half means west and its pronunciation *xi* calls up the ancient Western Chinese kingdom, *Xixia*. The lower half is woman. The image to be stored up then

Figure 2: Ideograph for *yao*.

becomes a woman from the west with an exotic look in the "vivid dress, the felt boots, the braided hair common to that region" (Spence, 95). Most of the Eastern Chinese would recognize her as Muslim, a *huihui*. Ricci knows that Jews and Christians were designated *huihui* as well, and connects this mnemonic of the exotic tribeswoman from the west with fundamental beliefs of Islam, Judaism, and Christianity. One *huihui* creates a multiplicity of potential explorations for the differences between Eastern and Western spiritual practice. Ricci places this image in the northeast corner of the reception hall close enough to the two warriors so that their stories can inform each other where, as Spence says, "She will stay…in the quiet light that suffuses the memory palace for as long as he chooses to leave her" (96).

In the ideographs Ricci gave new sense impact to traditional Chinese images; the Ink Garden illustrations help with remembering not so much the stories from the Gospels as the spiritual lessons Ricci wrote to accompany each picture. The actual depiction in the print matters less than the interpretation. This memory work itself confirms the disconnection between the memory of a thing and the thing itself as it filters through the screens of mnemonic encoding, confirming the distance between an item in a memory palace and its always absent dimensional reality. So far the images we have looked at underscore the temporal deferral of meaning in memory. In this next drawing we see both elements of Derridean *différance*.

This picture, *The Apostle in the Waves*, a drawing now in the public domain, (Fig. 3) is complicated on a number of levels.

The image itself is problematic in a universe embattled over subtle doctrinal distinctions as was Ricci's at the time. It claims to depict that story in Matthew's Gospel in which Christ walks on water. Ricci would have liked to have had Cheng Dayue engrave that image, but at the time that Cheng

Figure 3: Apostle in the Waves (Ricci).

agreed to work with Ricci on the project, the Jesuit had lent his copy of Jeronimo Nadal's *Images of Christ to* another missionary, to use as a means of evangelization. Ricci wrote in 1605:

> This book is of even greater use than the Bible in the sense that while we are in the middle of talking we can also place right in front of their eyes things that with words alone we would not be able to make clear (Spence, 62).

He used instead an illustration from the eight volume Plantin polyglot Bible that had just arrived in China. In it Christ is standing firmly on the shore. This image was actually an illustration of Christ appearing to his disciples after the Resurrection, with the stigmata of Crucifixion removed from the extended right arm. The real of the illustration changes signification at the behest of the desire of the user. Ricci's accompanying text suited his own views of what he thought might appeal to the Chinese. The last sentence of this story in Matthew says, "Truly you are the Son of God." Ricci focuses on the story as a means of comprehending the way one lives virtuously:

> A man who has strong faith in the Way can walk on the yielding water as if on solid rock, but if he goes back to doubting, then the water will go back to its true nature and how can he stay brave? (Ricci in Spence, 60)

The wrong picture did the right work by helping him inculcate a habit of Christian spirituality to a community whose belief system focused more on moral practice than belief, the *huihui* of the cross made amenable to the fundamental belief of Confucianism, Taoism, and Buddhism.

While Ricci is laboring in China, the intellectual climate in Europe is changing rapidly and the value of mnemonic technique is waning. In the early moments of *The Anatomy of Melancholy*, Robert Burton says, "there is no end of writing of books" (I, 22) and again, "who can read them? As already we shall have a vast chaos and confusion of books, we are oppressed with them..." (I, 24).[5] This 1620 manuscript marks the end of the wide use of mnemonic technique, especially the Erasmian program based as it was on a vast reading program, but I think that much the same case might be made for the memory palace with its need for ever-expanding buildings to house exponentially increasing information. The nascent episteme of the Age of Reason regards memory as immediate and erases, for the most part, the importance, or even the existence of, a temporal space between a thing and its inclusion in memory. Intellectual history has little to say about the

---

[5] Robert Burton, *The Anatomy of Melancholy*, edited by Holbrook Jackson, 3 volumes (London: Everyman's Library, 1932).

memory palace after the early sixteenth century. It itself has been placed in an attic for historical curiosities.

But I argue that the palace does not completely crumble as history might conjecture; rather it has migrated to the unconscious. What Renaissance rhetoricians and teachers did consciously over minutes, hours, sometimes weeks, happens instantaneously in the unconscious. In psychoanalytic practice the discursive *lapsus* of diurnal time provide entrances into its barred signification. That same bar creates a paradigmatic *differance*, and renders all unconscious memory different and deferred from any actual or real moment. In classic memory work, unconscious processes are invisible, as they often still are.

We still remember the mnemonics that get us through board examinations, and use repetition and visual or aural images to help us recall a name or a phone number. My cell number is mnemonically a picture of a leanish man who ate for one and worked for seven sick ones. In all it seems to come to little as a subject for serious analysis, but there is nothing simple about the simplest memory practice. What I cannot say easily is why I have chanced upon this set of images to recall the sounds of the numbers. And that matters. Our unconscious memory work is denser and more elusive than the earlier practices to which I suggest it bears much correspondence; this is especially true of the temporal aspect of memory.

Unconscious time bears little relation to the ticking of our watches, but that is not to say that it is atemporal. The time of memory functions constantly, not in discrete, linearly designated interstices, but always in its own synchronic dimension. Another time. By using the insights inherent in classical memory practice we might uncover some of its staggering complexity.

Jacques Lacan says that the unconscious is constructed like a language; the lingustic model he has in mind is Ferdinand de Saurrure's that sees all meaning as relational. The value of a sign, single word, or even a letter is empty until it is contextualized for each is arbitrary and unstable on its own:

> In the end, the principle it comes down to is the fundamental principle of the arbitrariness of the sign. It is only through the differences between signs that it will be possible to give them a function, a value. In language there are only differences. Even more importantly: a difference generally implies positive terms between which the difference is set up; but in language there are only differences *without positive terms*. . . . language has neither ideas nor sounds that existed before the linguistic system.[6]

---

[6] Ferdinand de Saussure, *A Course in General Linguistics*, edited by Charles Bally and Albert Sachehlaye in collaboration with Albert Heidinger, translated by Wade Baskin (New York, McGraw Hill, 1959), 120.

Lacan's unconscious recalls the memory palace: a crowded, arbitrary, but systemic construction. In the unconscious, however, the rooms are much denser. Unconscious memory might be more like the basement of the Metropolitan Museum or my attic, chock-a-block with stuff that might have some inherent, perhaps monetary, value but that ends up relegated to invisibility because another like it but judged better gains pride of place. Or because it is out of fashion, or vulgar or not vulgar enough by current standards or simply because or not because. To trope Pascal, the unconscious has reasons that reason does not know. But it is always there, and, unlike the objects in the memory palace or my attic, the things of the unconscious seem on a conscious level to move around on their own and can surface in dreams, slips, or accidents without being invoked. Why did I think of Barbara Sauerwald yesterday whom I have not seen nor heard much about in forty years? We were in college together but not especially close. Consciously it is a curiosity; unconsciously it is a mnemonic object that might uncover something meaningful for me if I am willing to scrutinize that memory in its unconscious fullness, an activity that exhausts more linear time than most of us are willing to give to it.

What gets saved and what gets forgotten in unconscious memory? It seems very little is forgotten, although most of our conscious experience might suggest otherwise. The plentitude of associations from the past or a feared or desired future cannot be measured. Every experience, feeling, or image remains in something more powerful than an unconscious trace. It is all there all the time and always shifting relationally as additional psychic material is added. What I might have to say about Barbara Sauerwald as I write will be substantially different from what I might say were you to question me after reading this. The work of teasing meaning out of memory objects adds another dimension to the time of memory.

Like the Saussurean linguistic chain, every memory object changes meaning as it is considered against everything else in the memory bank, an unstable construct that shifts and changes over the merest movement of measurable time. While this bank might not give interest, it is our unconscious psychic interests that make a relation important as well as overwhelming. It is difficult to grasp that so much is being processed in what seems so little time. Too much, I might be tempted to say with Burton if I insisted that the temporality of such a moment could be accessed consciously.

The unconscious is a vital force and not a place of sequestration, a dynamism, driving and driven by desire. Primarily, for me here the time of unconscious tropes from the niggardly cadged lacunae of the linear, into a synchronic counter measure capable of constructing complicated edifices that cannot be contained or contacted in ordinary, a word I use advisedly, discourse. And it never ceases. We are always remembering everything in an unconscious temporality that can accommodate the vastness of the past, the elusive, unavailable present as well as an imagined future.

Unlike classical memory construction, it seems to accomplish all this with little effort on the part of the individual. Cicero was not alone in thinking that he could not grasp the value of memory practice arguing that it required as much effort to encode and recall particulars mnemonically as it did to remember facts and ideas themselves (Yates, 19). Unconscious memory construction happens literally without thinking and exists in an autonomous loop constantly accumulating, redacting, and deploying memory images along the path of an always present but unarticulated desire.

Like the *huihui* woman from the west, some images have greater mnemonic intensity than others. The Freudian family romance posits a primal event that repeats itself endlessly because its initial construction derived from a shock that cannot be assimilated. Lacan addresses the transgressive origin of psychic structure in the more abstract notion of the Name of the Father.[7] In the end, perhaps, it comes to the same thing. All objects in the unconscious memory palace unfold the first shocking object placed there in the seminal emotional engagement with an alienating symbolic order. Because of this, unconscious moments are in important ways always the same moment. Every new imprint on the unconscious is determined by everything that is already there, a memory palace with a varied but repeated intense mnemonic object dressed, positioned, and lit differently in each of its incarnations, but, nonetheless, more alike than different from all the others. It takes the work of analysis to grasp these repetitions, an endeavor, I would argue, significantly more demanding than mnemonic construction for it requires that the conscious and the unconscious find a way to play nicely together. The conscious self depends on the functional utility of memory; the unconscious self insists on producing a chain of replication whose seeming caprice is frequently challenged and dismissed by that other self.

Two selves, two times, one quite literally faster than a speeding bullet, endowing objects with mnemonic value that connect them with sites in a memory palace—a process that often leaves the socially constructed self

---

[7] It is often difficult to know where to begin with Lacan. Most theorists know of the mirror stage and have read some seminars. If your interest is clinical I recommend *The Four Fundamental Concepts of Psychoanalysis*, edited by Jacques-Alain Miller, translated by Alan Sheridan (New York: Norton, 1977), as a fecund starting point.

breathless, confused, and alienated. While the relationship between both selves is vexed, it is not disconnected. Perhaps psychoanalysis is psychic marriage counseling that seeks to establish rules and boundaries for the frequently quarreling couple. Healthy integration of these two selves asks a person to accept that conscious choice of memory objects is always replete with unconscious resonance and that the time of memory of the latter is ungraspable in the order of the conscious.

The richness and complexity of unconscious time exhilarates me. Sigmund Freud was astonished "to find in it an exception to the philosopher's assertion that space and time are necessary forms of our mental acts."[8] My astonishment comes, however, from a conviction that there is space and time in the synchronic unconscious the understanding of which eludes our linear conscious comprehension. We know so much more than we know that we know. I take my leave as J. T. Fraser took his from *Of Time Passion and Knowledge*: Time "subsumes several levels of temporality each contributing some new and unique qualities to the noetic temporality of man."[9] The time of memory complicates things as it leads us reluctantly to some version of the truth about ourselves.

---

[8] *The Complete Introductory Lectures* (New York: Norton, 1966), 538.
[9] Princeton: Princeton University Press, 1990, 446.
I would like to thank Frances Chen, a tenth grade student at the Agnes Irwin School, for her graceful rendering of the Chinese ideographs that appear in this essay.

CHAPTER EIGHT

THE RADIANCE OF TRUTH:
REMEMBRANCE, SELF-EVIDENCE AND CINEMA

Heike Klippel

Summary

In the following essay I will discuss a certain characteristic of human memory, namely its self-evidence, as it has been theorized at the turn of the 20th century. I will then relate this discussion to the cinema in order to trace a quality of filmic images which can be considered analogous to the self-evidence of memories so often experienced when we remember. When speaking of the self-evidence inherent in memory, I refer to the truth of memories as it is subjectively perceived, regardless of whether they can be proven or not. The concept of self-evidence describes an experience that itself belongs to and emerges from the non- or pre-conceptual realm because such a spontaneous insight does not need a reason for how it came about. A phenomenon considered self-evident may have good reasons, but it also does without them because what is commonly called "self-evident" are such matters that are immediately obvious and experiences whose truth appears to be beyond doubt. When speaking of the self-evidence of cinematic images, I do not want to imply that the content and intended meaning of films is in itself obvious. Rather, I would like to address an intensity of subjective film perception as it emerges from the direct relation between the spectator and what is seen on screen. This relation allows for the cognition of a 'reality' which cannot be logically deduced, but which nonetheless does not necessarily escape objectification. Whether or not this realization coincides with the communicative intentions of a film is as irrelevant as is the factuality of experience with regard to the self-evidence of memories.

The attempt to present, by means of language, a phenomenon that itself escapes logical conceptualization carries certain risks which are mirrored in the form of this essay. Argumentation and reasoning collide with a subject that attempts to escape these forms of mastery. The self-evidence of cinematic images as well as of memory images is based on liveliness and plasticity, and their powers of persuasion are always also a seduction. Without rationalizing it away, this moment of seduction shall be taken seriously in the following essay.

Introduction

The period around 1900 is considered to have been one of cultural upheaval, during which the headlong progress of technical and media developments challenged traditional forms of thinking. In particular, the history of film, starting in 1895, is considered by Hans-Ulrich Gumbrecht as the beginning of an epistemological shift from conceptual thought to "the description of

Jo Alyson Parker, Michael Crawford, Paul Harris (Eds), Time and Memory, pp. 145–161
© 2006 Koninklijke Brill N.V. Printed in the Netherlands.

preconceptual layers of consciousness,"[1] because "filmic information cannot anymore be related through concepts"[2] but nevertheless creates a 'knowledge' whose relevance cannot be denied. According to Gumbrecht, the most important authors theorizing this "insight into the incommensurability between the moving images and the stability of concepts"[3] are Bergson and Freud, and the most important contributions to a theory of memory stem from them. On the other hand, attempts to freeze theoretically the dynamics of the moving image and the resulting liquefaction of conceptual thought, proved historically ineffective: "While the intellectuals worked on such attempts to save notions of experience, understanding and values as such, the transformation into a body-centered perception progressed in the everyday."[4] In the meantime, both realms have split, but, according to Gumbrecht, the problem is still current. "The epistemological situation of the present still seems to be dominated by the bifurcation of a sphere of conceptualization, stasis, and language and a sphere of perception, movement, and the body."[5]

Around the turn of the 20th century, descriptions of the pre-conceptual can be found in many publications relating to memory theories. In the following, I will focus on Bergson and Freud and read them together with August Gallinger.[6] Gallinger is unknown today, but a closer examination of his work is merited because he brought the issue of self-evidence to a head unlike any other author before or after him.

Around 1900, memory was an important theme in philosophy and psychology because it offered a terrain upon which a number of conflicts could be addressed: for instance, questions regarding the faithfulness of historical representations, questions relating to the interaction between body and mind, the problem of mental images and the element of imagination involved in the formation of judgments, questions concerning the ability to cope with a flood of information and the way in which this information is digested and integrated—to name but a few topics. During this time,

---

[1] Hans-Ulrich Gumbrecht, "Wahrnehmung vs. Erfahrung oder die schnellen Bilder und ihre Interpretationsresistenz," *Kunstforum* 128 (1995): 175.

[2] Ibid.

[3] Ibid.

[4] Ibid.

[5] Ibid., 176.

[6] August Gallinger (1872–1959) was a medical doctor and philosopher. He represented a phenomenologically oriented philosophy. His text "Zur Grundlegung einer Lehre von der Erinnerung [On the Foundation of a Doctrine of Memory]" from 1914 was part of a lively academic discourse before World War I which dealt with psychological and epistemological issues and questions. He returned from emigration after 1945, and was a professor of philosophy in Munich until 1952.

conceptual thought in the humanities suffered from a deep crisis while the new technologies of visualization were simultaneously hypostasized in a way which was to cover up doubts that thus became only more apparent.[7] Memory and remembering became of interest as a human capacity the results of which are, on the one hand, vitally necessary but which, on the other hand, were often held to be questionable. Moreover, the relative obscurity of the memory process complicated the investigations: neither neurological research nor biographical reconstruction could definitively show how, for instance, learning was achieved or memories were created and recalled.

The aforementioned questions equally relate to the simultaneously emerging mass media, above all the cinema. The cinematic experience presents something that seems immediately obvious, sensual-visual "truths", the status of which remains unclear: in the cinema, one sees 'things as they are'—this, however, is not the same as recognition in the way it has been traditionally understood. The complex relation between visibility and meaningfulness can, in the cinema, merge into a self-evident obviousness. Though 'judgements' formed in such a manner avoid intellectual reasoning, they cannot simply be dismissed as being irrational or invalid. They represent mass-medial transformations of Charcot's famous dictum that "seeing is believing," and they, together with judgments based on memory and remembrance, belong to the sphere of pre-conceptual cognition.

In the following, I will attempt to unfold the question of self-evidence inherent in selected theoretical texts concerned with memory dating from the turn of the century. Finally, these will be related to the medium of cinema as it presented itself in its beginning.

SELF-EVIDENCE IN MEMORY

Notions of self-evidence can primarily be found in the context of discussions of subjective remembrance, which represents but a part of memory activities.

---

[7] See Comolli's classic essay in which he shows how the "hype of visibility" prevalent in the 19th century, and, particularly, the invention of the photographic apparatus testify to the crisis in the supremacy of the look that had been established since the Renaissance: "At the very same time that it is thus fascinated and gratified by the multiplicity of scopic instruments which lay a thousand views beneath its gaze, the human eye loses its immemorial privilege; the mechanical eye of the photographic machine now sees in its place, and in certain aspects with more sureness. The photograph stands as at once the triumph and the grave of the eye." Jean-Louis Comolli, "Machines of the Visible" [1971/72], Teresa de Lauretis, Stephen Heath, Eds. The Cinematic Apparatus (New York: St. Martin's Press 1980), 123.

It is generally known that memory does not conserve the past as such, but rather compares and balances the past with the present, thereby proceeding selectively. Thus, subjective memories cannot be expected to offer conclusive evidence: "What matters [...] is not what my past actually was, or even whether I had one; it is only the memories I have now which matter, be they false or true,"[8] writes Henry Price. Memories point to a past the status of which always remains unclear: the historical context of its origin, that is the triggering event, mixes inseparably with its further evolution within the memory system, with the history of its modifications and restructurings; they are being "*usuriert*",[9] as Freud says, that is worn out. In the course of this history we always also find appropriations of "alien materials," i.e. the telling of events one had not been present at, so that the question of what has been truly experienced by oneself cannot anymore conclusively be decided. This position between validity and untruth, between unreliability and a subjective sense of conviction turned memory into an exemplary field of inquiry during this period. The question regarding the factuality of that to which memories refer is usually of subordinate importance. Accurate reproduction is more often expected from those memory activities that proceed automatically or from physical memory. Remembrance, on the other hand, is a special phenomenon the results of which are questioned in regard to their respective meanings rather than in regard to their verifiability.[10]

This is the case, for instance, with Bergson, who proposes a theory of consciousness emphasizing the qualitative properties of consciousness with regard to time. Thus, he develops a counter-position to concepts which attempt to describe human consciousness in spatial terms. What is characteristic for consciousness, according to Bergson, is duration (*durée*): it is essentially memory in which time periods pass into each other, moving from virtuality into actuality in a process of continuous differentiation. The present exists not only simultaneously with its own past, which it is constantly in the process of becoming, but also with the past in general. Consequently, duration is less determined by a sequence of events than by their simultaneity and co-existence. Consciousness is to be regarded as a quality, a merging of

---

[8] Henry H. Price, *Thinking and Experience* (London: Hutchinson 1969), 84.

[9] Josef Breuer and Sigmund Freud, "Studien über Hysterie" [1895], *Gesammelte Werke I* (Frankfurt/M.: Fischer 1999), 88.

[10] "The first [memory], conquered by effort, remains dependent upon our will; the second, entirely spontaneous, is as capricious in reproducing as it is faithful in preserving." Henri Bergson, *Matter and Memory* [*Matière et mémoire*, 1896], (Mineola, New York: Dover 2004), 102.

different complexes, not as a quantity that could be described in spatial terms. Bergson exemplifies his theory with regard to remembrance:

> I smell a rose and immediately confused recollections of childhood come back to my memory. In truth, these recollections have not been called up by the perfume of the rose: I breathe them in with the very scent; it means all that to me.[11]

The parts of the whole that emerge during recollection do not necessarily have to have been previously contained therein as individual components; the act of remembering only consists of a process of interpenetration.

Bergson very aptly grasps the particular fluidity of consciousness in the state of contemplation. However, he does not assume this quality to be the dominant feature of daily consciousness. For theoretical reasons, Bergson differentiates between consciousness and action; as soon as the need for action arises, the contents of consciousness are differentiated and ordered, in order for them to serve a potential activity. For him, remembrance is far removed from every form of action, and thus it is characterized by the virtual simultaneities of consciousness. It is not a return that could be described in any way by spatial metaphors, but a contraction of consciousness into layers of the past. Inhaling memories with the scent of the rose is an experience of self-evidence, and it is made possible because consciousness does not differentiate between different levels of time, but can span the past, present and future simultaneously. The question of truth and justification is irrelevant to this process, as the present, sensual perception triggers an immersion into past times and events, an immersion in which the truth emerges from a synthesis and not from the individual components involved in the process of remembering. It is also irrelevant whether the scent of the rose dates from the same time period as the childhood experiences; it is only significant that it is linked to them. This relation is "true" in the sense that the relation between this scent and this memory is unconditional. Such an experience does not function through a logic of representation which relies on demonstrable referents—if the scent "is" the memory, then there is no distance between the two requiring the construction of a signifying bridge.

An explanation of such coincidences and their meaning would certainly be possible, but not in categories that would satisfy the traditional requirements of constancy and reliability presumed by logical reason. Presentations of comparable structures can be found in Freud's writings. It is

---

[11] Henri Bergson, *Time and Free Will* [*Sur les données immédiates de la conscience*, 1899] (Mineola, New York: Dover 2001), 161.

widely known that very early in his work Freud gave up the assumption that the memories recounted to him by his patients were based on true events; instead, he analyzed their complex structures with regard to their inherent representational dynamics. This analysis involves a meticulous deciphering of the infinite displacements, condensations and disfigurations of the memory contents which continually take on new form each time they enter the realm of consciousness. The 'truth' of memories is subject to continuous negotiations.

A series of mostly shorter texts within Freud's work thematize memories which display, in a particularly stark manner, the disparity between the convincing presentation of a particular memory and its questionable basis in reality.[12] Here, he deals less with neurotic dispositions than with everyday phenomena in which the particularities of memory work manifest themselves. In the following, I will offer a discussion of an exemplary treatise dealing with "screen memories". The phenomenon of "screen memories," so vividly described by Freud, reminds one of film images due to its impressive visuality. Screen memories frequently date back to early childhood, and they mostly consist of indifferent or everyday impressions that often do not even refer to actual events. Usually, they are embellished with hyper-sensory elements (i.e. strong coloring, intensive perceptions of smell or taste); they seem to display hallucinatory qualities which, though seemingly exaggerated, make the respective memory very convincing. Related phenomena are memory illusions, sometimes referred to as *fausse reconnaissance* or *déjà vu*. These, too, are often impressively vivid, but they relate to occurrences that can be proven not to have taken place. 'Screen memories' owe their name to the fact that Freud interpreted them as a condensation of memory elements representing other, usually later, repressed contents which they simultaneously cover up, or screen. In *Screen Memories*[13] he describes the case of a young man who has merged two fantasies into one, both of them combining lost opportunities of the past with potential future developments: these fantasies were condensed into the memory of a slightly altered childhood scene dating from a much

---

[12] Sigmund Freud, "The Psychical Mechanism of Forgetfulness" [1898], *The Standard Edition of the Complete Psychological Works of Sigmund Freud*, Vol. 3, Ed. James Strachey (London Hogarth Press 1962, repr. 1995), 289–297; "Screen Memories" [1899], ibid. 303–322; "The Psychology of Everyday Life" [1901], ibid., Vol. 6; "Fausse Reconnaissance (Déjà vu) in Psycho-Analytical Treatment" [1914], ibid., Vol. 13, 201–207; "Remembering, Replaying and Working Through (Further Recommendations on the Technique of Psycho-Analysis II" [1914], ibid., Vol. 12.

[13] Sigmund Freud, "Screen Memories", ibid., Vol. 3, 303–322.

earlier time period that became highly symbolic as a result of this re-projection. The image of the past thus came to represent desires and hopes for the future. Freud comments on the example as follows:

> There is in general no guarantee of the data produced by our memory. But I am ready to agree with you that the scene is genuine. If so, you selected it from innumerable others of a similar or another kind because, on account of its content (which in itself was indifferent) it was well adapted to represent the two phantasies, which were important enough to you. Recollection of this kind, whose value lies in the fact that it represents in the memory impressions and thoughts of a later date whose content is connected with its own by symbolic or similar links, may appropriately be called a *screen memory*.[14]

Freud's analysis, explaining the scene to be based on later fantasy formations, lets the young man eventually doubt his very strong sense of the authenticity of this memory. Freud responds:

> It is very possible that in the course of this process the childhood scene itself also undergoes changes; I regard it as certain that falsification of memory may be brought about in this way, too. In your case the childhood scene seems only to have had some of its lines engraved more deeply: think of the over-emphasis on the yellow and the exaggerated niceness of the bread. But the raw material was utilizable.[15]

What is interesting in this passage is the use of the term "genuine" in contrast, for instance, to the attribution of "truth", which would have been inappropriate. Practically nothing is "true" about the scene described; on the contrary, it actually lies by pretending to be something else, masking that which is represented as an innocent childhood experience in order to show it and, at the same time, render it unrecognizable. "Genuine" only refers to the authentic core of the memory, a truth that has been whittled down to mere factuality; in this case, nothing more than the actual existence of the green meadow, the dandelion, the children's game. Here it becomes clear how little authenticity actually matters: it is turned into raw material in order to create meanings that maintain only superficial relationships with the manifest memory content. What is problematic is the immanent power of authenticity which seems to suggest that whatever is authentic cannot be wrong. This authenticity characterizes the experience of self-evidence and generates moments that ultimately escape analysis. More impressive than all the clarifying arguments in this case story are its strong sensual intensities:

---

[14] Ibid. 315f.
[15] Ibid. 318.

the yellow dandelion, the green meadow, the taste of bread, the innocently cruel game. Ultimately, it is this plasticity that has such a convincing effect on the young man, the one who remembers, and it is so strong that it even conveys itself to the reader of the text. One remembers the "screen memories" as if they had been one's own memories.

Although the experience of the self-evidence of memories is addressed by Bergson and Freud, it is a subordinate topic with regard to their actual theoretical interests which focus on the presentation of consciousness, on the one hand, and the effects of the unconscious on the other. Among the many, now forgotten treatises on memory from around 1900 there is one text that proves particularly relevant to our discussion. August Gallinger, in a paper titled "Grundlegung zu einer Lehre von der Erinnerung [Foundation of a Doctrine of Remembrance]" written in 1914, explicitly aims to show that the experienced self-evidence of memories is not a construction of the mind but occurs in an unmediated fashion.[16] Using a phenomenologically oriented method, Gallinger gives a detailed discussion of remembrance as a "fact of consciousness". He refers to a number of contemporary writers working with memory experiments, but above all to authors working in the fields of psychology and philosophy; he briefly mentions and quotes Bergson, but not Freud.

One of Gallinger's major concerns was to describe the act of remembering as an independent phenomenon and to differentiate it from the individual's memory as such. Despite all of the relations between remembrance and memory, the act of remembering remains, for Gallinger, an independent process of consciousness. For him, 'memory' refers to the reproductive activities of physical memory: memorizing, learning, and storing; as such, they are unconscious processes. The act of remembering, in contrast, is always coupled with an awareness of the past and thus is a form of experience. In his efforts to present remembrance with its particular qualities as an independent, mental activity, Gallinger hypostasizes the two forms of memory already differentiated by Bergson[17] and delineates an extreme profile of remembrance. Thus, he considers historical accuracy to be wholly irrelevant with regard to remembrance, a thesis which he repeatedly emphasizes: "What is meant here exclusively, is that something appears to the

---

[16] August Gallinger, *Zur Grundlegung einer Lehre von der Erinnerung* (Halle a. S.: Max Niemeyer 1914), 22.

[17] This corresponds to the presentation of different forms of memory in Bergson; however, Gallinger does not explicitly refer to Bergson's text. Cf. Henri Bergson, The two forms of memory, *Matter and Memory*, 86–105.

conscious subject *as* an earlier experience. Whether this subjective perception corresponds with the actual facts or whether it is just an alleged memory, a 'fausse reconnaissance,' remains to be seen".[18] Endlessly tracing and discussing the views of other writers, Gallinger gradually extracts what he considers to be the essence of remembrance: the act of remembering essentially consists in ascribing a subjective past to oneself which is not to be understood as a knowledge of actual events, but rather as a sense of conviction relating to their factuality. He rejects the possibility that these memories could be imaginative constructions, as imagination lacks a reference to the past and, relatedly, the moment of authenticity. This insistence on the independence of the experience of remembering gradually proves to be rooted in a question of identity; it becomes apparent that the act of remembering receives such detailed recognition from Gallinger because it is in the act of remembrance that the omnipresent identity of the Ego with itself is realized.

Apart from the idea that the act of remembering could unify the Ego(s) of different times and places and hence ward off fragmentation and self-alienation, Gallinger is, furthermore, greatly concerned with emphasizing the immediacy of remembrance. "'In remembering I am *immediately* convinced of the reality of my former experiences.'"[19] Consequently, remembrance is not a construction of the mind, as Gallinger repeats in various ways time and again; rather, it describes a capacity of the Ego to experience itself without differentiating between past and present. Remembering does not require "contemplation" or "reflection"; it is not a "reproduction" of the past, there are no "intermediate links," no references between present and past, no "attention [is paid] to the identical and reproducible in the present images. [...] The remembered image is underived and originally [...] accompanied by [...] validity and truth."[20] Remembrance contains recognition and knowledge which, without reasoning, can be grasped immediately. This becomes possible because the Ego can change its perspective and is thus able to readopt an earlier point of view. This capacity of the Ego to view experiences from an earlier perspective and its resultant sense of immediacy guarantees and justifies the certainty experienced in the act of remembering. For Gallinger, this process defies description. The sense of presence elicited by immersing oneself into former times and attitudes is an experience that

---

[18] Gallinger 1914, 22.
[19] Ibid. 113. Gallinger is quoting Johannes Volkelt, *Die Quellen menschlicher Gewißheit* [The Sources of Human Certainty] (Munich 1906) 9 et seq. remarking: "Jessinghaus (Psychol. Studien, Volume VII, P. 337 et seq.) also supports this view." Ibid. 114.
[20] Ibid. Gallinger is again quoting Volkelt.

cannot be verbalized. He can only describe it as an immediacy of existence, and his presentation culminates in the proposal of self-assurance in a dimension in which time has become irrelevant: "We [...] retreat [...] into ourselves, and without further ado, in a certain moment we are there. 'J'y suis maintenant.'—'Now I am there.'"[21]

The objection that the simultaneous experience of past and present would be contradictory in itself is dismissed by Gallinger with the argument that such simultaneity is an absolutely normal phenomenon of consciousness. "In principle, it is almost generally acknowledged that one can be one's own observer [...]".[22] In conclusion, Gallinger briefly mentions that the things viewed in this manner often display a special gleam, a "shimmer that is spread across the objects, adhering to them."[23] He explains this phenomenon with the change of perspective because, in his view, the former emotional value attached to the remembered experience is realized in such a positively colored atmosphere. A beauty is said to often reside in objects of remembrance "seen from the past, [...] which we cannot detect anymore when we actually look again or which appears to be hugely exaggerated. Thus, many people avoid a repetition, 'in order not to destroy their memories.'"[24] The shining reflection of the past in its wondrous presence is only possible by excluding every perception of the time that has passed in the meantime.

Gallinger's writings lack differentiation, but what they do possess, in excess, is an insistence that sometimes borders on obsession. This proves to be an advantage insofar as it clarifies the problematics related to phenomena from which my essay took its start, namely the diminishing belief in the possibilities provided by a conceptual explanation of the world and the interest in a "description of pre-conceptual layers of consciousness" common at the turn of the century. As soon as the complexity of Bergson's and Freud's approaches gives way to an affirmative, non-analytical stance, a wishful thinking comes to the fore that, although already irreversibly anachronistic around 1900, still remains virulent today. Gallinger and the circle of authors to whom he makes reference attempted to formulate new ways of gaining insight, without acknowledging any relationship to the new visual technologies and mass media. Consequently, they ignored the uncertainties regarding the ability of traditional epistemology to adequately address reality, and by

---

[21] Ibid., 132.
[22] Ibid., 142.
[23] Ibid., 143.
[24] Ibid., 144.

focusing on the presentation of the experience of self-evidence, they tried to resolve contradictions and control the fear of a loss of identity. The form of remembrance thus described is a form of recognition intended to meet the challenges of the epistemological transformation around 1900; however, it simultaneously covers up the related dispersion of the Ego in the masses and the new media by proposing remembrance as a form of contemplative self-assurance.

The specific characteristics of this notion of self-evidence now become obvious. To begin with, it contains a strongly subjective affirmation, as the subject (re-)assures itself by judging an experience as self-evident. Consequently, potential factualities are ignored and remain of lesser significance. What is experienced as self-evident cannot be wrong, even when it contradicts blatant facts, as it always has authenticity on its side. In this innermost sphere, in which the Ego coincides with its object, there is no representational gap, no pain of reconstruction. Still, one should not just dismiss it as a regressive waking dream of paradise before the fall of the symbolic systems, before differentiation sets in; eventually, the replacement of identity by only an impression of itself, no matter how convincing, amounts to abandoning the concept of identity as such. In the experience of self-evidence, disparate elements come together: the presence of a past that as such has never been real and a sensual shimmer that owes its existence to precisely this unreality. Apart from the certainty inherent in self-evidence and the negation of temporality, the experience of remembrance displays a third characteristic: it exerts a certain fascination in which sensuality and emotionality meet, whereby the remembered is often colored in a positive light. Placing the experience of remembrance, in which the subject brightens its own past and watches it in projection, "on the side of perception," Gallinger unintentionally conceives his "mode of intuition" as cinema.[25]

CINEMA AND SELF-EVIDENCE

Before I relate the experience of self-evidence in the act of remembering to the cinematic experience, I would first like to recall some of the characteristic features of early cinema. In the beginning of the 20th century, cinema was

---

[25] Ibid., 135. Translator's note: Gallinger's use of the term "Anschauung" is based on the Kantian notion of "Anschauung" which denotes awareness, or direct knowledge, through the senses, and is translated as 'intuition' in English versions of Kant.

an unusually fast-spreading industry; to illustrate its expansion, I would like to quote some figures referring to Berlin's early cinemas. Between 1905 and 1907, the number of stationary cinematographs rose from 83 to 300, and since the early 'teens there was a choice of theaters ranging from the grimy storefront cinema to the ritzy film theater; about 500 new releases were offered per week.[26] The show usually consisted of a diverse mix of short films, with occasional *variété* numbers presented in between, and these programs were shown throughout the day, starting in the morning. The cinema could only become mass entertainment because it attracted the lower classes of society. Only the best balcony seats in well-equipped film theaters were as expensive as regular theater seats, and even the poorest could afford the entrance fee to a storefront cinema only equipped with uncomfortable wooden benches. The programs shown in the expensive theaters were the same as those shown in the cheap ones. What was characteristic for the cinema, however, was that its offering suited the taste of the working class, the lower middle class and, generally, women;[27] thus, it inevitably carried the stigma of being uncivilized, and it was said to lack culture. It quickly became an economic branch; its goods were measured in meters and classified according to production companies. It served no other goal but entertainment, and it had a mass audience—people whom the bourgeoisie did not even consider to be an audience.

The choice of a cinema was not only determined by the film program, but the surroundings were at least as crucial; some cinemas were preferred because they were the darkest ones, others because of their comfort, or because of the quality of the music or the refreshments that were offered. The effect of the films changed with the surroundings even if the same program was shown; the experience created by a particularly elegant film theater is wholly different from the situation in a poorly equipped place because "a film story is a stretchable thing, and the happenings hinted at can be fashioned individually."[28] This multiplicity is supported by the incoherence of the film program changing between the most diverse genres: "10 minutes [...] about the Tsetse-fly or the brooding process within a chicken's egg or the fabrication of shoes,"[29] before or after a grotesque, images of nature, a melodrama and the

---

[26] Cf. Helmut H. Diederichs, *Anfänge deutscher Filmkritik* (Stuttgart: Fischer und Wiedleroither 1986).

[27] Cf. Emilie Altenloh, *Zur Soziologie des Kino* [On the Sociology of Cinema] (Jena: Verlag Eugen Diederichs 1914), 63–92.

[28] Ibid., 19.

[29] Ibid., 38.

news. "You come and go whenever you want, you watch the closing sequence of a melodram, followed by the latest news, two bad jokes and then Act I and II."[30] Even if one is overcome by emotion, it often remains unclear which quality of the film exactly this emotion is rooted in: "The final image of 'The Detective's Death at the Hospital' is deeply moving in its versatility, affecting the audience in an undefinable way."[31] I provide this brief sketch of the atmosphere surrounding early cinema in order to point to the continuity of the forms of experience, perception and recognition between that period and today's cinema even if it rarely shines through.[32] Early film was, on the one hand, characterized by a certain freedom because the code of narrative cinema had not yet been established;[33] on the other hand, it featured an often mechanistic repetition of themes and structures—which created, within early cinema, a tension between aesthetic innovation and stereotype. Though it has been defamed for its "lack of soul", for its exuberant materialism, Georg Lukács ascribes to cinema a tendency towards metaphysics especially because of its strong sensual presence.

> The temporality and flow of the 'cinema' are completely pure and unspoilt: the essence of 'cinema' is movement as such, eternal variability, the unresting succession of things. The different fundamental principles of composition as applied on stage as well as in the cinema correspond to these different

---

[30] Heinrich Stümcke, "Kinematograph und Theater," *Bühne und Welt*, 15 (1912): 89–94. Cit. op. Jörg Schweinitz, *Prolog vor dem Film*, (Leipzig: Reclam 1992), 244.

[31] -n., "Wovon man spricht," *Der Komet*, 1121, (1906), cit. op. Diederichs 1986, 64. The comment refers to the Gaumont-film *La Pègre de Paris* (1906) directed by Alice Guy, the "best film to be released to date". Ibid.

[32] The limited space of this essay does not allow for further consideration of these early testimonies. However, the contemporary writings on cinema available today are of special significance for the history of early cinema, as they provide the only testimonies aside from the films still preserved. Caution and carefulness provided, these materials allow for gaining an impression of the state of cinema before World War I and, moreover, an understanding of how much early cinema differed from what we mean by cinema since the establishment of the narrative code. Writing film history, one necessarily is forced to reconstruct and thus is always confronted with the danger to obstruct the view onto the particularities of this era. One of the most important sources in this regard is Emilie Altenlohs investigation from 1914, ibid. With regard to the presentation of experiential reports on early cinema and their relation to memory theories see Heike Klippel, *Gedächtnis und Kino* [*Memory and Cinema*], (Frankfurt/M.: Stroemfeld 1997). The most comprehensive, not only historical but also theoretical study of early cinema remains Heide Schlüpmann's *Unheimlichkeit des Blicks* [*The Uncanniness of the Look*], (Frankfurt/M.: Stroemfeld 1990). Though it focuses on Germany's social-historical context, its conclusions are in large parts relevant for early cinema in general; to date, there is no comparable study available that would equal its depth.

[33] See, for example, Tom Gunning, "*An Unseen Energy Swallows Space: The Space in Early Film and ist Relation to American Avant-Garde Film*," John L. Fell, *Film before Griffith*, (Berkeley, Los Angeles: Univ. of California Press 1983), 355–366.

concepts of time: one is purely metaphysical, the other is so strongly, so exclusively empirical, life-like, unmetaphysical that, through its most extreme intensification, another, completely different metaphysics is generated. [...] The single moments which merge and thus create the temporal sequence of cinematic scenes, are only connected by the fact that they follow each other immediately and without transition. There is no causality which would link them; or, more precisely, the causality is not restricted by or bound up with any content. 'Everything is possible': this is the world view of 'cinema,' and because its technology expresses the absolute (if only empirical) reality in every single moment, the validity of 'potentiality' as a category opposed to 'reality' is abolished; both categories become [...] identified. 'Everything is true and real, everything is equally true and equally real': this is what the sequences of images called 'cinema' teach us.[34]

This direct relation to the material world ascribed to the cinema was, especially in early cinema, bound up with an insufficient integration of narration, thereby releasing filmic images from any representational duties. As such, they were free to transcend their profane, reproductive character and lend themselves to the sense of a self-evidence that, in the eyes of the spectator, seems to come across without mediation. The less the object is symbolically charged through the narrative plot, the more opportunities its concrete materiality gets to manifest itself. Images of nature, in particular, support this kind of spectatorial appropriation; the images testify to an alien reality, they do not serve as a bridge of understanding. Emilie Altenloh, for instance, who took extensive polls for her sociological investigation of the cinema in 1914, emphasizes the particular attraction moving water held for the spectator; filmic presentations of water were particularly impressive due to the brilliant quality of the nitrate-based film stock. Other writers as well testify to the interest in water imagery: "We all know the delightful beauty of cinematic images showing calmly sparkling water surfaces as well as breaking waves," writes Ferdinand Avenarius.[35] Another author picks out this type of scenery in *Dante's Inferno* in 1913; he enjoys himself watching "Charon [...] rowing the damned across wonderfully sparkling and rippling water,"[36] even

---

[34] Georg Lukács, "Gedanken zu einer Ästhetik des Kino," *Frankfurter Zeitung* 25 (1913). Cit. op. Fritz Güttinger, ed., *Kein Tag ohne Kino*, (Frankfurt/M.: Deutsches Filmmuseum 1984), 11.

[35] Ferdinand Avenarius, "Vom Schmerzenskind Kino," *Der Kunstwart* 19 (1918): 2. "A breaking ocean wave presented in images of light belongs to the most beautiful one can see." Alfred Baeumler, "Die Wirkungen der Lichtbühne'" *März* 6 (1912): 339. Both cit. op., Fritz Güttinger, *Der Stummfilm im Zitat der Zeit*, (Frankfurt/M.: Deutsches Filmmuseum) 11.

[36] Ludwig Volkmann, "Dante im Kino," *Der Kunstwart* 15 (1913): 213. Cit. op. ibid.

though he otherwise hated the film. In documentary recordings of images of nature, water, due to its indeterminate motion, was especially favored by women: "They [...] often name [when asked about their preferences, H.K.], for instance, images of water and the ocean, like 'Italian waterfall' or simply 'waterfalls and the movement of waves' and 'drifting icebergs' without adding any more specific descriptions."[37]

The active appropriation of filmic images by the viewer is facilitated even more by the silence of early film as well as the musical score. Emilie Altenloh sees therein the reason for why some people equally enjoy the opera and the cinema; they both "affect exclusively the senses and not the mind", their "unspecific way of expression" allows for "potentially more interpretations".[38] What is important in this respect, is the balance between determination and indetermination also held by the music. It is supposed to be of rather stereotypical character, and it should not be tuned too much into the individual film or even individual scenes, in order not to direct the imagination of the viewer too strongly: "because the music inspires other visions, which only interfere with those elicited by the film when they are too close together."[39]

Though always bound to a material referent, film is thereby nonetheless able to release the imagination: the cinematic experience enables the viewer to return to himself and, simultaneously, makes an emotional perception possible. In this quality of cinema we find realized what Gallinger had tried to grasp with such great effort and still could not quite put into words. His concept of recognition is based on remembrance, which does not focus on detecting historical truth but is rooted in the experience of knowledge emerging from a decentered view of the Self, a view that could not be achieved through any reasoning. In the cinema, this experience extends beyond the limited realm of pure subjectivity. Based on the mixing and interpenetration of cinema and spectatorial consciousness, respectively remembrance, Gallinger's construct releases the reactionary element that characterizes a concept of unquestionable self-evidence. Film images are self-evident for the viewer because they show him the objects and, simultaneously, himself.[40] Thus, the cinema offers a space for self-recognition.

---

[37] Altenloh 1914, 89.
[38] Ibid., 102.
[39] Bela Balázs, "Der sichtbare Mensch" [1924], *Schriften zum Film*, (Budapest, München, Berlin: Akadémiai Kiadó, Hanser, Henschelverlag 1982), 130.
[40] With regard to the significance of this moment of self-recognition in early cinema, especially as it relates to female perception, see Schlüpmann 1990, 8–183, especially her

Experiences of self-evidence are moments in which the decentered, dissolving subjectivity of the fading 19th century realizes itself: not as an illusory identity, but as an authentic non-identity which confirms itself in and through its own heterogeneity. The cinema offers a privileged space for these moments in which imagination and perception are bound together and draw on the material of the past. Film offers to the spectator a plastic, moving past, a strange life that he can imitate and make into his own. In the cinema, he does not watch his past self from without, but he sees other people, if only in fragmented images, from within. Conveyed through the actors' bodies, the materiality of the places shown, the writing, and the music, the spectator experiences something that he will only be able to understand if he connects it with his own life. What is important here is not the understanding of any significance or that he gains access to a "work," but that his own experiences become activated.

The stimulation of imaginative powers used to be a privilege of the bourgeoisie; with the spread of the cinema, in an unprecedented way, it has become a mass phenomenon. Recognition and knowledge do not anymore confirm the unifying identity of the subject through a reflection of "truthful" objectivity; rather, recognition and knowledge are only possible by crossing (out) the demarcation between the Self and its alien Other. Thus, recognition turns into an experience mediated through bodily affection and remembrance, which leads the subject to leave its singular existence behind and which releases the object from the meaninglessness of objectivity by placing it into a realm of multiple subjective perspectives. In this regard, the quality or value of a film is irrelevant; film achieves this epistemological transformation by merely being film: life caught in silver, made to move and conveyed by means of an eye-catching materiality. The wondrous sparkling of the ocean waves reminds the viewer of the ocean, regardless of whether he had been there before or not. Inner and outer images interpenetrate. The image of the ocean as projected within the conscious mind of the spectator is neither the photographic image nor his own memory; it is a new one that carries with it a specific quality of recognition and enables the viewer to make an aesthetic experience which is related to the world in a very concrete way.

The self-evidence experienced in remembering as it relates to the cinema has two aspects: it confirms the Self in the spontaneous judgment "that's

---

"Einleitung: Zur heimlichen Komplizenschaft zwischen Kinematographie und Frauen-emanzipation in der wilhelminischen Gesellschaft" [Introduction: On the secret complicity between cinematography and the emancipation of women in the Wilhelminian society], ibid., 8–23.

how it is," and it charges presence with temporality. Temporality is doubly important, as the temporal distance between experience and memory, respectively between the production of filmic images and their reception, is constitutive for the immediacy of an experience of self-evidence. The coincidence of image and knowledge is only possible when the gap between the 'object' and its 'representation' is sufficiently large. However, temporality is not acknowledged, as past and history are ignored when something is experienced as being 'self-evident'. But this is why temporality is constitutive for the experience of self-evidence because the past is undeniably present in this experience and thus exerts pressure onto the present. By perceiving cinematic moving images as present ones, and not as a reproduction of former movements, the energy of their history is condensed by charging the present. To look for places known from intense memories or film scenes is often not worth the effort because, at most, they can be real, but they can never be truth and dream at once. Their past can neither be denied, nor can the past be turned into a surplus of presence.

Translation: Astrid von Chamier

CHAPTER NINE

TONES OF MEMORY: MUSIC AND TIME IN THE PROSE OF YOEL HOFFMANN AND W. G. SEBALD

MICHAL BEN-HORIN

SUMMARY

This article examines the relationship between music and time in contemporary Hebrew and German literature that creates poetic memories. A poetic memory in this context refers to a mode of representing the past that is consciously shaped within literary narratives. I claim that the transformation of musical-time concepts into literature is crucial for the formation of alternative temporalities to historical ones.

The article focuses on *The Christ of Fish* (1991) by Yoel Hoffmann, and *The Emigrants* (1992) and *Austerlitz* (2001) by W. G. Sebald. These prose texts demonstrate new modes of poetic documentation on one hand, while challenging the very essence of testimony and commemoration on the other. Therefore, the article also includes a critical discussion on the problem of representing the past, while dealing with cases that manifest the affinity of literature with music.

In his book *The Time of Music* Jonathan Kramer writes:

> Under the rigidity of absolute time, past-present-future is governed by memory-perception-anticipation, while in music absolute time does not reign solely: music's earlier-simultaneous-later unfolds in absolute time while its past-present-future can occur in gestural time. Music thus frees us from the tyranny of absolute time. In its ability to create unique temporalities, music makes the past-present-future exist on a plane other than that of the earlier-simultaneous-later. (163)

How does music free one from the "tyranny of absolute time", and in what sense can such a process be translated into a literary context? This article attempts to answer these questions by studying the relationship between music and time in those works of literature that establish poetic memories. A 'poetic memory' in this context refers to a mode of representing the past that is consciously shaped within literary narratives. I claim that the transformation of 'musical time' concepts into literature is crucial for the creation of an alternative time structure to the historical one. Literature, therefore, deliberately borrows the time concepts of music in order to establish narrative representations of the past that may escape the

Jo Alyson Parker, Michael Crawford, Paul Harris (Eds), Time and Memory, pp. 163–175.
© *2006 Koninklijke Brill N.V. Printed in the Netherlands.*

tyranny of absolute time and enable different modes of documentation.[1]

In an attempt to demonstrate this proposition, the article focuses on three main aspects of music-literature relations (Scher, *Literatur und Musik* 10) regarding the question of time: first, the thematic use of intertextual elements such as names of composers, musical pieces, or acoustic images which are associated with specific time structures; second, the analogy between narrative and musical forms that manifest certain time concepts; and third, the configuration of phonetic components and prosodic patterns from which unique modes of temporality are generated.[2]

Since the second half of the twentieth century the problem of representing the past has occupied both German and Hebrew literature. A major reason for this phenomenon, which stands at the center of many works and analyses, was the need to find modes of expression and documentation to deal with the 'crises' of the Second World War and the Shoah.[3] Such a need, however, does not relate only to past discussions, but rather has been an inevitable part of actual cultural discourse in both Germany and Israel.[4] This article considers two contemporary literary works that may be placed within these contexts, and which demonstrate the complexity of representing the past. More precisely, I will attempt to show that specific musical time concepts play a central role in the 'memory poetics' of two authors, Yoel Hoffmann and W. G. Sebald, and that musical temporality can therefore be a fruitful means for analyzing their works.[5]

Both Hoffmann and Sebald have dedicated their major works to the twentieth-century world of Jews from German-speaking countries, a world that was destroyed during the Second World War. Their texts thus constitute representations of vanished and forgotten being. They tell the stories of anonymous immigrants, lost figures that reappear only in voices and tones

---

[1] On the potential of music for documentation see Adorno, *Philosophie der neuen Musik* 47.

[2] For further discussion of these aspects as part of musico-literary study, i.e. the form-analogies, 'word music', as well as 'verbal music', see Scher 10–15.

[3] A key essay that questioned the very possibility of dealing with these past events within the literary medium and aroused ambivalent responses from both authors and researches is: Adorno, "Kulturkritik und Gesellschaft" 7–31. For further critical discussions, see Parry 109–124; Hell 9–36.

[4] On the German discourse, see, for example, Brockmann 18–50; Shapira (40–58) discusses the construction of representations of the past in Israel.

[5] Relatively recent attempts to inquire into the role of audible components within narratives of memory are found in: Morris 368–378; Schmitz 119–142. Yet both authors stop at this point without analyzing further aspects of the musical system in relation to the question of representing the past.

that have become an immanent part of a poetic language. These narrative representations of the past are facilitated by a unique elaboration of musical time concepts. Indeed, the question of time, in this case the possibility of alternative temporalities, functions as a central motive in Hoffmann's novel *The Christ of Fish* (1991). Sebald also uses a-historical temporalities to shape a new authenticity of testimony in two of his dominant works, *The Emigrants* (1992) and *Austerlitz* (2001). In what sense, however, can these temporalities be considered as specifically musical ones?

In attempting to answer this question, I return to Kramer's musical analysis, including his fruitful dialogue with Fraser. In dealing with the relationship between time and music, Fraser claimed that music is unique among the arts because "it can determine, define, an audible present with respect to which expectations and memories may be generated" ("The Art of the Audible Now" 181). However, it is precisely the musical system that can also blur the distinctions between past, present and future, by placing a higher priority on perception than on memory. Using Fraser's hierarchical theory of time levels (*Voices of Time* xxv–xlix), Kramer metaphorically defined structures of musical time that contradict the structure of 'absolute time' (*The Time of Music* 165). More precisely, Kramer's structures subvert orders that depend on 'absolute time', such as mental processes of memory, perception and anticipation.[6]

The first of these structures is 'moment time' in music, which is associated with the 'prototemporal level'. Moment-form compositions were first composed by Stockhausen, a contemporary German composer, who defined them as "forms in a state of always having already commenced, which could go on as they are for an eternity" (201). Moment music thus has neither conventional 'beginnings' nor 'endings'; rather it starts and stops as if it has already been played for hours and could have continued being played for many more. Its most striking feature is discontinuity and fragmentation, in that it presents sections that are never outgrowths of earlier ones, as well as moments that only occasionally repeat other moments (219). Moment music, therefore, radically expresses the structure of the fragment, while challenging the very nature of linear progression. The second structure is 'vertical time' in music, which manifests the 'atemporal level'. Vertical music demonstrates a temporal continuum of the unchanging in which there are no separate events and everything seems part of 'eternal present'. It thus evokes

---

[6] Despite his disagreement with Fraser's interpretation of certain temporal levels in specific musical pieces, the crucial point for Kramer (*The Time of Music* 451) is that art and music do propose metaphors for all temporal levels of the hierarchical model.

a sense of vertical accumulation instead of horizontal expansion. This results from the extreme and systematic use of repetitions (much more extreme than in moment music), which avoids offering cues of musical phrases, but rather suggests 'horizons' that extend to infinity (374). Furthermore, the extended present of vertical time can evoke a different reality, distinct from, yet similar to the social unreality of mental schizophrenia (376). The third structure, 'multiply-directed time' in music, is analogous to the 'eotemporal level' in which the order of events is important, but where several different sequences, moving in different directions, are presented simultaneously in the same composition (396). Due to this simultaneous motion of different sequences as well as the ambivalent formation of 'beginnings' and 'endings' in multiply-directed music, backwards and forwards in some sense become the same. This kind of music may, therefore, interrupt or rather reorder the sense of linear motion through time (Kramer, "Postmodern Concepts" 23).

The three temporalities associated with moment music, vertical music and multiply-directed music, can be contrasted with the 'nootemporal level' (*Time of Music* 396), namely, the fully linear level of the 'individual' in which personal identity exists, and in which beginnings and endings, as well as memory and anticipation, are fully developed. I intend to elucidate below how both Hoffmann and Sebald elaborate these musical temporalities, which interfere with linear time, while transforming them into literary manifestations of 'memory poetics'.

1

In his novel *The Christ of Fish*, Yoel Hoffmann, an Israeli author born in 1937, describes the worlds of immigrants who came from Europe to Palestine in the nineteen-thirties. The narrator tells the story of his aunt Magda, interweaving it with other stories of family members and acquaintances. From a position that blurs the distinction between children's and grownups' perspectives, the narrator follows the characters in their daily encounters, documents their foreign languages (German, Hungarian, Rumanian, Yiddish), and listens to their desires and fears.[7] More unconventionally, however, Hoffmann's novel also portrays the return of dead figures to the world of the living.

---

[7] Similar to his two previous books, *The Book of Joseph* (1988) and *Bernhard* (1989), here as well Hoffmann's poetics seeks the limits of language as a representative medium, by using various techniques of estrangement and by interweaving unique textures of perspectives. Herzig considers this a possible explanation for the ambivalent reception of Hoffmann's

The novel begins with a short prologue, namely an image of uncle Herbert, a cembalist, arriving at night and speaking to the narrator. A brief dialogue follows this visit, as the narrator addresses his father: "But uncle Herbert is dead. Am I dreaming?" And his dead father answers: "He is alive".[8] By introducing a 'dialogue with the dead', this prologue thus demonstrates a narrative form that requires an alternative time conception to the historical one. This conception is poetically shaped due to the intense affinity of Hoffmann's text with music; precisely in the ways it borrows and develops the time structures of music.

The novel's initial paragraphs recount the death of aunt Magda, the death of the narrator's father: "When my father died [...] it seemed at first as though I still had my Father, but my father was a Dead Father" (*The Christ of Fish* 2); and the death of the narrator's mother: "My mother gave birth to me and died (in this sequence of events there is evidence of perfect order) so my father searched in the books until he found me a name you can pronounce backwards, and remained [...] alone" (9–10). It transpires, however, that NATAN, the name given to the narrator (6), embodies within its phonetic composition a central motif that dominates Hoffmann's 'memory poetics'.

Reading the name backwards and forwards generates two identical semiotic successions. This bi-directional motion suspends the immediate semantic operations of language (i.e. the denotative value of the Hebrew word Natan is "gave"), and thus enables other modes of signification.[9] The ascribed meaning of "giving birth" (life), which might be activated in this context, is neither primary nor exclusive. Instead, a potential mode of 'returning' appears that is generated from the reversibility of sound succession. In this sense, the semiotic configuration manifests an essential interference with the time concept of daily life, namely nootemporality; one that depends on 'absolute time' and is reflected in the "perfect order" of events as birth precedes death (10). And indeed, Hoffmann's novel explores the ability of 'returning' not only in terms of space, but also of time:

> Many ships set sail from the port of Constantsa. Some came back there again and again. But no ship ever set sail a second time on its first voyage. How can ships break the time barrier? The clocks show a different time and the

---

literature, which evoked in the readers either fascination or extreme antagonism ("Hoffmann's Poetics of Perspectives" 115–116). Compare also with Openheimer 26.2.1988.

[8] My translation. These lines were not included in the English translation.

[9] In his essay on the poetry of Paul Celan, Szondi ("Celan Studien" 362–263) discussed modes of signification that escape description or representation, by stressing the composition principle, i.e. the syntax, in favor of the sense, i.e. the semantics.

wood rots. Maybe one should spin the great wheel on the bridge of the ship simultaneously forwards and backwards and then, in a flash, *what was will be again, for the first time once more*. (61, italics mine)

The figurative solution of the wheel's simultaneous motion (forwards and backwards) allegorically reveals the desire to break the tyranny of 'absolute time' within which past events are passé, fixed in 'history', and the dead are forgotten. Yet another answer to the time dilemma is to be found in the structure of music, namely the concept of multiply-directed time that simultaneously comprises two opposite motions: backwards and forwards. This is demonstrated by Hoffmann's use of specific musical allusions such as Bach's suite. Listening to music that embodies the multiply-directed time structure, Mr. Moskowitz, one of Hoffmann's characters, becomes a "resonating body" (128), thus giving voice to those who can no longer speak:

> When Casals played his cello by virtue of such love, he was ninety-six years old. In the middle of Bach's C Minor Suite, his dim eyes opened wide and it was clear he was looking past death at life. Mr. Moskowitz remembered those notes as though he himself were the cello. His eyes surveyed his inner organs. He heard the dead breathing. From that day on all the winds that blew were for Mr. Moskowitz internal winds. (127)

The reversible motion of the gaze which, due to the 'power of music', turns from outside to inside, enables us to "look past death at life," instead of looking past life at death. This capability, which is soon formulated in terms of time, suggests a poetics that seeks to musically 'resonate' the past within the present:

> It was a dreadful shock when his hair began to grow inward, into his body. This was the external sign of the breakdown in the flow of time. *His hair returned, as it were, to that other time* that Casals had seen with blind eyes when he was ninety-six and playing Bach's Suite in C Minor, almost with no body. (127–128, italics mine)

The confrontation between the absolute "flow of time" and the "other time that Casals had seen [...] when playing Bach's suite" (128) demonstrates the attempt to emancipate oneself from the tyranny of ordinary time. More precisely, this is the ability of music to divorce the past-present-future succession from the earlier-simultaneous-later that progresses in fixed 'absolute time'. Thus, a 'gestural time' future, i.e. a musical cadence, can exist earlier in 'absolute time' than a present, just as a past may succeed a future. The musical time structure that penetrates into Hoffmann's fictive world with the allusion of Bach's suite becomes an integral part of the characters' experience.

A similar experience to that of Mr. Moskowitz, whose future (death) precedes his present (life) while hearing Bach's music, characterizes uncle Herbert. On the Day of Atonement, Herbert, the musician, is rising out of a glass bowl to the sound of his harpsichord, as if "he was doomed [...] by some strange karma, to come back from the realms of the dead (like a Christ of fish) in times of woe" (115–116). It seems that here as well the musical image resonates in the reappearance of the dead in the world of the living, a reappearance that is also a way of remembering those who have passed away.

However the return of uncle Herbert reflects another central feature of Hoffmann's poetics, namely the fragmented structure of a text within which the characters repeatedly appear and disappear. The narrative structure of *The Christ of Fish* is analogous to the fragmentary form of 'moment music' in that it has no beginning or ending, but rather 'starting' and 'stopping', in which poetic sections are not necessarily an outgrowth of previous sections. Nevertheless, because of narrative repetitions, the text evokes an impression of ongoing present, one that reflects forgotten being and reverberates with echoes of vanished worlds. Transposing these temporal concepts from the world of music by shaping simultaneous as well as reversible narrative motions, Hoffmann's novel thus confronts presence with absence while proposing new modes of representing the past. It is not, however, only the allusion to musical intertexts, or the form analogy, but also the unique organization of semiotic elements that emphasize the language's phonetic level, which play a central role in Hoffmann's poetics of memory.

Julia Kristeva, who studied the relationship between poetic language and psychic mechanisms, described a situation in which "language tends to be drawn out of its symbolic function [...] and is opened out within a semiotic articulation"; and concludes that "with a material support such as the voice, this semiotic network gives 'music' to literature" (Kristeva, "Revolution in Poetic Language" 113). The 'semiotic' as opposed to the 'symbolic', therefore sets in motion subversive manifestations that undermine the determined signifying practices.[10] Kristeva analyzed phonetic configurations in literature, such as rhythmic and metric sequences or tonal repetitions (i.e. alliterations, rhyme) that subvert grammar and syntax, as well as conceptualization practices, and she thus allegorically reveals repressed elements of the 'unconscious'. One may conclude that the musicality of a poetic language gives expression to that which is inexpressible and doomed to elimination and oblivion.

---

[10] The 'semiotic' is a pre-linguistic level that exists prior to the logical and grammatical structures of the 'symbolic'. It may be identified with primary processes that fit the Freudian concept of 'unconscious', and is characterized by tonal as well as vocal and gestural rhythmicality. For further discussion of the semiotic versus the symbolic, see Kristeva 90–93.

The question still remains of how exactly the 'semiotic', namely a musical structure presented in language, could be of value for creating new modes of representing the past. A preliminary answer is that semiotic expression allegorically exposes literature's timeless dimension, namely its 'unconscious',[11] in that it presents elements that resist immediate semantic construction. Nevertheless, the phonetic unfolding of these elements within the literary text, characterized as it is by tonal repetitions and acoustic similarities, enables other modes of signification. Proposing a psychoanalytical model for narrative structures, Brooks argued that certain repetitions create a *return* in the text, a doubling back, that might manifest either "a return to origins or a return of the repressed. Repetition through this ambiguity appears to suspend temporal processes, or rather, to subject it to an indeterminate shuttling or oscillation that binds different moments together" (Brooks, "Freud's Masterplot" 100). Similarly, the semiotic expressions in Hoffmann's novel that are based on tonal repetition and similarity, might bind earlier and later as part of an attempt to convey an impression of 'eternal present' in which nothing can get lost.

Hoffmann's literature is, indeed, semiotic, requiring the reader to listen, and demanding from him, as an integral part of the reading process, to hear its rhythmic and tonal configurations. This happens, for instance, in the case of the narrator's name (NATAN), or with various wordplays based on transcriptions of foreign words, such as emphasizing the phonetic similarity between the German words "Tante" (aunt) and "Tinte" (ink), as well as with semiotic flows such as the presentation of foreign verbs (pasian, srinwan, sprisan, swapan etc.) without translating them—thereby suspending immediate signification. By undermining determined practices of signification, Hoffmann shapes poetics of memory that avoid conventional schemes of commemoration.

Another example of semiotic flow is onomatopoeia, such as the repetitive morphemic pattern "Tweetweetweet" heard with minor variations over a whole page by Dr. Staub, "who could tell Bachbirds from Mahlerbirds" (*Christ of Fish* 143–144). And he does this with the "ears he had brought from Braunschweig [that] revolved like saucer antenna" (143). Emphasizing the acoustic images alongside the audible organs reveals a central meta-poetic notion: Hoffmann attempts to shape poetics that listens to voices from the past, such as the voices that were once heard or played in Braunschweig.

---

[11] 'Timelessness' (Zeitlosigkeit) is one of the features used by Freud ("Das Unbewusste" 146) to define the psychic processes of the unconscious.

This is his attempt to talk about lost worlds and absent beings in order to repeatedly free them from the irreversibility of "passing away".

2

W. G. Sebald, a German author born in 1944 who came to England in the nineteen-sixties, yet continued to write in German, poetically reconstructs the atemporal level of an 'eternal present'. Like Hoffmann's work, Sebald's literature demonstrates an attempt to break the linear flow of ordinary time in order to represent what was excluded by it and which got lost in history. Atemporalirty is manifested in the mental context by extreme schizophrenia, namely a situation of a 'split mind' that might simultaneously function in different time modes. A schizophrenic subject may, therefore, experience his internal 'present' without being aware of the external time succession of daily life.

Sebald's novel *Austerlitz* seems to explore the schizophrenic experience in order to escape the repressive determinations of earlier-simultaneous-later as they unfold in 'absolute time'.[12] This he does not only by poetically using the medium of photography, but also by a stratified application of musical temporalities such as vertical time as well as multiply-directed time.[13]

Describing his experience of listening to vertical composition that corresponds to atemporality, *Pages mystiques* (1893) by Erik Satie, Kramer writes:

> For a brief period I felt myself getting bored, becoming imprisoned by a hopelessly repetitive piece. Time was getting slower and slower, threatening to stop. But then I found myself moving into a different listening mode. I was entering the vertical time of the piece. My present expanded, as I forgot about the music's past and future [...] the music was not simply a context for meditation, introspection, or daydreaming. I was listening [...] I became incredibly sensitive to even the smallest performance nuance. (379)

Austerlitz, Sebald's protagonist, seems to convey similar impressions while trying to find traces of his dead mother. In 1938, at the age of four, he was

---

[12] On this point, see also Eshel's argument that "Sebald's 'time effects' model a modern postcatastrophic temporal consciousness, one that reflects the loss of a sense of successiveness, chronology and coherence" (93).

[13] Most research on Sebald's literature with regard to the problem of representing the past has focused on his use of photographic images. See, for instance Harris 379–391; Long 117–136. This article, on the contrary, tries to shed light on Sebald's use of acoustic and musical images, which are nonetheless presented in his prose.'

put on a train carrying Jewish children from Europe to England. Growing up away from his biological parents and his birthplace (Prague), Austerlitz experiences himself as a stranger. The identification with 'outsider' figures reflects his compulsive desire "to be outside time" like the "dying and all the sick at home or in hospitals", like those people who actually "are not the only ones, for a certain degree of personal misfortune is enough to cut us off from the past and the future" (*Austerlitz* 101). Only after long hours of conversation, formally characterized by repetitive transmit-patterns indicating that Austerlitz is the one who speaks while the narrator listens, does Sebald's protagonist reveal the consequences of his arrival in England to the narrator (and to the readers as well). Uncovering this information is, however, a slow process of self-revelation.

Sebald uses narrative strategies in order to give poetic form to a psychic mechanism of recollection. The whole scene is configured within an allegoric context of exposure: Austerlitz talks about a planned project—describing the history of civilization—that consists of integrating historical perspective with archeological insights, and notes with photographs. This project, however, will never be realized since Austerlitz buried his drafts in order to get rid of an unknown burden. At that moment another story begins of uncovered stratums as Austerlitz speaks about lost images and impressions that he suddenly became aware of. The slow exposure of repressed events is analogous to the novel's form of narration: the information about early life experience that initially arises within Austerlitz's involuntary confrontation with his own past is suspended due to narrative regressions. Sebald also shapes such regressions that interfere with processes based on 'absolute time' by poetically elaborating concepts of musical time.

The repressed information that gradually comes to the surface, is indeed, thematically connected to the question of time. In this sense, the clock metonym demonstrates Austerlitz's desire to resist the clock-time within which history unfolds:

> I have never owned a clock of any kind [...] perhaps because I have always resisted the power of time [...] keeping myself apart from so-called current events in the hope [...] that time will not pass away, that I can turn back and go behind it, and there I shall find everything *as it once was*, or more precisely I shall find that *all moments of time have co-existed simultaneously*. (101, italics mine)

Similar to Hoffmann's attempt to narrate a 'returning' in terms of time by borrowing the time concepts of music into his literary fragment, Sebald examines the possibility of blending the past within the present by combining musical images within a narrative of a paradoxical encounter. Austerlitz's longing for a simultaneous co-existence of "all moments of time", while experiencing "everything as it once was" (101), is realized, albeit in a weak form, within the novel's fictive world by the use of musical intertexts. He finally 'meets' his mother, who was murdered by the Nazis, in a propaganda film about Terezin. However, identifying the mother's face is only possible under certain conditions. In the following passages the narrator describes Austerlitz watching a slow-motion copy of the propaganda film as it is extended to four times its original length. Yet the technical extension of the documented fragment grotesquely distorts its sound track—one that included popular musical pieces by European composers that was supposed to give a vivid and lively background to the filmed characters from Terezin:

> Strangest of all [...] was the transformation of sounds in this slow-motion version. In a brief sequence at the very beginning [...] the merry polka by some Austrian operetta composer on the sound track of the Berlin Copy had become a funeral march dragging along at a grotesquely sluggish pace, and the rest of the music pieces accompanying the film, among which I could identify only the can-can from *La Vie Parisienne* and the scherzo from Mendelssohn's *Midsummer Night's Dream*, also moved in a kind of subterranean world, through the most nightmarish depths [...] to which no human voice has ever descended. (247–250, italics in origin)

The estrangement of the time manipulation is manifested in tones and sounds of musical pieces. Despite the fact that these musical pieces are not based on vertical time structure, the time sensation evoked by their tonal distortion is similar to that evoked by vertical compositions, and soon connects with topographic images of depth. In this sense, the deformed musical images metaphorically reflect the non-conceptual "subterranean world" of the unconscious. Moreover, this transference of vertical time sensation is followed by that of multiply-directed time, which happens when Austerlitz focuses on a sequence showing the first performance of a piece of music composed in Terezin, Pavel Hass's study for string orchestra (250). Only when running the tape back and forth repeatedly and confronting clock-time with its alternative temporalities—these are indicated "in the top left-hand corner of the screen, where the figures [...] show the minutes and seconds"—does he discover the face of his own mother (251). This crucial moment of decoding past existence is narratively interwoven with musical images, as well as with technical manipulation of rhythms, sounds and repetitive back-and-forth motions—these are poetic strategies used by Sebald to create modes of documentation.

Other instances of musical time transferred into literature in a way that creates poetic memories can be found in Sebald's collection of stories entitled *The Emigrants*. At the end of the last story the narrator hears the voice of a tenor that evokes recollections of photographs from the Ghetto:

> And now, sitting in the Midland's turret room above the abyss on the fifth floor, I heard him again for the first time since those days. The sound came from so far away that it was as if he were walking about behind the wing flats of an infinitely deep stage. On those flats, which in truth did not exist, I saw, one by one, pictures from an exhibition that I had seen in Frankfurt the year before. They were color photographs [...] of the Litzmannstadt ghetto that was established in 1940 in the Polish industrial center of Lodz. (235)

Here, as well, musical images are used to shape certain temporalities that enable the emergence of the past within the present. However, it is clear that in contrast to Hoffmann, Sebald's use of music is metaphoric (combining musical images) rather than literal (shaping phonetic configurations). Thus it appears that the tones heard by the narrator trigger forgotten knowledge. Indeed, these sounds are poetically connected with specific images of time, distance and depth, and indicate involuntary mental processes of recalling: due to the sounds of music heard a long time ago, the narrator now "sees" the pictures he has seen before. These pictures that are imagined in the narrated 'now' present traces of lost existences. And yet, Sebald's characters, similar to Hoffmann's, do not remember anything. Far more than remembering the past, they function as poetic agents of alternative temporalities in which past is experienced and represented.

In conclusion, both Hoffmann's and Sebald's reasons for translating the time structures of music into literature, point to their need to find new modes in dealing with the past and the problem of representation. By alluding to musical images (Sebald) as well as by emphasizing phonetic elements (Hoffmann), the authors seem to create alternative temporalities for either daily time, or for linear, traditional and historical time concepts. However, their literature does not only resist 'absolute time' processes such as memory, perception, and anticipation, but also redefines the very act of testimony or commemoration. Interfering with the tyranny of 'absolute time' might challenge practices of recalling the past as blocked events that passed away. Instead, these works of literature propose different modes of memory that enable continuous, open dialogue with the past. They are necessary, therefore, for a critical discussion on memory processes of poetic, socio-cultural or historiographical discourses. The critical approaches thus become an integral part of the 'poetics of memory', one which borrows the time structures of music in order to endow the 'absent' with a voice.

REFERENCES

Adorno, Theodor W. *Philosophie der neuen Musik*. Frankfurt am Main: Suhrkamp, 1975.
———. "Kulturkritik und Gesellschaft." In *Prismen*, ed. Rolf Tiedemann, 7–31. Frankfurt am Main: Suhrkamp, 1976.
Brockmann, Stephen. *Literature and German Reunification*. Cambridge: Cambridge University Press, 1999.
Brooks, Peter. "Freud's Masterplot: A Model for Narrative." In *Reading for the Plot*, 90–112. New York: Vintage, 1985.
Eshel, Amir. "Against the Power of Time: The Poetics of Suspension in W. G. Sebald's *Austerlitz*." *New German Critique* 88 (2003): 71–96.
Fraser, J. T. *The Voices of Time*. Amherst: University of Massachusetts, 1981.
———. "The Art of the Audible 'Now'". *Music Theory Spectrum* 7 (1985): 181–184.
Freud, Sigmund. "Das Unbewusste." In *Studienausgabe*. Vol. III: *Psychologie des Unbewußten*, 121–173. Frankfurt am Main: Fischer, 2000.
Harris, Stefanie. "The Return of the Dead: Memory and Photography in W. G. Sebald's *Die Ausgewanderten*." *The German Quarterly* 74 (2001): 379–391.
Hell, Julia. "Eyes Wide Shut: German Post-Holocaust Authorship." *New German Critique* 88 (2003): 9–36.
Herzig, Hanna. "Hoffmann's Poetics of perspectives. (Hebrew)" In *First Name: Essays on Jacob Shabtai, Joshua Kenaz, Yoel Hoffmann*, 115–133. Tel Aviv: Hakibbutz Hameuchad, 1994.
Hoffmann, Yoel. *The Christ of Fish* (Hebrew). Trans. Edward Levenston. New York: New Directions, 1999 [1991].
Kramer, Jonathan D. *The Time of Music: New Meanings, New Temporalities, New Listening Strategies*. New York: Schirmer Books, 1988.
———. "Postmodern Concepts of Musical Time". *Indiana Theory Review* 17 (1996): 21–62.
Kristeva, Julia. "Revolution in Poetic Language." Trans. Margaret Waller. In *The Kristeva Reader*, ed. Toril Moi, 89–136. Oxford: Basil Blackwell, 1986.
Long, J. J. "History, Narrative, and Photography in W. G. Sebald's *Die Ausgewanderten*." *The Modern Language Review* 98 (2003): 117–136.
Morris, Leslie. "The Sound of Memory." *The German Quarterly* 74 (2001): 368–378.
Openheimer, Yochai. "An Author's Poetry. (Hebrew)" *Davar* (26.2.1988).
Parry, Ann. "Idioms for the Unrepresentable: Postwar Fiction and the Shoah." In *The Holocaust and the Text. Speaking the Unspeakable*, eds. Andrew Leak and Goerge Paizis, 109–124. Houndmills: Macmillan Press, 2000.
Scher, Steven P. "Einleitung: Literatur und Musik—Entwicklung und Stand der Forschung." In *Literatur und Musik: Ein Handbuch zur Theorie und Praxis eines komparatistischen Grenzgebietes*, ed. Steven P. Scher, 9–25. Berlin: Erich Schmidt, 1984.
Schmitz, Helmut. "Soundscapes of the Third Reich. Marcel Beyer's *Flughunde*." In *German Culture and the Uncomfortable Past*, eds. H. Schmitz, 119–142. Aldershot: Ashgate, 2001.
Sebald, W. G. *The Emigrants* (German). Trans. Michael Hulse. London: Harvill Press, 1996 [1992].
———. *Austerlitz* (German). Trans. Anthea Bell. Toronto: Alfred A. Knopf, 2001.
Shapira, Anita. "The Holocaust: Private Memories, Public Memory." *Jewish Social Studies* 4 (1998): 40–58.
Szondi, Peter. "Celan Studien." In *Schriften II*, ed. Jean Bollack et al., 321–397. Frankfurt am Main: Suhrkamp, 1978.

RESPONSE

David Burrows

Ben-Horin shows how both Hoffmann and Sebald play with everyday expectations for time. In the normal course of things, we expect our lives to give us a varied succession of events. Each event is followed by a new one, with each event implying a past and a future proper to it, the whole possessed of an overall direction of flow. These authors, however, give us a time shaped in ways that Jonathan Kramer has described for some twentieth-century music. Some music exists in moments that have no past or future. Other music stacks and accumulates instances of the same moment repeated over and over so that time seems frozen. In yet other music independent discourses are yoked together, their conflicting agendas undermining any sense of an overall direction. Music has of course always made it its business to play with time. In fact, any of the arts that take time—film and dance for example—can use the time they take to play with duration, repetition, and order of succession, and one effect in all these cases is a pleasureful sense of mastery over time.

"Ordinary time" is conflated with "absolute time" in Ben-Horin's essay, but she no doubt realizes that this ordinary time is quite different from the "absolute time" described by Newton. Ordinary time is a construction imposed on happening by nervous systems, conspicuously including the human ones, in the interest of their self-regulation. In ordinary time, happening comes with a surround of alternative states, the past and the future. The brain clings to and fixes occurrence and creates a file, the past, where hypostatized occurrence can be stored. The past is partnered by a zone where the disequilibrium of the present is hypothetically resolved: the future (and the future, where hope flourishes, is conspicuous by its absence from the texts Ben-Horin reports on). So time as we find it in these writers, and in music, is a construction superimposed on a construction.

Music shares a liberated approach to time with another area of experience that stands to one side of everyday experience, dreaming, mentioned in passing in Ben-Horin's essay. When we dream we are notoriously out of our ordinary minds, and this applies inevitably to ordinary time. In fact, the time-play in both Hoffmann and Sebald, for all their differences, is accompanied

Jo Alyson Parker, Michael Crawford, Paul Harris (Eds), Time and Memory, pp. 176–177.
© *2006 Koninklijke Brill N.V. Printed in the Netherlands.*

by a dream-like affective quality tinged, especially in the case of Sebald, by a melancholic lyricism.

Through the use of techniques that include time-play, Hoffmann and Sebald create a distance between their readers and the shattering historical displacements and genocide that gave rise to their narratives. As is the case with listeners to music and dreamers of dreams, these readers exist in a buffer zone. In their case the buffer zone recalls the Shoah and the war while insulating readers from their brutal impact. Sebald and Hoffmann sing their dreams of the horror, which is thereby aestheticized into a sad pleasure that we experience as wise and knowing.

CHAPTER TEN

ONCE A COMMUNIST, ALWAYS A COMMUNIST: HOW THE GOVERNMENT LOST TRACK OF TIME IN ITS PURSUIT OF J. ROBERT OPPENHEIMER

KATHERINE A. S. SIBLEY

SUMMARY

The government's denial of a security clearance to atomic scientist J. Robert Oppenheimer in 1954 has always carried the taint of McCarthyism. Less often recognized is that Oppenheimer's hearings were also linked to events eleven years earlier, when the physicist had learned of Soviet intelligence agents' interest in his nuclear research. That this episode continued to be central at the time of his security hearings resulted from officials' construction of a powerful memory about him which had become institutionalized, despite agents' awareness of its flaws.

This narrative of disloyalty persisted in large part owing to the moment in which it first surfaced in 1943. Although Oppenheimer had by then left the Communist Party, he was evasive about his awareness of Soviet spying even as the FBI learned of a separate Russian atomic espionage operation linked with a friend of the physicist. The attempt to develop a case against Oppenheimer then was only stopped by his importance to the successful development of the atomic bomb.

The affair never died down completely, and in 1954, it resurfaced with a vengeance in the wake of debate over the hydrogen bomb, which Oppenheimer had opposed. Though investigators soon conceded that he had no espionage contacts, he lost his clearance anyway. His fate suggests the way in which memories are constructed and concretized at a particularly charged moment, with devastating results.

The Atomic Energy Commission (AEC)'s denial of a security clearance to Manhattan Project pioneer J. Robert Oppenheimer in 1954 has long carried a strong association with the polarized political atmosphere of its time. Referring to the lengthy transcript of the deliberations that led to this decision, historian Richard Polenberg writes that no other recorded document "better explains the America of the cold war—its fears and resentments, its anxieties and dilemmas."[1] Oppenheimer, too, identified "the evil of the times" in which his fate was decided.[2] Senator Joseph McCarthy can hardly

---

[1] Richard Polenberg, ed., *In the Matter of J. Robert Oppenheimer: The Security Clearance Hearing* (Ithaca: Cornell University Press, 2002), xiv.

[2] Dr. J. Robert Oppenheimer: Summary for May 7, 1954. FBI Security File: J. Robert Oppenheimer, cited in Polenberg, *In the Matter of JRO*, xv.

Jo Alyson Parker, Michael Crawford, Paul Harris (Eds), Time and Memory, pp. 179–196
© *2006 Koninklijke Brill N.V. Printed in the Netherlands.*

be ignored in any discussion of the physicist's treatment; he was all too eager to conduct an investigation of the scientist, a probe prevented when AEC head Lewis L. Strauss pointed out that the Wisconsin Republican's handling of such a delicate matter would be "most ill-advised and impolitic."[3] Strauss, a convinced Oppenheimer critic, preferred to handle the matter himself.

Yet, though the political culture of the early to mid 1950s is key in understanding the fate of Oppenheimer, so too are events eleven years earlier, and it was the memory of those distant events, as this article will argue, that was chiefly responsible for the physicist's fate. In 1943, Oppenheimer learned of Soviet intelligence efforts to garner information on the bomb at the Berkeley Radiation Laboratory where he worked, and was less than forthcoming about them with his Manhattan Project superiors. Though the physicist kept his leading role as an advisor in nuclear matters throughout the war and well into the Cold War, this only murkily understood episode continued to resonate in the collective memory of security officials. By 1954, when a new presidential administration exposed government advisors to greater scrutiny, Oppenheimer's wartime history re-emerged as instrumental in the intelligence community's campaign against him. The kind of "scriptlike plotline" with which officials conveniently lined up past events in Oppenheimer's life has been well described by sociologist of memory Eviatar Zerubavel. Zerubavel writes that such a practice bestows "historical meaning" on otherwise unconnected episodes, which thus lend themselves to "inevitably simplistic, one-dimensional visions of the past."[4]

This constructed narrative of disloyalty persisted in large part owing to the moment in which it had first surfaced during World War II. Oppenheimer had been evasive about what he knew about Soviet intrigue at the same time that the FBI learned of a separate Russian espionage operation against the Berkeley Radiation Laboratory linked with California Communist leader Steve Nelson, who himself claimed a friendship with Oppenheimer.[5] This coincidence was especially searing for officials, making Oppenheimer's

---

[3] Quoted in Polenberg, *In the Matter of JRO*, xvii.

[4] Eviatar Zerubavel, *Time Maps: Collective Memory and the Social Shape of the Past* (Chicago and London: University of Chicago Press, 2003), 19.

[5] See Comintern Apparatus (COMRAP) Report, May 7, 1943, in Steve Nelson file 100–16847–112, 12, Federal Bureau of Investigation, Washington, D.C.; Recording No. 3, Nelson and "X", (Soviet representative Vassili Zubilin), April 10, 1943, attached to D.M. Ladd to Director, Communist Infiltration at Radiation Laboratory (CINRAD) report, Nelson File 100–16847–201; FBI Report, "Soviet Activities in the United States," July 25, 1946, 21, in Clark Clifford Papers, box 15, Harry S. Truman Library, Independence, Mo. (hereafter HSTL).

evasions more "eventful" than they otherwise might have been, and imparting them with heightened significance.[6] Oppenheimer and Nelson also shared a personal connection, since the scientist's wife, Kitty, had been married to Nelson's late friend, Joseph Dallet, a casualty of the Spanish Civil War. Nelson, moreover, visited Oppenheimer on at least five occasions in 1941. Still, the physicist claimed that "Nelson never approached him directly or indirectly to obtain information regarding the experimentation ... at the Radiation Laboratory, and ... never ... contacted him at the University."[7]

Despite such concerns, World War II has often appeared to be the golden age of Soviet-American relations: Lauchlin Currie, an aide to FDR during the war, declared that "we were all united ... the atmosphere of suspicion and caution only arose after I left the government" in 1945.[8] Currie's views may be less than convincing, owing to his secret role as a Soviet spy;[9] yet even a skeptic of all things Soviet like George F. Kennan was convinced that "in 1943 the Soviet Union was hardly regarded by our top people in our government as an enemy."[10] Indeed, a "wholly different atmosphere" then prevailed, as Oppenheimer's counsel Lloyd K. Garrison contended in 1954: "The whole attitude toward Russia, toward persons who were sympathetic with Russia, everything was different from what obtains today."[11] Nevertheless, inside security agencies like the FBI and military intelligence, a much more critical outlook prevailed, and the discovery of Soviet espionage in World War II only confirmed this posture. For these reasons, Oppenheimer's lack

---

[6] For a fascinating discussion of how past time is divided in our memories into either intensely "memorable" periods or eminently "forgettable" ones, see Zerubavel, *Time Maps*, 25–34.

[7] Harry M. Kimball, SAC San Francisco to Director, FBI, September 19, 1946, Nelson file 100–16847–354; July 4, 1946, July 19, 1946, Surveillance log, Steve Nelson, Nelson File 100–16847–394 Volume 5, 67. At his 1954 security hearings, however, Oppenheimer said he knew that Nelson was "an important Communist." See U.S. Atomic Energy Commission, *In the Matter of J. Robert Oppenheimer: Transcript of Hearing before Personnel Security Board and Texts of Principal Documents and Letters* (Cambridge, Mass.: MIT Press, 1971), 195 (hereafter *IMJRO*). Despite abundant evidence, Nelson consistently denied his espionage role during his lifetime, enabled by a long-held scholarly reticence on the subject in the post-McCarthy era. See Steve Nelson, James R. Barrett, and Rob Ruck, *Steve Nelson: American Radical* (Pittsburgh: University of Pittsburgh Press, 1981), 294; author's note from Maurice Isserman, January 3, 2005.

[8] Quoted in Roger Sandilands, *The Life and Political Economy of Lauchlin Currie: New Dealer, Presidential Adviser, and Development Economist* (Durham: Duke University Press, 1990), 149.

[9] R. Bruce Craig, *Treasonable Doubt: The Harry Dexter White Spy Case* (Lawrence, Kan: The University Press of Kansas, 2004), 96.

[10] *IMJRO*, 98.

[11] *IMJRO*, 358.

of candor became indelibly stamped in the minds of counterintelligence agents, who seized the moment in 1954 to address their memories of his earlier obfuscations. The fate of J. Robert Oppenheimer offers an instructive example of how memories can be constructed and corresponding narratives hardened at a particularly intense moment, and with highly damaging consequences.

The FBI itself had closely monitored Oppenheimer as early as 1941, seeking to determine the nature of his numerous Communist connections. Agents kept track of his conversations, magazine subscriptions, activities, and lovers, as well as his involvement in the Party, which included meetings with party members and generous donations. Bureau and military intelligence officials also watched his wife, brother, and former girlfriend, all of whom had ties to the party.[12] The Bureau learned that up to the early 1940s Oppenheimer was regarded as a Communist by party members in Berkeley. The physicist, while acknowledging membership in "practically every known front group," always denied any Communist affiliation.[13] However, as newly-unearthed documents reveal, Oppenheimer's relationship with the party was more than

---

[12] Julius Robert Oppenheimer, February 17, 1947, FBI background, attached to Hoover to Harry Vaughan, February 28, 1947, President's Secretary File box 167, HSTL; Communist Infiltration at Radiation Laboratory (CINRAD) Summary Memo #1,100–190625–2017, March 5, 1946, 8. I am indebted to Gregg Herken for a copy of the CINRAD report. Oppenheimer's brother, sister-in-law, and wife were all former Communists, as was former paramour Jean Tatlock. On monitoring of Kitty Oppenheimer, see FBI Report, November 27, 1945, 39, box 97, Record Group 233, Records of the U.S. House of Representatives, Special Committee on Un-American Activities (Dies), Exhibits, Evidence, etc. re Committee Investigations: Communism Subject Files 1930–1945, National Archives, Washington, D.C. In July 1943, the War Department's Military Intelligence Division had opened an investigation on Frank Oppenheimer, in response to his attendance at Communist meetings in 1940 as well as his brother's leadership of the Manhattan Project. According to Lt. Col. Pash, who instigated the probe, Frank Oppenheimer was "reliably reported to be a Communist and possibly engaged in espionage." Pash called for an investigation of "Oppenheimer's associates, activities, and political sympathies." But one of his former teachers, G.H. Dieke of Johns Hopkins, saw Frank as a shy and retiring type, and dismissed his political leanings, only to say that as a "Hebrew", Frank Oppenheimer would not have any warm feelings about Nazism. Boris T. Pash, T. Col, Military Intelligence, Chief, Counter Intelligence Branch, to Major Geroge B. Dyer, Director, Intelligence Division, headquarters Third Service command, Baltimore, July 10, 1943, Richard Connolly, Special Agent, CIC, Memo for the Officer in Charge, July 19, 1943, *Washington Post*, June 15, 1949, all in Frank Freidman Oppenheimer File, E2037207, box 51, RG 319 Records of the Army Staff, Records of the Office of the Assistant Chief of Staff, G-2, Intelligence, Records of the Investigative Records Repository, Security Classified Intelligence and Investigative Dossiers, 1939–1976 (hereafter IRR), National Archives, College Park, Md.

[13] Memo to Boardman, April 16, 1954, FBI Oppenheimer file 100–17828–1208; Comintern Apparatus (COMRAP) Summary Report, San Francisco, December 15, 1944, 100–17879, 38–39. I am grateful to John Earl Haynes for a copy of the COMRAP report.

theoretical: he gave $150 monthly donations as late as 1942, and was almost certainly a member.[14]

Though the atomic physicist's party membership could not be proven during the war, this did not stop Lt. Col. Boris T. Pash, who headed the Manhattan Project's investigative arm, from asserting that Oppenheimer "is not to be fully trusted and that his loyalty to the Nation is divided." Given the physicist's dissembling on the issue of his party membership, the government's "considerable doubts" about him are not without foundation.[15] They do not, however, justify assertions that Oppenheimer was a spy, as Pash claimed he was, providing material to persons who "may be furnishing... [it] to the Communist Party for transmission to the USSR." Though this security officer urged that the Los Alamos leader be replaced, Manhattan Project manager General Leslie R. Groves, convinced that the physicist was essential to the bomb project, was able to foil such attempts to remove him during the war.[16]

The FBI, meanwhile, was learning that Oppenheimer's party connections were, if anything, increasingly attenuated. In a 1943 conversation recorded by the Bureau, Steve Nelson informed Radiation Laboratory scientist Joseph Woodrow Weinberg that while he had talked to Oppenheimer about the project, he had not pressed him, since "I didn't want to put him in an unfriendly position. I didn't want to put him on the spot, you know." Weinberg agreed, noting that Oppenheimer would not think Nelson the "proper party" for gaining any materials.[17] Nelson acknowledged that he and the party made the physicist "uncomfortable," even "jittery." Weinberg complained that as for himself, Oppenheimer had "deliberately kept" him from the project, in part because of Weinberg's politics. Weinberg believed

---

[14] Gregg Herken cites a letter that Haakon Chevalier wrote Oppenheimer much later, discussing their membership in the same unit of the party from 1938 to 1942. See Chevalier to Oppenheimer, July 13, 1964, as well as further substantiation in Gordon Griffiths' unpublished memoir, *Venturing Outside the Ivory Tower: The Political Autobiography of a College Professor*, on deposit at Library of Congress, Washington, DC; both of these may be found also on the website for Herken's *Brotherhood of the Bomb*: www.brotherhoodofthebomb.com.

[15] *IMJRO*, 260, 270; Pash to John Lansdale, September 6, 1943, cited in *idem*, 273.

[16] Pash to Lansdale, June 29, 1943, cited in *IMJRO*, 822; Oppenheimer was considered a "calculated risk" owing to his past political associations. The notion of "calculated risk" allowed for giving clearance to a candidate who otherwise had security issues, but who "is a man of great attainments and capacity and has rendered outstanding services." The Hearing Board felt that such a standard only applied during times of "critical national need." See *IMJRO*, 15/1013, 43/1041. Also see Leslie Groves, *Now It Can be Told: The Story of the Manhattan Project* (New York: Harper Brothers, 1962), 63.

[17] Steve Nelson File, August 20, 1945, 100–16847–NR, 15.

this was "a strange thing for him to fear. But he's changed a bit." Both men agreed that Oppenheimer was "just not a Marxist" and that while he "would like to be on the right track," he was being pushed away by his wife, who was "influencing him in the wrong direction." Her ambitions for him, as well as his own, were leading him away from his old associations.[18]

Nelson and Weinberg's association, however, definitely included espionage. On the night of March 29, 1943, the Communist leader was overheard at his home securing "some highly confidential data regarding the nuclear experiments" from Weinberg.[19] Informing the party leader that the separation process for uranium was the Lab's chief problem at the moment, Weinberg noted that he was "a little bit scared" to hand over a published document on the project. Nelson, nevertheless, pressed him for material, but did not attempt to assuage Weinberg's fear that his espionage could lead to him being put under investigation.[20] With some trepidation, then, Weinberg provided Nelson a formula of about 150 to 200 words which concerned the "calutron", a separator that would enrich uranium. Most of it the Bureau's listeners found unintelligible.[21] Soon after this meeting, an FBI tail saw Nelson passing materials to Soviet vice consul Peter Petrovich Ivanov on the grounds of nearby Saint Joseph Hospital.[22]

Ivanov was also working other avenues. He had contacted George C. Eltenton, a sympathetic chemist at a nearby oil company, and complained to him that "the Russians . . . needed certain information and . . . for political reasons there were no authorized channels through which they could obtain such."[23] Eltenton knew that left-leaning professor of French Haakon Chevalier was a friend of J. Robert Oppenheimer at Berkeley, and got in

---

[18] Steve Nelson File, August 20, 1945, 100–16847–NR, 4–7, 15. Lansdale agreed with this; Kitty Oppenheimer's "strength of will was a powerful influence in keeping him away from what we would regard as dangerous associations." See *IMJRO*, 266.

[19] See U.S. House, Committee on Un-American Activities (HUAC), *The Shameful Years: Thirty Years of Soviet Espionage in the United States.* (Washington, D.C., 1951), 31; FBI Report, September 27, 1945, 40. On the surveillance of Nelson, also see D.M. Ladd to the Director, April 16, 1943, Re: Cinrad, FBI File 100–16847–201; "Communist Infiltration of Radiation Laboratory, University of California, Berkeley, California," FBI File 100–16980, July 7, 1943, 1; FBI Steve Nelson File, 100–2696, June 25, 1945, 28.

[20] Steve Nelson File, August 20, 1945, 100–16847–NR, 8–20 *passim*.

[21] Steve Nelson File, August 20, 1945, 100–16847–NR, 18; COMRAP Report May 7, 1943, 100–16847–112, 5; FBI Report, "Soviet Activities in the United States," 26.

[22] CINRAD Summary Memo #1 100–190625–2017, 7; Ladd Memorandum for the Director, May 5, 1943, 100–16847–104; FBI Report, San Francisco, May 7, 1943, 100–16847–112.

[23] FBI background on Julius Robert Oppenheimer, February 17, 1947, attached to Hoover to Vaughan, February 28, 1947, President's Secretary's Files, box 167, HSTL.

touch with him. Eltenton pressed Chevalier "to find out what was being done at the radiation laboratory, particularly information regarding the highly destructive weapon which was being developed." He told the professor that since "Russia and the United States were allies, Soviet Russia should be entitled to any technical data which might be of assistance to that nation."[24]

In Chevalier's later version of the story, the whole suggestion was "preposterous," and Oppenheimer would be "horrified" to hear of it. Yet he nevertheless decided to tell the physicist, whom he considered his "most intimate and steadfast friend," owing to his professed concern that Eltenton's might be only the first attempt to get more information about the secret project.[25] Chevalier reported that he had informed Oppenheimer at a visit to the physicist's home in the winter of 1943, whereupon Oppenheimer flatly slammed the entreaties as "treason."[26] But Oppenheimer did not alert Army security to the episode until some six months later—and the story he told then was largely fabricated. So too, as it turned out, was Chevalier's portrayal of events.[27]

In August 1943, Oppenheimer informed his Army interviewers that sometime the previous winter he "had learned from three different employees of the atomic bomb project... that they had been solicited to furnish information, ultimately to be delivered to the USSR." None of the men had cooperated, according to Oppenheimer, so he refused to name

---

[24] Quoted in HUAC, *Hearings regarding Communist infiltration of the Radiation Laboratory and Atomic bomb project at the University of California, Berkeley, California,* (Identification of Scientist X), December 22, 1950, 81st Congress, second sess. (Washington, 1951), 3495; also see HUAC, *Report on Soviet Espionage Activities in Connection with the Atom Bomb; Investigation of Un-American Activities in the United States,* September 28, 1948, 80th Congress, second sess. (Washington, 1948), 182; John Lansdale, Jr. to J. Edgar Hoover, August 27, 1943, December 13, 1943, RG 319, 383.4 USSR; FBI Report, November 27, 1945, 39–40; David Holloway, *Stalin and the Bomb: The Soviet Union and Atomic Energy, 1939–1956* (New Haven: Yale University Press, 1994), 103. Eltenton's argument was also made by spies like Ted Hall. See Albright, Joseph and Marcia Kunstel. *Bombshell: The Secret Story of America's Unknown Atomic Spy Conspiracy* (New York: Times Books, 1997), 89–90.
[25] Haakon Chevalier, *Oppenheimer: The Story of a Friendship* (New York: George Braziller, 1965) 53–54. Chevalier's discomfort was not so apparent to Eltenton, however. See, Gregg Herken, *Brotherhood of the Bomb: The Tangled Lives and Loyalties of Robert Oppenheimer, Ernest Lawrence, and Edward Teller,* (New York: Henry Holt and Co. 2002), 92.
[26] Chevalier, *Oppenheimer,* 19.
[27] Oppenheimer's reply to Nichols, March 4, 1954, cited in *IMJRO,* 14. Despite his sympathy toward better relations with the Soviet Union, scholars have persuasively dismissed Oppenheimer's links with espionage. See, for instance, Allen Weinstein and Alexander Vassiliev, *The Haunted Wood: Soviet Espionage in America—The Stalin Era* (New York: Random House, 1999), 183–185; Herken, *Brotherhood,* 93. On Oppenheimer's "passivity" in taking action on what he did know, see Haynes and Klehr, *Venona,* 330.

them. He also refused to name the intermediary who had been assigned by Eltenton to approach them, declaring this person "innocent."[28] This was, of course, a cryptic reference to Chevalier. Authorities were alarmed: even if no espionage had taken place, an official at the Soviet consulate stood ready "who was said to have had a great deal of experience with microfilm and who was in a position to transmit the material to Russia."[29] Oppenheimer's story led to much fruitless work for intelligence agents, who attempted to find the three men, spurred on by their knowledge of Nelson's activities. Eleven years later, memories of that looming espionage threat remained vivid in the hearing transcripts.[30]

In December 1943, after having been repeatedly badgered about it, Oppenheimer at last revealed to General Groves that his intermediary had been Chevalier.[31] Then, demanding he keep it a secret, Oppenheimer told Groves that the French professor had contacted not three people, but only one—his brother, Frank.[32] Robert apparently told his convoluted story of Chevalier's approach to himself in order to protect his brother, who had been a member of the Party and was also a Radiation Laboratory scientist. Thanks to Groves' discretion, this more accurate story remained largely unexplored until 1954.

Meanwhile, the FBI, which had been kept off the case by the Manhattan Project security team during the war, eagerly resumed its investigation of Oppenheimer in 1946.[33] Agents continued to attempt to demonstrate

---

[28] Report on J. R. Oppenheimer, February 5, 1950, Summary Brief on Fuchs, February 6, 1950, FBI Klaus Fuchs file, 65–58805–1202, 3.

[29] "Chevalier Conspiracy," Part II, August 20, 1945, Steve Nelson FBI file, 100–16980, 34–35.

[30] *IMJRO*, 148, 815.

[31] "Chevalier Conspiracy," Part II, Steve Nelson file, August 20, 1945, 100–16980, 36; Julius Robert Oppenheimer, February 17, 1947, FBI background, attached to Hoover to Vaughan, February 28, 1947, PSF box 167.

[32] CINRAD Summary Memo #1, 100–190625–2017, March 5, 1946, 14–15, *IMJRO*, 264–265. This story has been most recently and convincingly told in Herken, *Brotherhood*, 113–114. Oppenheimer may have sworn Groves to secrecy about Frank, but the general told Colonel Lansdale, who informed the FBI. As Herken relates, Oppenheimer's reporting on Chevalier was "comeuppance" for the French professor's attempt to recruit his brother. See *idem*, 120.

[33] See Hoover's request for a wiretap, April 26, 1946, memo to the Attorney General, FBI Oppenheimer File 100–17828–29, where he justified the request based on Oppenheimer's "contact with individuals reported to be Soviet agents." Also see George M. Langdon Report, Oppenheimer file, 100–17828–16; San Francisco Office to Washington Director, May 8, 1946, Oppenheimer file, Internal Security 100–17828–33; SAC San Francisco to Director, May 14, 1946, Internal Security 100–17828–33. The FBI took over the Manhattan District files on Oppenheimer in the summer of 1946. See *IMJRO*, 417.

Oppenheimer's party membership, explore his potential espionage links, and study his positions on international control of atomic armaments and later, the H-bomb.[34] Frank Oppenheimer, who had just taken a job at the University of Minnesota, was also wiretapped by the FBI. The Bureau reported both men's alleged Communist connections to Truman in 1947, though to little immediate effect.[35] Robert Oppenheimer, who had been named Director of the Institute of Advanced Study in Princeton that year, maintained a security clearance for his position as chairman of the Atomic Energy Commission's General Advisory Committee (GAC), which was consulted on scientific and technical matters.[36] The case against Robert Oppenheimer was weak in the early Cold War, as even the FBI acknowledged, noting "there is no substantial information of a pro-Communist nature concerning Oppenheimer subsequent to 1943."[37] In fact, the FBI's most compelling evidence on Oppenheimer's party membership showed that his last monthly payment was made in April 1942.[38] As a result, the Bureau focused on such tenuous links with Communism as his wartime relationship with Jean Tatlock, a former Berkeley student and Communist who later committed suicide.[39]

The FBI's allegations largely drew from their "confidential sources"— wiretaps and informants. At a November 1945 meeting of the North Oakland Club of the Alameda County Communist Party, for example, Jack Manley declared that "Oppenheimer told Steve Nelson several years ago that the Army was working on the bomb."[40] But these reports and allegations of espionage were simply unconfirmed gossip, as the Bureau knew, and its agents were driven to more aggressive measures to find additional evidence. An obliging priest, Father John O'Brien, attempted to influence Oppenheimer's

---

[34] Oppenheimer had been an adviser to Bernard Baruch in the latter's eponymous plan for internationalizing atomic energy, and contributed to the Acheson-Lilienthal Report for the control of atomic energy. Julius Robert Oppenheimer, February 17, 1947, FBI background, attached to Hoover to Vaughan, February 28, 1947, PSF box 167.

[35] Hoover to Vaughan, February 28, 1947, PSF box 167; see discussion of Frank Oppenheimer's investigation in Herken, *Brotherhood of the Bomb*, 181.

[36] Memo on Oppenheimer, April 16, 1954, 100–17828–1208.

[37] W. A. Branigan to A. H. Belmont, April 7, 1954, 100–17828–1055.

[38] See FBI memo, April 21, 1947, cited in Jerrold and Leona Schecter, *Sacred Secrets: How Soviet Intelligence Operations Changed American History* (Washington, D.C.: Brassey's, 2002), Appendix 1, 314.

[39] Memo on Oppenheimer attached to Ladd to Director, March 18, 1946, Oppenheimer file 100–17828–51, 6.

[40] Memo on Oppenheimer attached to Ladd to Director, March 18, 1946, 100–17828–51, 6; also see Belmont to Boardman, April 15, 1954, 100–17828–1159; Memo to Boardman, April 16, 1954, 100–17828–1208.

secretary, whom the Bureau hoped "on the basis of her religious convictions and patriotism" might become a source for the Bureau. The secretary would not be so typecast.[41] Even the fact that Oppenheimer admitted to the FBI after the war that he "'concocted a completely fabricated story'" about the three contacts of Chevalier, and that he "failed to recognize the potential threat to the nation's security which was present in the incident," did not affect him adversely.[42]

Thus the new surveillance campaign ended in 1947, as AEC chairman David M. Lilienthal reported the consensus of his advisors that "Dr. Oppenheimer's loyalty was an accepted fact."[43] Called to testify in front of HUAC two years later, while it was investigating wartime espionage at the Radiation Lab, Oppenheimer was treated with kid gloves—even by Richard Nixon. Speaking at the same hearings, however, his brother ended up losing his job when it was revealed he had kept his Communist past a secret.[44]

The renewed emphasis on Robert Oppenheimer's danger as a "security risk" that emerged less than five years later shows the tenacious grip which memories of his earlier misstatements had on the minds of Washington's intelligence community. As Zerubavel observes, "memory often schematizes history," applying an overarching narrative to it. This "editing" process, as he notes, "mentally transform[s]… noncontiguous points in time into seemingly unbroken historical continua."[45] Government officials would once again proclaim their suspicions of Oppenheimer's disloyalty despite a clear understanding that his Communist associations were more than a decade old by 1954—a mustiness confirmed by the FBI's having dropped his surveillance for the previous seven years. Oppenheimer's lack of support for the hydrogen bomb no doubt played a significant role in the newly critical scrutiny he received, but the powerful memory of his past political associations and obfuscations appear to have been most central in the new charges.

Resistance to Oppenheimer's continued advisory position had arisen even before former President Harry S. Truman returned home to Missouri. In June 1952, Oppenheimer was essentially forced off the GAC when his term ended, and here his opposition to the H-bomb was key. The FBI was busily marshalling reports from interviews with critics like physicist Edward

---

[41] Strickland to Ladd, March 26, 1946, 100–17828–24.

[42] Report on J. R. Oppenheimer, February 5, 1950, Summary Brief on Fuchs, February 6, 1950, 65–58805–1202, 4–5.

[43] "Scientist Upheld Despite FBI File," *New York Times,* April 13, 1954, 19.

[44] Herken, *Brotherhood,* 196–97.

[45] Zerubavel, *Time Maps,* 25, 40.

Teller, who complained that Oppenheimer had "always swayed opinions against the Super." Nevertheless, Oppenheimer had soon been named an adviser to the Atomic Energy Commission. At first, it seemed as if he and the new president, Dwight D. Eisenhower would get along; Oppenheimer appreciated Ike's openness to discussing the arms race, and he and other scientists hoped that the Eisenhower Administration would give them more of a say in the military buildup. Instead, though, the growing animosity to the physicist in the military and intelligence communities, as well as in Congress, contributed to ending Oppenheimer's access to nuclear secrets.[46]

This process was accelerated when Eisenhower named Lewis L. Strauss to head the AEC in June 1953. Strauss, who had once been so favorably inclined toward Oppenheimer that he recommended him for his job at the Institute for Advanced Study, had by this time developed "an almost paranoid obsession" with the scientist.[47] He immediately had classified papers in Oppenheimer's office at Princeton removed and taken to an AEC facility. It was too late to keep Oppenheimer away from everything, however; the consultant's contract had just been renewed for another year.[48]

Meanwhile, a similarly fixated government servant was busily assembling the weaponry for Strauss to permanently get rid of Oppenheimer. Fed up with his former Congressional colleagues for lack of action, William Liscum Borden, who had been executive director of the Joint Congressional Committee on Atomic Energy from January 1949 to May 1953, stewed over Oppenheimer's continuing presence in government the following summer and fall, and in November 1953 sent a lengthy missive about him to J. Edgar Hoover.[49] Borden's claims went further than the government was prepared to; for example, he charged that Oppenheimer "more probably than not ... [is] functioning as an espionage agent."[50] Borden's vehemence was connected with a common memory practice, that of constructing a "continuous biography." When applied to the self, this narrative production generally highlights the most favorable aspects of one's identity while masking less attractive attributes, but here Borden did precisely the opposite—emphasizing Oppenheimer's

---

[46] Herken, *Brotherhood*, 249–64 *passim*.
[47] Herken, *Brotherhood*, 265–67. The scientist had made fun of Strauss at a 1949 Congressional hearing. When Strauss pressed him as to the "dangers" of radioactive isotopes being used to make bombs, Oppenheimer compared their explosive potential to shovels or bottles of beer.
[48] "President barred data to scientist," *NYT*, April 14, 1954, 18.
[49] *IMJRO*, 832–833.
[50] See Borden's letter to Hoover, November 7, 1953, cited in *IMJRO*, 837–838.

perceived past transgressions as the synecdoche of his life, despite his heroic status and his long years of government service. Indeed, the physicist's very presence in the government now carried the weight of "historical analogy"; to Borden, Oppenheimer's status as a consultant to the AEC was a glaring example of replicating the errors of the past, when blinkered officials had allowed Communists into the federal ranks.[51]

Borden's letter reached Strauss on November 30, and Eisenhower on Dec. 2. The president, after speaking to several top officials, decided the following day to construct a "blank wall" between Oppenheimer and secret information, suspending his clearance, although not yet informing him about it.[52] Neither Hoover nor Strauss recommended pursuing the matter in the halls of Congress, to be subjected to the treatment of Senate investigative pioneers like Democrat Pat McCarran and Republican Joseph McCarthy. At the same time, however, General Kenneth D. Nichols of the AEC prudently worried that a separate hearing of the AEC's personnel security board would distance other scientists from the Commission, especially since the charges prominently featured Oppenheimer's opposition to the hydrogen bomb, an issue that seemed to suggest that scientific advice was subject to a "political correctness" standard.[53]

Then, just as Eisenhower announced the erection of his wall, further discussions among the AEC, White House, and military officials stirred up the story of the old Chevalier incident—and the even more effectively suppressed tale that it was Oppenheimer's brother Frank, and not Oppenheimer himself, who had been the French professor's contact in 1943. The AEC Chairman decided that a hearing of the personnel security board would be necessary after all. Borden had not even mentioned the Chevalier affair. Now, it seemed that the "decade-old approach by Chevalier would be key to any legal case that might be made against Oppenheimer," Gregg Herken notes. An already outraged Eisenhower snarled that "Oppenheimer is a liar," as officials quickly seized upon the fact that the physicist's past misstatements, rather than his present views on the "Super", would be the more compelling hook on which to hang him. Oppenheimer was not informed of the various charges against him until after he had returned from a trip to Europe some two weeks later—a trip that ironically included a visit

---

[51] See discussion of these concepts in Zerubavel, *Time Maps*, 48–53.
[52] "Eisenhower Order Barred U.S. Data to Oppenheimer," *NYT,* April 14, 1954, 1, 18; Herken, *Brotherhood*, 268.
[53] Herken, *Brotherhood*, 269–270.

to Paris where he actually saw Chevalier and intervened on his behalf in a passport matter! General Nichols inventoried the AEC's verdict in a letter announcing Oppenheimer's suspension as a consultant, a document that highlighted the scientist's earlier obfuscations about the French professor and also asserted that he had delayed development of the H-bomb.[54] A good chunk of the charges against Oppenheimer, however, were connected with his former links with Communists in the early 1940s and before, such as the allegation that "he fell in love with one Communist and married another former Communist," or that he "hired Communists or former Communists at Los Alamos."[55] Deciding to fight the suspension, Oppenheimer retained New York lawyer Lloyd Garrison to defend him. He also wrote a 43-page letter about himself, acknowledging his "fellow traveling" in the 1930s but denying his party membership.[56] The entire matter remained unknown to the public until April 1954, when the personnel security board's hearings began.

By then, the government had been bugging Oppenheimer for five months, including the dubious practice of listening to his conversations with counsel.[57] In January 1954, Agent Alan Belmont defended the wiretapping as a way to detect "any indication that Oppenheimer might flee"—despite the scientist's professed plans to defend himself at the hearings three months later. The FBI even entertained a rumor that Soviet intelligence "had reportedly contacted Oppenheimer to arrange his disappearance behind the Iron Curtain."[58]

The hearings, eventually bound in lengthy proceedings entitled *In the Matter of J. Robert Oppenheimer*, focused in detail on the physicist's having told two different stories about Chevalier, evasions and untruths that he did not deny, and this wartime case became crucial to undermining Oppenheimer's credibility. He admitted that the version he gave Groves in 1943 about Soviet efforts to reach three atomic researchers through Chevalier was a "cock and

[54] Herken, *Brotherhood*, 270–74; "AEC Suspends Dr. Oppenheimer," *NYT* April 13, 1954, 15. The Eisenhower quote is in A. Branigan to AH Belmont, March 31, 1954, FBI Oppenheimer File, Internal Security 100–17828–1279. Branigan was quoting Lewis Strauss, Chairman of the Atomic Energy Commission, via David Teeple, Lewis' assistant, on Eisenhower's view of Oppenheimer.

[55] "Dr. Oppenheimer Suspended by A.E.C. in Security Review; Scientist Defends Record," *NYT*, April 13, 1954, 1; "Strauss, Oppenheimer Serve Same Institute," *NYT*, April 13, 1954, 21.

[56] See *IMJRO*, 7–20.

[57] Herken, *Brotherhood*, 275.

[58] Belmont to Ladd, January 5, 1954, 100–17828–587; W. A. Branigan to A. H. Belmont, April 7, 1954, 100–17828–1055.

bull story," completely at odds with what he had said to the FBI in 1946, when he had dismissed the earlier tale. He conceded, "I was an idiot." The government did not attempt to illuminate the more dramatic reality of the 1943 affair, since it would only make Oppenheimer look like "a hero" who had tried to protect Frank.[59] That particular memory would be blotted out in favor of a more incriminating scenario, one that placed the scientist in the "villain" category—and Oppenheimer wasn't about to name his already beleaguered brother to thicken the plot.

The board's three members voted 2–1 on June 1, 1954, to affirm the suspension and deny the physicist a position as a consultant. Despite its members' professed belief in Oppenheimer's loyalty, the board was troubled by Oppenheimer's "continuing conduct and associations." Harking back to his record with Chevalier, the majority report noted that the scientist "has repeatedly exercised an arrogance of his own judgment with respect to the loyalty and reliability of other citizens to an extent which has frustrated and at times impeded the working of the system." Moreover, they pronounced that "his conduct in the hydrogen bomb program [was] sufficiently disturbing as to raise a doubt as to whether his participation... in a Government program relating to the national defense, would be clearly consistent with the best interests of security."[60]

Oppenheimer's supporters strongly disagreed. John J. McCloy, for instance, pointed out that "We are only secure if we have the best of brains and the best reach of mind in the field."[61] Well aware, too, of the scientific community's distaste for the hearings, the majority on the board pointedly defended themselves against the charge of "anti-intellectualism."[62] They needed to, since many scientists were furious that Oppenheimer's "enthusiasm" (or lack thereof) for the H-bomb was used as a criterion in the case. Nichols thus urged that the AEC drop the issue of the hydrogen weapon as a consideration in its upcoming hearing of Oppenheimer's appeal of the security board's decision, focusing on the matter of his history and obfuscations instead.[63]

The AEC then rejected Oppenheimer's appeal on June 29, with Strauss joining three other commissioners against him in a 4–1 decision

[59] Herken, *Brotherhood*, 287–88; *IMJRO*, 137.
[60] "Dr. Oppenheimer is Barred from Security Clearance, Though 'Loyal,' 'Discreet,'" *NYT*, June 2, 1954, 1, 13.
[61] "Counsel Defend Dr. Oppenheimer," *NYT*, June 16, 1954, 18.
[62] "Scientists Views Stir Panel Worry," *NYT*, June 2, 1954, 17.
[63] Polenberg, *In the Matter of JRO*, xxv.

which declared that "his associations with persons known to him to be Communists have extended far beyond the tolerable limits of prudence and self-restraint."[64] Along with Chevalier, these Communists included two lab employees (Weinberg and Giovanni Rossi Lomanitz) and a Communist functionary, Rudy Lambert, all of whose party connections the physicist had been slow to elaborate. As Nichols had recommended, the H-bomb issue was played down. Instead, the focus was on Oppenheimer's "character": his "falsehoods, evasions, and misrepresentations."[65] The Commission insisted that Oppenheimer's "early Communist associations are not in themselves... controlling" in denying him a clearance. Instead, "they take on importance in the context of his persistent and continuing association with Communists," the Commission alleged, creating in their patterning of Oppenheimer's behavior—"mnemonic pasting," as Zerubavel calls it—a selectively constructed memory of his life.[66]

Certainly Oppenheimer's recent activities were not immaterial in the verdict against him: in 1950, Oppenheimer had helped Chevalier to get a passport, and hosted the former Berkeley professor at his home; he had visited Chevalier twice in Paris, all while he was on the AEC payroll. Such contact, authorities feared, could have allowed the scientist to be whisked to the Soviet bloc! All the same, the narrative of Oppenheimer's disloyalty that emerged from the hearings was based far more on his inconsistencies of 1943 than his friendships in 1953. Oppenheimer's admission that "I was an idiot" and Eisenhower's conviction that the physicist was a "liar" had little to do with his postwar visits to France, and everything to do with his actions eleven years earlier—and the powerful memory of those events had most influence in shaping the outcome of the hearings.

The physicist returned home after his ordeal, but his life was hardly private. Six G-men watched Oppenheimer in his "exclusive" Princeton neighborhood each night, including a posse a half mile away on Mercer Road with a three-way radio, who were on the alert for possible train trips to New York. Another FBI vehicle monitored automobile travel north of the residence. Such fears were excited by a lengthy vacation Oppenheimer took in the Virgin Islands, which prevented surveillance. Yet the FBI conceded

[64] *IMJRO*, 1049.

[65] "Oppenheimer Loses Appeal to AEC, 4 to 1," *NYT*, June 30, 1954, 1, 9.

[66] "AEC Vote Bars Oppenheimer, 4–1," *NYT*, June 30, 1954, 9; Zerubavel, *Time Maps*, 40.

that "no information was developed by this method indicating his contacts with espionage agents or Soviet officials."[67]

The lack of results did not stop the FBI from doggedly applying for yet another wiretap authorization of Oppenheimer from the Justice Department two months after the hearings had ended; the Department, meanwhile, was considering the Bureau's reports for possible action against the scientist. Alas, sighed Belmont, "Oppenheimer suspects that his telephone conversations have been monitored," thus eliminating much chance for interesting information. Moreover, "the residential nature" of his neighborhood hindered effective surveillance, and monitoring Oppenheimer on Princeton's campus was no easy trick either. At last, he admitted, "the Bureau cannot anticipate any great results from either the technical or physical surveillance." The FBI discontinued surveillance in October 1954.[68]

While the memory of Oppenheimer's wartime behavior and even earlier associations stubbornly persisted for officials well into the Cold War, the Institute for Advanced Study's board was happy to keep him as director until Oppenheimer, by then afflicted with cancer, resigned in 1966. Joseph McCarthy, ironically, had been disgraced the same month that Oppenheimer's hearings took place, and the Cold War had rapidly cooled off. Indeed, the very year that Oppenheimer lost his security clearance, he was honored by a group of black businessmen from the Pyramid Club in Philadelphia, who gave him an achievement award.[69] Despite the powerful narrative of disloyalty the U.S. government created and nurtured about him, therefore, Oppenheimer would remain a source of inspiration for many Americans who had constructed different memories of his contributions.

---

[67] FBI Newark to Director, FBI, July 13, 1954, 100–17828–1880; Belmont to Boardman, August 24, 1954, 100–17828–1974; see also Hoover's earlier request for a wiretap, April 26, 1946, memo to the Attorney General, 100–17828–29, where he justified the request based on Oppenheimer's "contact with individuals reported to be Soviet agents."

[68] Belmont to Boardman, August 24, 1954, 100–17828–1974; Hoover memo, October 15, 1954, Oppenheimer, Internal Security 100–17828–2072.

[69] "Club to Honor Dr. Oppenheimer," *NYT*, September 15, 1954, 19.

REFERENCES

Primary Sources

Independence, Missouri

  Harry S Truman Library

    President's Secretary's File
    Clark Clifford Papers

Washington, D.C.

  Federal Bureau of Investigation Files

    Communist Infiltration of Radiation Laboratories (CINRAD)
    Klaus Fuchs
    Steve Nelson
    J. Robert Oppenheimer

  National Archives

    Record Group 233, Records of the U.S. House of Representatives, Special Committee on Un-American Activities (Dies), Exhibits, Evidence, etc. re Committee Investigations: Communism Subject Files 1930–1945.
    Record Group 319, Records of the Army Staff. Army Intelligence Project Decimal File, 1941–45.
    Record Group 319, Records of the Army Staff. Records of the Office of the Assistant Chief of Staff, G-2, Intelligence, Records of the Investigative Records Repository, Security Classified Intelligence and Investigative Dossiers, 1939–1976.

National Security Agency

  Venona cables, 1942–1946

Congressional Hearings

  U.S. House. Report on Soviet Espionage Activities in Connection with the Atom Bomb; Investigation of Un-American Activities in the United States. Eightieth Congress, second session. Washington, D.C., 1948.
  —— Hearings Regarding Communist Infiltration of Radiation Laboratory and Atomic Bomb Project at the University of California, Berkeley, Calif. (Identification of Scientist X). Eighty-first Congress, first and sessions. Washington, D.C. 1949–1951.
  ——. The Shameful Years: Thirty Years of Soviet Espionage in the United States. Washington, D.C., 1951.

Newspapers

  New York Times, 1954.

Books

Albright, Joseph and Marcia Kunstel. *Bombshell: The Secret Story of America's Unknown Atomic Spy Conspiracy*. New York: Times Books, 1997.
Chevalier, Haakon. *Oppenheimer: The Story of a Friendship*. New York: George Braziller, 1965.
Craig, R. Bruce. *Treasonable Doubt: The Harry Dexter White Spy Case*. Lawrence: University Press of Kansas, 2004.

Groves, Leslie. *Now It Can be Told: The Story of the Manhattan Project*. New York: Harper Brothers, 1962.

Haynes, John Earl and Harvey Klehr. *Venona: Decoding Soviet Espionage in America*. New Haven: Yale University Press, 1999.

Herken, Gregg. *Brotherhood of the Bomb: The Tangled Lives and Loyalties of Robert Oppenheimer, Ernest Lawrence, and Edward Teller*. New York: Henry Holt and Co., 2002, and website: www.brotherhoodofthebomb.com

Holloway, David. *Stalin and the Bomb: The Soviet Union and Atomic Energy, 1939–1956*. New Haven: Yale University Press, 1994

Nelson, Steve, James R. Barrett, and Rob Ruck. *Steve Nelson: American Radical*. Pittsburgh: University of Pittsburgh Press, 1981.

Polenberg, Richard, ed. *In the Matter of J. Robert Oppenheimer: The Security Clearance Hearing*. Ithaca: Cornell University Press, 2002.

Sandilands, Roger. *The Life and Political Economy of Lauchlin Currie: New Dealer, Presidential Adviser, and Development Economist*. Durham: Duke University Press, 1990.

Schecter, Jerrold and Leona. *Sacred Secrets: How Soviet Intelligence Operations Changed American History*. Washington, D.C.: Brassey's, 2002.

Sibley, Katherine A. S. *Red Spies in America: Stolen Secrets and the Dawn of the Cold War*. Lawrence: University Press of Kansas, 2004.

U.S. Atomic Energy Commission. *In the Matter of J. Robert Oppenheimer: Transcript of Hearing before Personnel Security Board and Texts of Principal Documents and Letters*. Cambridge, Mass.: MIT Press, 1971.

Weinstein, Allen and Alexander Vassiliev. *The Haunted Wood: Soviet Espionage in America—The Stalin Era*. New York: Random House, 1999.

Zerubavel, Eviatar. *Time Maps: Collective Memory and the Social Shape of the Past*. Chicago and London: University of Chicago Press, 2003.

RESPONSE

DAN LEAB

Katherine Sibley's arguments about Robert Oppenheimer's ties to the less than democratic Left in the 1930s and 1940s cogently and convincingly demonstrate that the physicist was not the innocent naif that his apologists have presented over the years. Indeed, as one observer has noted, Oppenheimer "may not have actually obtained a Party card but... did everything short of it." Oppenheimer handled his problems with the government in the mid-1950s with great aplomb and considerable dignity, but his was not, as the *New York Review of Books* titled a September 2005 article on a spate of books about him, "An American Tragedy." Perhaps "An American Melodrama" would have been a more accurate and revealing heading.[1]

As Professor Sibley's article points out Oppenheimer benefited from the U.S. government's limited and faulty security apparatus to do properly the job it initially should have. As has become clear through publication of various memoirs, intelligence histories (drawing on hitherto classified information from both Western and Communist sources), and the VENONA excerpts, "the penetration of the Manhattan Project by Soviet spies" (to use the respected intelligence analyst Thomas Powers' words) "is not in dispute." One does not have to wholly accept the more melodramatic charges of Jerold and Leona Schechter in their 2002 account of "How Soviet Intelligence Operations Changed American History" that Oppenheimer became "a willing source for the Soviet Union of classified military secrets" to appreciate the scientist's involvement with the espionage directed against the Project.[2]

Oppenheimer without any doubt whatsoever obviously enhanced significantly the development of a nuclear bomb by the United States. And

---

[1] James Nuechterlein, "In the Matter (Again) of J. Robert Oppenheimer," *Commentary*, October 2005, 55; Thomas Powers, "An American Tragedy," *The New York Review of Books*, 22 September, 2005, 73–79.

[2] Thomas Powers, *Intelligence Wars: American Secret History From Hitler To Al-Quaeda* (New York: The New York Review of Books, 2002), 62–63; Jerrold and Leona Schechter, *Sacred Secrets: How Soviet Intelligence Operations Changed American History* (Washington, D.C.: Brassey's Inc., 2002), 205.

however one may morally view the destruction wrought on Japan by the use of nuclear weapons, the response to the reasons for the dropping of the atomic bombs by such gifted and articulate polemicist scholars as Gar Alperovitz and Martin Sherwin (which it seems to me had more to do with anti-Cold War politics) is one with the discussion of Oppenheimer's later supposedly tragic situation; as Professor Sibley points out its roots lay in his political feelings, however complex these may have been. Sherwin and Kai Bird in their compelling, comprehensive biography of Oppenheimer maintain that the scientist knew that "the atomic bombs" he had "organized into existence were going to be used. But he told himself that they were going to be used in a manner that would not spark a postwar arms race with the Soviets." She foregoes that tendentious after-the-fact insight.[3]

If Professor Sibley fails to address thoroughly any aspect it is that she did not follow through on the public relations side of the debate over Oppenheimer, and his relations with Communism and the Soviets. "Oppie" in the many images of him which appeared in the press and on TV after the AEC hearings until his death looked ascetic, frail, and worried—pipe in hand or mouth. He did not present the stern visage of a Lewis Strauss or the rough-hewn features of a Edward Teller, his prime opponents (neither of whom even Bachrach could have made attractive) at the time when Oppenheimer lost his clearance. Perhaps if he had more of their physical attributes his standing as an icon would have been less likely. In the TV parlance of the day Oppenheimer was "hot"; they were not.

---

[3] Gar Alperovitz, *The Decision to Use the Atomic Bomb and the Architecture of an American Myth* (New York: Alfred A. Knopf, 1995); Martin Sherwin, *A World Destroyed: Hiroshima and Its Legacies*, 3rd ed. (Stanford, CA: Stanford University Press, 2003); Kai Bird and Martin Sherwin, *American Prometheus: The Triumph and Tragedy of J. Robert Oppenheimer* (New York: Alfred A. Knopf, 2005), 314.

CHAPTER ELEVEN

TEMPORALITY, INTENTIONALITY, THE HARD PROBLEM OF CONSCIOUSNESS AND THE CAUSAL MECHANISMS OF MEMORY IN THE BRAIN: FACETS OF ONE ONTOLOGICAL ENIGMA?

E. R. DOUGLAS

SUMMARY

This essay concerns the intricate and multifaceted metaphysical relationship between memory, intentionality, physical time, psychological time, causality, and the arrow of time. Specifically, I argue that they have resisted scientific explication for very similar reasons, and I conclude that natural philosophy can come to terms with them once several fundamental notions, including directedness and transience, are clearly articulated and formalized. The implications are significant for the philosophy of mind, the metaphysics of time, the cognitive sciences, as well as contemporary cosmology and natural philosophy generally. This work moreover highlights an important difference in the operative assumptions respectively grounding the humanities and sciences.

1. INTRODUCTION

The three greatest metaphysical questions to confront humanity at the beginning of the twenty-first century are arguably: What is time? What is mind? What is the physical origin of the material universe? Each of these perennial issues has generated a colossal body of philosophical, literary and scientific investigation. However, as is so often the case in our overly complex epoch, an excessive emphasis on problem analysis, the methodological hallmark of the previous century, has also led to a paucity in the rigorous synthesis of such seemingly disparate issues. In this essay, I argue that these three questions have proven especially difficult to answer on account of a single underlying enigma, a riddle that turns on the nature of temporal transitivity and direction.

Investigations of such expansive questions are traditionally more customary in the humanities than in the sciences, but it is a central tenet of this essay that both kinds of methodologies are necessary to their possible resolution(s). In particular, I adopt the analytically motivated rigor of the sciences, maximizing

Jo Alyson Parker, Michael Crawford, Paul Harris (Eds), Time and Memory, pp. 199–222

the clarity of the disparate issues and concepts at hand, but also motivated by the humanities' focus on the function of human agency in the pursuit of natural truth, I claim rational grounds for drawing a synthesis among the corresponding problem sets. So on the one hand, I attempt to explicate such diverse terms as, for example, *time, intentionality, causality, memory* and *directedness*—whose commonly understood definitions leave them too ambiguous and/or obscure for an analytically empirical methodology. On the other hand, while they prove under inspection to have a common conceptual thread, its empirical credentials depend for merit on the acceptance of the phenomenological relationship between the scientist and nature, or the philosopher and philosopheme, as itself legitimately evidential. While this step does not require an endorsement of Protagoras's relativist claim that *man is the measure of all things*, we must nonetheless accept that he remains the *measurer* of all that we know.[1]

From the vantage of this interdisciplinary perspective, it becomes increasingly evident that physics, and the scientific *Weltanschauung* generally, is coming to a crossroads. While there remain many questions without definitive answers, science as most generally construed now comprises a patchwork of models and theories that collectively can speak to almost every conceivable issue in natural philosophy. Yet, internal inconsistency abounds, and I would further contend that at least mainstream science has overlooked a very important aspect of nature. Illustrating for the nonce with a metaphor that may yet prove more than mere metaphor, science has sketched the design of the cosmos, but the blueprints reproduced remain monochromatic; whereas nature in all its wonder remains demonstrably colourful.[2] Nonetheless, one can inscribe all the colours with a single ink—so the metaphor continues—and though it is disputable whether one can thus entirely characterize their natures, we should at least attempt such 'scientific' descriptions as completely, accurately and truly as possible.

In particular, I nominate three classes of 'empirical' phenomena that mainstream science has not sufficiently explored: consciousness, causality

---

[1] Many scientists of course do tacitly accept this point, even celebrating science as a product of a humanist cultural revolution against the medieval metaphysics that preceded it. Nevertheless, there are no mainstream scientific methodologies that legitimize themselves as objects of their study. Indubitably, the sophists would not approve.

[2] This metaphor of colour may be more than a metaphor. In his argument against physicalism, the philosopher of mind, Frank Jackson proposes a celebrated thought experiment: Mary is a woman of the future who knows all there is 'scientifically' to know about colour, but she herself is colorblind, and so limited to a physicalist vocabulary to explain the phenomena. (Jackson 1986)

and psychological time. With respect to the first, it has been widely objected that purely physicalist, materialist and even cognitivist accounts can suffice to explain away (mental) phenomena qua phenomena (and in particular for our purposes, *intentionality*, but also notably, *qualia* or indeed the experience of colour itself). To the second, I ask: does causality as we perceive it actually equate with a statistical correlation between events as stipulated in most contemporary analytic accounts and, if so, how did the universe come to be, and what would that even mean? Finally, the third begs whether 'time' really reduces accurately and completely in all its forms to a static tableau, in spite of our manifest experience of temporality as profoundly transient and directed.

Each of these three problem sets is marked by several possible solutions, theories or philosophical approaches that one might adopt, some easier, some 'harder' or, indeed, some considerably more 'colourful' than others. The first and 'easiest' solution falls under the name of eliminativism in the philosophy of mind, but it generalizes to all three of the problem domains we are considering here. This view relegates all mental phenomena to instances of mechanistic brain chemistry, but we may regard its cousin theories equating cause with correlation and time with space respectively. The more difficult, but more colourful—i.e., I contend, more 'empirical'—approaches involve wrestling with the underlying hard problems of explicating the emergence and nature of consciousness and intentionality, temporal transience and direction, as well as the metaphysics of being, becoming and causality. Although I discuss some of the 'colourful' theories available to explain the emphasized phenomena endemic to each domain of inquiry, my focus throughout this essay remains primarily on distinguishing eliminativist solutions—which at root simply expurgate their respective ontologies of the problematic phenomena—from those that do not.

The ultimate thesis of this essay asserts that the apparent quandaries of these three introduced problem sets are really each facets of a single, underlying enigma, but I make this case through several lemmas or staged arguments. First, I link these disparate concepts (time, mind, cause) by first explicating their meanings and then identifying how directionality plays a central role in the interpretation of all three; this comprises the remainder of section one. Second, I remonstrate something of a collective bad faith marked by a tendency in the philosophical and scientific literature to explain the most difficult aspects of each of the problem sets by smuggling implicit assumptions from the other two respective problem domains. This discussion strongly emphasizes the concept of directionality and so also significantly informs yet a quasi-fourth quandary of import to the natural philosophy of time: the meaning, origin and nature of the arrow of time. This last matter figures prominently in the larger problematic, for, in contrast, none of the other 'empirical' phenomena I endorse (consciousness, apparent causality or temporal transience) has aggravated especially physical scientists so much.

From here, I contend that while there may not necessarily be clear *a priori* grounds to choose one class of explanation over another—i.e., 'monochromatic' universe over 'colourful' cosmos—to explain the three (or four) aforementioned quandaries, insofar as metaphysics motivates arguments introduced in natural philosophy, the principles invoked for one solution should be employed consistently across the other two problem domains. In this way, I maintain an all or nothing approach to eliminativism in natural philosophy; the cosmic models we endorse are either very colourful or very monochromatic, there is little room for compromise. However, the stage is thus set to revisit the issue of whether and how science should attempt to explain the 'colourfulness' of the cosmos, and from this more global, human perspective, I maintain a considerably stronger case presents itself for the affirmative on the first question through a kind of logical inter-corroboration of empirical evidence.

As for 'how' science should proceed, both the principle of parsimony and this weave of logical association between problem domains support theories able to resolve all the aforementioned difficulties synchronically. Although I mention a few promising scientific attempts to come to terms with these issues, a complete review of contemporary research lies beyond this essay's scope; notably, I have had to omit due consideration to ideas stemming from Bohm's implicate order—I could not do it the justice here it deserves. Finally, I would conclude this introduction with a small warning: although rigor is desirable, *natural philosophy* is neither mathematical nor empirical per se, and the discourse herein is less an analytical proof than a metaphysical sketch stencilled with plausibility arguments, all hopefully inducing further dialectical investigation.[3]

---

[3] Schulman analogizes speculative philosophy as "scouting" new terrain, leading the way for the "heavy-tanks" of physics (his discipline) to follow. (1997) While I regard the military metaphor as a trifle extreme, there is much to be said for philosophy that does not limit its scope only to what can be proved.

1.1　*Chronological and Rhealogical Time*[4]

It is no great exaggeration to say 'time' has as many meanings as token applications, and yet they do share a common essence—sufficiently, that most disciplines tacitly assume that their 'temporality' is similar, if not identical, to those of every other field. J. T. Fraser has made great inroads in explicating this semantic jungle of *times*, and his characterization of six sorts of temporality proves an excellent point of departure. (1975; 1987; 1996; 1998; 1999) Each is distinguished by the accumulation of novel properties and qualities which emerge in increasingly complex natural systems to form a nested hierarchy of times. This scheme has proven very useful and influential, so that one finds its trace etched across the sciences and the humanities; in particular, the physicist, Rovelli, employs it to argue that a final theory of cosmology will have 'no time,' since the most fundamental levels of reality lack temporal structure, congruent with Fraser's *atemporality*. (Rovelli 1995) In most respects, their schemes agree; direction corresponds respectively with thermodynamics and life (*biotemporality*), and both tacitly regard the *present* or 'passing now' as a feature properly attributed to the time of (human) intentional agents (*nootemporality*).

However, I must begin by taking issue with these two properties, *directedness* and *transience*, as characterized above. They are in fact both present in Fraser's lowest echelon, *atemporality*, and so by evolutionary extension pervade all of his levels. In contrast, Rovelli's scheme highlights the formulae and physical models of his (meta-)physics, emphasizing the static and permanent, relegating transience to illusion. Moreover, direction finds no place in his scheme whatsoever, for the asymmetry and direction of time are not equivalent, as I show in the next section. Thus, all other similarities aside, Rovelli plays Parmenides to Fraser's Heraclites, and to celebrate this difference, I introduce a category distinction between species of time to cleave the distinction sharply: any model or idea of time that invokes *transience*, even qua illusion, is *rhealogical*, and those that do not, but are *static*, are termed *chronological*. Moreover—and I argue this further in the

---

[4] The term, 'rhealogical,' is a neologism I introduce to characterize a class of models of time that compliments especially Bergson's use of 'chronological.' It avoids many semantic pitfalls prevalent amongst temporal terminology by addressing the essential property of authentic change (which I discuss in this essay). The etymology is both the Greek word, *rhein*, meaning to flow (as in 'time flows'), but it also nicely dovetails with the name of the titan-goddess Rhea, who was sister and wife to Kronos, god of time. For further discussion on the nature and significance of rhealogical models of time, see Douglas (2006).

next section—directedness appears to be a feature of all and only rhealogical models of time, whereas chronological time is devoid of such qualities. Thus, Fraser's nested times are paradigmatically rhealogical, whereas all orthodox physical models are chronological.[5]

However, what precisely is this mysterious *transience*? This proves to be a surprisingly subtle question whose answer quickly extends beyond the scope of this essay, but a brief digression is in order. First, we may recall Taylor's 'Doctrine of the Similarity of Space and Time,' wherein he claims everything that can be expressed about time can be transposed about space. (1955) While I do not agree with the thesis, it serves as a useful rule of thumb: if a particular characterization of time can be 'spatialized,' to use Čapek's term, then it is chronological, but if not, then it is rhealogical. (1961) Second, and a little more precisely, transience is any instance of *authentic change*, which is to be distinguished from Newton-Smith's 'Goodman changes.' (Newton-Smith 1980; Goodman 1977) Interestingly, Webster's dictionary defines such arguably counterfeit 'change' as: '...concomitant variation in time and some other respect—Nelson Goodman.' (1993) The problem with such a definition becomes evident in McTaggart's celebrated analysis of time, wherein '...it would be universally admitted that time involves change.' (1908) Thus, the Goodman definition proves circular and so permits a deflationary interpretation of *change* that puts it on par with such concepts as *difference* or *variation*.

The following example illustrates the problem. Consider two bottles of Chianti-Rufina standing next to one another on a table, one empty, the other full; there is a *difference*, but it would be obscure to claim there lies an authentic *change* between them. Yet, suppose the sense in which one finds them 'next' to one another is temporal, not spatial, for example, as seen in two still-lifes painted by an artist with a penchant for *veritas*—such 'change' is common fare. So, what qualitatively distinguishes these two cases—i.e., we say 'spatial' *difference* but 'temporal' *change*—or is the distinction purely nominal?[6] I identify at least two explicit features: *authentic change* and

---

[5] Cf. Park (1996; 1972). Similarly, in the only article to address transience in Scientific American's most recent issue dedicated to time, Davies summarily dismisses it as an illusion: 'Nothing in known physics corresponds to the passage of time. Indeed, physicists insist that time doesn't flow at all; it merely is.' (2002)

[6] The view I am confuting here is nicely summarized in Oaklander's conclusion to his refutation of A-series interpretations of time: 'What distinguishes greater than among numbers from later than among events? ...only the relation itself... [which] is a simple and unanalyzable relation. Thus, there is nothing that we can say about temporal succession that would distinguish it phenomenologically from other relations that have the same logical

the *direction* of process. However, not wishing to beg the question before discussing *direction*, allow me to propose another definition of *authentic change*: a single *token* entity manifesting as two or more inconsistent *types*. In the philosophy of mind, a 'token' denotes a unique and particular or singular entity, whereas 'type' refers to a mould or form, such as the class of full wine bottles with 'Chianti-Rufina-1998' etched onto them. In the example above, the *token* is '*this* bottle of Chianti,' and the two *types* separated by time are 'full' and 'empty' respectively.

While I choose to accept transience as inevitably involving an (onto)-logical contradiction, there are many who impugn such impossibility as absurd and, to their credit, not without good cause. In particular, a topic or object of discourse only makes (scientific) sense if one can characterize its nature in a formal language, whereas from an inconsistent phrase, anything and everything can be logically adduced, with all the pragmatic inadequacy this 'explosiveness' entails. Thus, the law of non-contradiction is a first law in logic that one only perilously transgresses, and I suspect this accounts for a good part of the reason why analytic philosophy has had such a difficult time coming to terms with temporal transience (and perhaps love).

However, recent work in 'paraconsistent logics' suggests this law is not inscribed in stone, though I have yet to find any of the developed logics satisfactory for the characterization of transience.[7] Nevertheless, there are grounds for hope, for the contradictions in question are marked by the order they are presented: i.e., while a full bottle may become empty, or an empty bottle might become full—both of which are only expressible in most logics as 'bottle is full *and* empty' or, equivalently, 'bottle is empty *and* full'—we only wish at most one of these contradictions to be *true*. Thus, it may prove possible to control the aforementioned 'explosiveness' of our logic of transience by introducing a non-commutative operator—e.g., an ampersand with an arrow—that delimits truth or satisfaction to some, but not all, inconsistent expressions: e.g., in this instance, 'Full(Bottle) &► Empty(Bottle),' corresponding to 'the full bottle *becomes* an empty bottle,' could be true, but 'Empty(Bottle) &► Full(Bottle)' would remain false.

If this digression into formal logic seems a trifle pedantic to some readers, I would assure them that it is difficult to overestimate its importance here.

---

properties. Nevertheless, succession is something more than its logical properties and we all know what more it is although we cannot say… There is no further basis for the difference between temporal and non-temporal relations with the same logical properties, they are just different.' (Oaklander 1984: 17)

[7] For a good introduction to paraconsistent logics and the problems facing their construction, see Beall (2004) and Priest (2004).

The scientific community can only investigate those 'colours' of nature that it can articulate, disprove and corroborate clearly. Furthermore, with respect to temporal transience, something else very interesting appears in the logical arrangements I have been considering. It may be the case—possibly *provable*—that any logic capable of expressing transience in a useful manner will intrinsically comprise features that correspond in a natural way to what I introduce in the next section as *directionality*. If correct, it follows that all rhealogical models of time manifest direction. Moreover, such 'direction' may prove ineffable in purely chronological logics—i.e., those that do not permit sentences expressing the kind of inconsistency introduced earlier—in which case, *only* rhealogical models of time will be authentically directed. Such a mutual logical entailment remains conjecture for now, since the full arguments range beyond the scope of this essay. On the other hand, the views espoused here are hardly revolutionary; indeed, Kant long ago recognized the logical relationship implicit between direction, transience and inconsistency.[8]

1.2   *Directedness and the Arrows of Time*

This brings us to the issue of the arrow of time or, more precisely, the *directedness* of time. Fraser maintains that a 'short arrow' first appears in *biotemporality*, and Rovelli similarly attributes such to thermodynamics. (ibid.) However, there is already a direction intrinsic to all of Fraser's nested hierarchy defined by the generative progression of evermore complex, nested models of time. This is thus well in accord with our working hypothesis that temporal directedness and rhealogical time mutually entail one another. Furthermore, I maintain there is in fact no directedness in any of Rovelli's models of temporality; his arrow of time is only an asymmetry. To appreciate this, a clearer distinction between *directedness* and *asymmetry* should be drawn. Consider the following image: Does it define a *direction*?

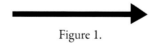

Figure 1.

---

[8] Kant identifies both the issue of inconsistency and that of direction or order: "...the concept of motion, as alteration of place, is possible only through and in the representation of time... render[ing] comprehensible... a combination of contradictorily opposed predicates in one and the same object, namely, *one after the other*. Thus our concept of time explains the possibility of... motion..." (Kant 1998: B48–49)

Anyone licensed to drive an automobile probably feels it does *point* to the right, but this is a cultural artefact, a mere convention. On the other hand, the property of *asymmetry* is rather intrinsic to the geometry of the shape itself, independently of our interpretation. Thus, while it may be that all instances of directedness may logically imply a measure of asymmetry, the inverse is apparently not the case.[9] This property of directedness begs further explication, but let us first consider it in the context of the arrow of time.

There is a rather extensive literature concerned with the so-called 'arrow of time,' but from the beginning, much of it fails even to recognize the term is a misnomer.[10] Really, there is a quiver of double-headed arrows, which are typically, but inexplicably, presumed to aim together at a single target, namely, *the end of time*, thereby supporting the contention that there is some intrinsic *directedness* to time. However, most of these putative arrows are mere asymmetries, as illustrated in the table below.[11]

1. *Psychological arrow*: a *directed* asymmetry reflecting the 'feeling of relentless forward temporal progression, according to which potentialities seem to be transformed into actualities.'

   1a. *Ontological asymmetry*: a determinate past contrasted with an open future.

   1b. *Epistemic asymmetry*: exemplified by memory, that we know more about the past than the future.

   1c. *Causal arrow*: the *direction* in which events as effects *seem* to follow causes.

2. *Entropic asymmetry*: associated with second law of thermodynamics.

3. *Electromagnetic asymmetry*: radiation never 'converges' on an antenna in phase, but frequently 'emits' thus: only *retarded* solutions of Maxwell's equations, and not *advanced* ones, are real.

---

[9] Logically, let direction be a relation, $D(\ ,\ )$, of elements, e_1 and e_2. All we can infer about the finite models that satisfy it is *asymmetry*, because we can construct an inverse expression, $D^{-1}(\ ,\ )$, representing the opposite direction, such that: $\models D(e_1, e_2)$ if and only if $\models D^{-1}(e_2, e_1)$.

[10] Although the 'arrow of time' has entered the common vocabulary, its meaning remains very disputatious in both philosophy and physics. (Hawking 1995, Price 1996; 1995; Savitt 1995; Schulman 1997; Sklar 1974) It originally appears as 'Time's Arrow' on p. 28 of Eddington's 1928 Gifford Lectures. (Savitt 1995: 1)

[11] Roger Penrose distinguishes between three aspects similar to 1a–c. (1979: 591) Also, if the seventh, quantum asymmetry should prove genuine, then it may well be moreover directed, depending on whether one interprets the collapse of the wave function as an ontological event—cf. (Penrose 1994).

4. *Neutral Kaon asymmetry*: the neutral K meson anomalously decomposes into pions at a rate that varies with the temporal direction of the process.
5. *Relativistic asymmetry*: some world lines end (begin) at singularities in black holes.
6. *Cosmological asymmetry*: corresponding to the expansion or contraction of the universe—i.e., the *big bang* vs. the *big crunch* (or its lack, as the case may be).
7. *Quantum asymmetry*: disputatiously resulting in quantum *measurement*.
8. *Biological asymmetry*: corresponding to evolutionary features of biological systems and organisms.

Naturally, some of these arrows may reduce to or derive from one another, and this list is not intended to be complete; our concern here is with *directedness*, or its lack, and this becomes yet more transparent in the following thought experiment. Suppose that after much laudable science, seven independent asymmetries were discovered to remain: Which direction does time tend? If each is thought to point in one of two directions, then there would be $2^7 = 128$ possible combinations, and in only two would all seven converge. The moral of the story is that 'forward' is trivialized as conventional, and so most research on the arrow of time only concerns the asymmetric distributions of properties across space-time; it says nothing about temporal *directedness*. (Price 1996; Sklar 1974: 355–60) The only feature of time that clearly incorporates an intrinsic *direction* is its perceived flux! Thus, we should proceed carefully, for much like Augustine's celebrated dictum about time, that we only know it when we do not ask too much of it, closer inspection similarly reveals *directedness* to be a surprisingly Janus-faced concept.

1.3 *Causality, Intentionality and the Mathematical Infinite*

Thus, let us broaden the scope of our investigation and consider all possible instances of directedness in our experience. In the first category, I take it as given that we (human beings) all *perceive* the passage of time as directed. We may designate a second class, if we accept a generative theory of causality into our physics (cf. 1c in the previous table), for then the cause-effect relationship is also directed; i.e., causal events may exist independently of their corresponding effectual events, but the latter only exist (are generated) through the action of the former. However, the vast majority of instances of directedness fall under the third grouping, *intentionality*, either explicitly or implicitly.

The term 'intentionality' finds most of its employment in the cognitive sciences and the philosophy of mind, and its definition and nature are highly disputatious. Nevertheless, 'about-ness' is an approximate synonym, and we are referring to intentional relationships when we say a word is *about* its denotation, a thought is *about* its content and an image is *about* whatever it reproduces. In each case, there is nothing *physical*, but the metaphysical significance of language, signs, and thoughts begs further explanation. Although intentionality is generally regarded as an essential component of *consciousness*, its etymology suggests a more general interpretation: the Latin word, *intendo*, means 'to point at,' which is rather how Aquinas and his medieval, scholastic colleagues employed it in their metaphysical musings. Such directedness remains in the signification of contemporary 'intentionality,' for the *about-ness* relationship is intrinsically uni-directional: e.g., the word 'wine' may be *about* the preferably red liquid often found in glass bottles, but it is nonsensical to suppose such a beverage is *about* a four-letter concatenation.

Now, thinking back to the right-handed arrow that figured earlier, a little reflection finds almost all instances of perceived or posited direction are conventional. This means they derive their putative direction from the way we, as intentional agents, interpret them; they are *secondary qualities*, to use Locke's nomenclature. Thus, most empirical instances of direction vicariously reduce to such a phenomenological subclass subsumed under intentionality, without which all conventions would cease to exist. One might well argue that all instances of directedness are dependent on intentionality in this way, but that presupposes directedness is always a secondary quality, hence extrinsic to the natural order, and this is refuted by the intrinsic directedness of intentionality, qua *natural kind*. Nevertheless, the onus remains on us to demonstrate which instances are intrinsic to nature, where directedness proves to be a *primary quality*. Besides intentionality itself, the only physical possibilities seem to be time and causality, and only provided they respectively prove to be rhealogical and generative.

This completes my inductive argument with respect to 'physical' instances of direction, but there still remains a putative fourth category that bears mentioning. Certain mathematical structures intimate direction, at least as symbolically expressed. Prima facie, an infinite sequence of points with a proper limit form a set that can only be articulated in one direction; i.e., the sequence $(1, \frac{1}{2}, \frac{1}{3}, \frac{1}{4}, \ldots)$ with 0 as a limit cannot be expressed beginning with its end point. Yet better examples may be found in logic, such as recursively constructed Gödel-sentences. However, even if there exist mathematical objects as natural kinds (Platonism), it is still not self-evident in what sense the referents of mathematical language are directed, for such appearances may result from the choice of mathematical language employed in their description, in which case the directedness would be a secondary quality and so subsumed under intentionality. On the other hand, this seems less plausible with respect to the Gödel sentences noted above, whose construction might well demonstrate an ostensible intrinsic directedness that belongs to some kind of existent symbolic domain—if Platonism proves correct in this way. However, mathematical realism can take still several forms, many of which do not affect the pursuant arguments of the next section significantly.

1.4 Memory and Time

Memories are variously the thoughts, feelings, signifiers and other instances of manifest information that refer to events properly attributed to the past. As such, they compose a subspecies of *intentions*, but they have a double relationship with time. Firstly, as *intentions*, they are intrinsically directed, and I will argue that this entails an intimate relationship with rhealogical temporality. Secondly, they comprise an important subset of intentional vectors that refer to (define?) the chronological past.

Thus we find a curious metaphysical circularity in the way memory and time support and refer back to one another. Past events and entities prove the most ambiguously existential (actual) of all potentialities, since their status in this respect is largely derived from memory. Yet, the liminal reality of that-which-has-gone-before is essential to the causal determination of all present identities, including the very *intentional* agency that creates those memories. Thus, the following three questions are posed: How does a human create memories? How do memories and the past correspond? How does the past determine a person? Such circularity suggests that memories, along with their symmetric counterparts, *expectations*, play an especially interesting role, both in the marriage of rhealogical and chronological temporalities—whose union we may call 'time'—and in the metaphysical integration and explication of mind and body. Moreover, because memories are arguably ostensibly empirical, they demonstrate the deficiency of contemporary scientific theories to account in 'monochromatic' terms for the spectrum of 'colourful' mental phenomena evident in nature.

2. SYNTHESIS OF RHEALOGICAL TIME, INTENTIONALITY AND CAUSALITY

I have thus far argued inductively that these three categories are sufficient to account for all physical instances of directedness, and I have conjectured with some supporting rational that transience and the (directed) arrow of time mutually entail one another. We may now turn to three further methodological arguments, each respectively linking our conceptions of time, causality and intentionality together into a larger logical-conceptual synthesis. The first shows a general correspondence between models of generative causality and rhealogical temporality; the second correlates such causality with intentionality; and the third argues that our ontological commitments to intentionality and time are strongly interdependent. This sets the stage for the last section's conclusions about the natures of memory, agency, physical reality and time.

2.1 *Temporality and Causality*

If the model of time employed to describe a set of events is genuinely directed, this suffices to introduce directedness into a causal scheme on those events. Similarly, a genuinely directed theory of causality suffices to construct a directed model of time.

Causal theories come in two shades: either they are generative (*colourful*) or correlative (*monochromatic*). Aristotle's views exemplify the former, whereas Hume's paradigmatizes the latter; the key difference lies in that correlative, 'scientific' causality does not qualitatively differentiate between counterfactual causes and causes simpliciter, undermining the important intuition that causes *precede* effects, but not visa-versa.

The inability of such 'scientific' models of causality to demonstrate intrinsic direction appears to result from their underlying assumptions, which allow them in general to be expressed in a first order language. For if we suppose T_C is such a causal theory with axioms, events $\{e_i\}$ and n-relationships $C(causes; \, effect)$, in general, an inverse theory T_C^{-1} is constructible that is *satisfied* by the same *structures* as T_C; i.e., $Mod(T_C) = Mod(T_C^{-1})$. This is accomplished by defining $C^{-1}(effects; \, causes)$ so that $\exists e_{1 \leq i \leq n} C \, (e1, \ldots \, ;e_n) \Leftrightarrow \exists e_{1 \leq i \leq n} C^{-1} \, (e_n \ldots \, ;e_{i<n})$ is *satisfied*. Such formalism is simply a long-winded way of articulating how the causal arrows between events can be reversed without injury to the expressiveness of a first order causal theory, because the direction of the relations between events are superfluous to the correlations, as illustrated below:

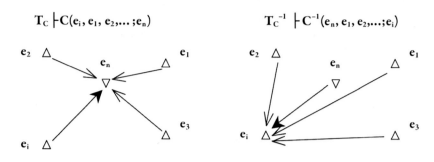

Figure 2.

Yet, we wish to explain the origin of the apparent direction and order of the many causal relationships we commonly experience. Hume grounded such structural features in temporality as he experienced it, which was psychological, hence transient and rhealogical. (Hume, 2000) Similarly, most contemporary theories follow suit, for as Reichenbach observed, 'Time ... represents not only an ordered series generated by an asymmetrical relation, but is also unidirectional. This fact is usually ignored. We often say simply: the direction from earlier to later events, from cause to effect is the direction of the progress of time.' (1958: 138–39) And, indeed, if time describes a manifold with an engraved direction and includes every physical event, it follows trivially that the said direction maps onto any subset of those of events characterized in a causal theory or relationship. Thus, rhealogical time suffices to account for the directedness of causality.

This state of affairs would all be fine and well, but the problem deepens considerably when an origin for temporal direction is sought. As discussed earlier, the physical arrows of time provide no answers, and a suitable scientific account founded on a chronological conception of time has proven sufficiently vexing, that several thinkers have rather attempted to derive the latter's structure from the 'natural' order and direction of causality. (Bunge 1959; Tooley 1997; 1999) Indeed, if we suppose the cosmos is composed of some set of events arranged as a directed causal *lattice*, its order and direction can be *homomorphed* back onto those events, introducing a directed structure to time. In this case, two events are said to be 'simultaneous' if-and-only-if they are identical or do not belong to any common causal *chain*, and an event is said to 'precede' another if-and-only-if it precedes that other causally on such a *chain*. Finally, the temporal orientation of an event corresponds to the direction of causal relations to which the event belongs. Thus, the directedness of an ontological account of causality suffices for a corresponding structure of time.

However, the articulation of a directed causal theory is fraught with the same logical difficulties facing the explication of rhealogical time. As the logical argument and illustration above make abundantly clear, any meaningful, generative causal theory will require new methods and formalization. It is perhaps for this reason that, by and large, generative theories of causality found their greatest circulation in the pre-Leibnizian—and so pre-symbolic and pre-scientific—era of the medieval scholastics. And overlooking these factors in would-be generative theories of causality has tended to undermine some of the most interesting attempts to explain our experience of time.

Finally, let me conclude here with a brief note on so-called 'backward causation,' which is code in much physical literature for time travel into the past, and which also indicates how physics largely understands the structure of time to derive from the causal relationships between events. Leaving the issues associated with causal paradoxes aside for now—the rhealogical-chronological modelling of time may shed much light here, but the discussion would take us too far a field in this essay, since it begs important questions about freewill—there are no significant problems introduced to this account of the relationship between time and causality. If an event in a directed causal sequence of events is also its own cause, then the structure introduced onto time will be circular and closed, but directed all the same.

2.2   Causality and Intentionality

Naturally occurring causal directedness suffices to account for the directedness of intentionality. Similarly, the directedness of intentionality suffices to explain the appearance of causal directedness in nature.

Intentionality is a primary property of consciousness and, adopting a naturalist position, is generated from a composite of processes and elements in the (human) brain. Explaining the emergence of consciousness from matter has a long history as the mind-body problem in philosophy, and it has shown itself remarkably resilient to resolution. However, if we accept that mind *supervenes* on the physical body and brain, then the web of experiences, feelings, memories and thoughts must map into some corresponding set of embodied processes. If those underlying parts then interact in a genuinely-directed causal manner, it follows that such interactions will similarly transpose onto the elements of mentality, providing a plausible account of the directedness of intentionality.

Let us consider an example that does not stray too far from a number of current cognitive theories. Let M_1 be an experienced memory that reminds us of another memory M_2, and each intentionally signifies some objects,

O_1 and O_2, which in turn likely signify other memories or experiences, which in turn refer to yet other objects, O_{11} and O_{21}, etc. Ultimately, I postulate such a chain of signification yields a web of objects, which at certain loci start to take on the characteristics of physical events, E_1 and E_2, the putative referents of memories, M_1 and M_2. When these events actually occurred—so the theory goes—they left causal impressions on the brain, affecting brain states, B_1 and B_2, though both are also affected by all the elements of the chain of objects. Per our assumptions above, these are the two physically instantiated brain states upon which our two respective memories putatively *supervene*. There is then a web of relationships—which, *in vivo*, would be complex to a degree beyond description—but isolating an absurdly simplified example, I illustrate below the different kinds of directed relations to consider, namely, intentional (unbroken lines), causal (dashed) and mixed:

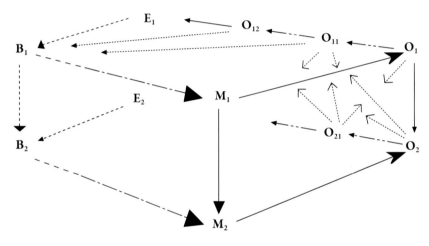

Figure 3.

I propose here that the origin of the direction of intentionality naturally obtains from the directedness of physical causality. This follows directly for the M_1-M_2 mental relation of remembering, since it is isomorphic to the B_1-B_2 physical causal relation. Similarly, the E_1-E_2 *about-ness* relationship, for example characterizing the signifier and signified of a word and its imagined association, may similarly derive its direction as an analogue of the original B_1-B_2 relationship. The M_1-O_1 and M_2-O_2 intentional relationships, which seem to tend against the grain are more subtle to explain. However, all of the objects, O_{xy}, synchronously represent both physical brain states and quasi-mental states (many perhaps not properly conscious as such), and so there are ample resources to sum over the many interrelated vectors to produce the requisite directedness. Nevertheless, what is most significant to appreciate from this illustration is that the directedness of intentionality need not be miraculously generated *ex nihilo*. Thus, a genuinely directed theory of causality suffices to explain the origin of intentional direction.

We might pause to consider here whether the reverse does not also obtain, that intentionality might similarly account for the apparent directedness of causality. Indeed, the logic appears symmetric. However, it resembles the highly mystical metaphysics of many pre-scientific cultures, for the causal interactions of physical processes then result from the dynamics of purely intentional, spirit-like entities, manitous. On the other hand, just because an idea is old does not mean it is irrelevant.

Alternatively, there is an epistemic correlate to the previous ontological argument. Since all theories of nature are conceived through the lenses of intentional minds, it is not such a stretch to claim we anthropomorphize and baptize *a* direction as *the* direction. This remains especially plausible so long as physics remains ambiguous on the matter, which indeed it does:

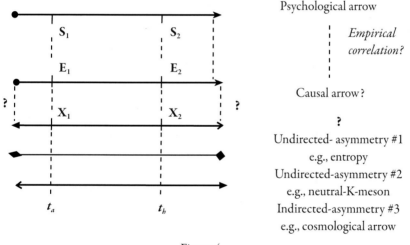

Figure 4.

Above, 'causally' related events (E), experienced states (S) and physical phenomena of some asymmetric class (X) are all respectively simultaneous at times (*t*). Such asymmetric 'arrows' are the closest contemporary science comes to explicating causality, but they all fail to account for its apparent directedness, whatever may be claimed.[12] The alternative anchors our

---

[12] Hawking writes, 'The psychological arrow, our subjective sense of time, the fact that we remember events in one direction of time but not the other,... [and] the electromagnetic

experience of causality in our psychology, though the ontological origin of its directedness is then begged of intentionality itself. Nevertheless, whether it is physics appealing to intentionality and psychology for the apparent directedness of causality, or psychology referring back to physical causality to ground the directedness of mental processes, both demonstrate an intimate relationship between domains.

2.3   *Intentionality, Consciousness and Rhealogical Time*

The genuine, intrinsic directedness of intentionality is sufficient to account for the apparent directedness of time. On the other hand, if time is genuinely directed, it suffices to account for the directedness of intentionality.

Cartesian dualism splits mentality and physicality in a way that almost perfectly parallels the schism between rhealogical (psychological) time and chronological (physical) time. However, here it becomes important to distinguish between consciousness and intentionality, which is arguably only one important aspect of the former; others include *qualia*, agency, freewill, and self-consciousness. Similarly, the directed arrow of time is only one aspect of rhealogical temporality, which also comprises such other aspects as transience and the distinction or dimension of *actua et potentia*. Now, a satisfactory analysis (and subsequent synthesis) of these two domains is beyond the scope of this essay, but a few remarks may suffice to sketch the argument.

Mentality derives much of its physical significance from its capacity to realize the future and past in the present, through memory, prediction, differentiation, association, identification and other intentional actions. How does it do this? Indeed, the physical significance of one workman calling out to another, 'slab!'—to recall Wittgenstein's example—correlates strongly with the likelihood that the other will in fact give the first a slab in the future. (1958) Without wishing to oversimplify a matter that is really considerably more complex, an important way intentionality obtains physical significance lies in connecting agency, identity and signs in the present moment with events of the past and future and, moreover, predictably affecting the constitution of that future.[13]

---

arrow… can be shown to be consequences of the thermodynamic arrow, which says that entropy is increasing in one direction of time.' (1993: 3) However, he never does tell how this is done, as Price notes. (1995, 1989)

[13] Freeman maintains psychological time is rooted in future-directed intentionality. (2000) As Modell writes, 'The emergence of a goal thrusts the organism's past into its future.' (2002:

In sum, I claim thus that a conscious agency without an intrinsic capacity to refer or associate in a unidirectional manner would, nevertheless, be able to recreate that directedness if it were subject to rhealogical process in time. Here, the identities of all elements, physical and mental, are subject to a uni-directional evolution—i.e., they either persist or they change—provided we accept the conjecture introduced in section 1.1. However, it then follows that an association between two possible identities belonging to different moments of time are distinguishable in that the one can become the other, but not vice versa. From here, all species of directed intentions, including memories, can be created and recreated through the faculties of creative imagination and identification. Again, the difficult step is to derive an initial directedness ex nihilo, and the posited directedness of time provides the needed pigment to colour the remainder of the ontology.

Suppose on the other hand we imagine the reverse scenario: a universe devoid of rhealogical temporality, but inhabited by intentional agency. Interestingly, Weyl describes such a world when he famously writes,

> The objective world simply is, it does not happen. Only to the gaze of my consciousness, crawling upward along the lifeline of my body, does a section of this world come to life as a fleeting image in space which continuously changes in time. (Weyl 1963: 116)

Here, time is characterized as purely chronological, but can consciousness survive embedded in such a spatialized block-universe?[14] I contend that it is not possible and so conclude that rhealogical time is essential to consciousness and intentionality.

Assuming the postulates of cognitivism for the nonce, mentality derives from *information* instantiated in some physical material, whether that be carbon-based biology, silicon-based technology or wood-based tinker-toys. Since the choice of material is inconsequential, I will follow Leibniz's example and choose wood on aesthetic grounds to construct a hypothetically sentient machine.[15] Indubitably, such a 'consciousness' would be massive and complex, but engineering issues aside, its design falls within the parameters of cognitivism, the popular view today in those academic circles dedicated

---

22) Similarly, Peirce writes, 'One of the most marked features about the law of mind is that it makes time to have a definite direction of flow from past to future.' (Øhrstrøm 1995: 133) For further research that intimates a deep connection between time and mind, see Penrose (1994), Hameroff (2001).

[14] Cf. Schlesinger writes, 'a spatial world in which time did not exist would be entirely stripped of capacity to contain the basic ingredients of a viable universe.' (1980: 18)

[15] Cf. Leibniz's comparison of the mind to a windmill. (1714)

to creating artificial intelligence and explaining the evolution of mind in the cosmos. Thus, let us baptize *him* 'Tinky,' the timeless, tinker-toy boy. Now, what environmental conditions are required for Tinky to tick?

Presumably, at each chronological moment, Tinky realizes a corresponding physical structure, so that his *genidentity* is composed of a series of states foliated along a 'temporal' dimension in space-time. However, as time is chronological, it has nearly equivalent properties with space.[16] Thus, there is no reason to believe Tinky's psychological constitution would be affected by transposing his physical parts onto spatial dimensions only. Admittedly, some contortion may be necessary, but we may assume enough 'temporal' thickness to maintain his structural integrity. Now, the problem is to identify, in principle, any indication of Tinky's sentience in this static, spatialized state; even a hint of the directedness intrinsic to his putative intentionality would suffice.

However, isolating his intentionality's physical origin in such a crystallized, inert state is absurd. His genidentical body is composed of alternating regions of space and material at best asymmetrically distributed. What geometric form can, even in principle, signify directedness? Any mark or pattern like '$\longrightarrow$' that we might discover uncovers nothing because, as discussed earlier, such symbols do not intrinsically signify anything, least of all direction. Whatever ecstasy Tinky experienced during his short lived temporal interval, there remains no corresponding physical trace of it or his *intentionality*, contretemps with cognitivism. Thus, I conclude that the constitution of memory and intentionality apparently require more than material and information; some tertiary ingredient is required, and the most plausible candidate is the directed process of rhealogical time.

3. CONCLUSIONS

Time, causality and intentionality intersect in many ways, sharing structural symmetries and ontological-existential similarities, but no scientifically established theory is available to explain them. However, their global and synthetic consideration together suggests a deeper ontological relationship, as illustrated below:

---

[16] This assumes a classical space-time. However, the relativity theories do not appear to undermine my argument, for they do not address either directionality or transience explicitly. The key difference is the Lorentz metric, which can be interpreted to mean time adopts an imaginary value, and the fact that the "proper time" of a world line is determined by velocity and accelerative forces.

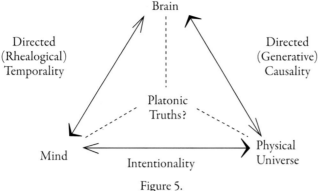

Figure 5.

I have considered how these three doubled arrows coalesce and how each finds resolution in the other. Thus, each individual plausibility thesis contributes to a collective corroboration of the others, and it suggests strong grounds that the directedness of causality, time and intentionality all share a common ontological origin. This view has been espoused by others, such as J. R. Lucas, albeit argued differently, but it has yet to be aggressively pursued.[17] This results in great part, because so little is really understood about the brain, mind and time.

Nevertheless, there are indications that a new philosophical approach may be gaining momentum to uncover these mysteries. At this point, we are still in the dark, and the best we can do is ask questions. However, these questions are now being posed:

> Time is the brain's glue… The experience of time is a neurophysiological construction that is generated actively within our brain, but how this is accomplished no one really knows… the sense of time follows from intentionality but the links that translate such a neural process into a subject experience are unknown. (Modell 2002, 34)

Indeed, there is growing recognition that these 'perennial' mysteries are related.

So, where does that leave us in this search? Firstly, I have argued that we are in fact searching for a single underlying principle. Secondly, this suggests refining the methodological approaches adopted in disparate domains, that they better cohere and so support interdisciplinary dialogue. Thirdly,

---

[17] Lucas similarly argues for an ontological integration of time, consciousness, and causality, though not via directedness. (1999)

and most central to my thesis, each of the three aforementioned domains bears sufficient similarity to one another that we should adopt an analogous epistemic attitude toward them all.

Thus, if one feels compelled to adopt mental eliminativism, which is akin to claiming there is no intentionality to be uncovered, then the same should hold for time and causality. Similarly, a purely chronological modelling of time or a blanket prohibition on directed causality will undermine the other two, respectively. On the other hand, if one finds the experience of self-consciousness, time or physical causality as directed more convincing, then it behooves one to meditate carefully on the natures of all three. In this way, a stronger empirical case is presented for each on the grounds that evidence for any one of the three problem sets is amplified as evidence for all three, very much undermining such eliminativist views to my mind. Not only does the world appear in spectacular Technicolor, 'monochromatic' theories that deny it are now trebly wrong.

In tracing the ontological and theoretical relationships between these three puzzling domains, more than such a denial of denial follows. The similarity between these quandaries suggests that whichever approach one adopts in one domain, methodological consistency would compel a similar attitude towards the other two. Thus, if a *mysterion* view of mind seems closest to correct in the case of consciousness, corresponding antirealist accounts would best serve in explaining the natures of time and causality. Similarly, *functionalist* interpretations of any one of the three problem sets should stand or fall together with corresponding theories in the other domains. By extension, we may then include such paradigms as mechanism and materialism.

I remain tentatively optimistic about the potential of science to uncover much of the underlying nature of temporality, causality, and intentionality. However, this will only become possible once its institutionalized monochromatic glasses are removed and some of the most difficult questions resolutely posed: e.g., what is directedness and where does it come from? Although its definition is subtle, its empirical merit and pragmatic value are not easily denied. Moreover, as noted, mathematics indirectly expresses directedness in some of its formalisms, providing an objective form with which to characterize it. Thus, features of the universe necessitating models with such formalisms suggest themselves for further research with respect to the origin of directedness.

REFERENCES

Beall, J. C., 2004. Introduction: At the Intersection of Truth and Falsity. In *The Law of Non-Contradiction: New Philosophical Essays*, edited by Priest, Graham, Beall, J. C. and Armour-Garb, Bradley. Oxford: Oxford University Press.

Bunge, Mario, 1959, 1979. *Causality and Modern Science, 3rd revised edition.* New York: Dover Publications Inc.

Čapek, Milek, 1961. *The Philosophical Impact of Contemporary Physics.* Princeton; NJ: Princeton University Press.

Davies, Paul, 2002. That Mysterious Flow. Scientific American (September, 2002).

Douglas, E. R., 2006. Rhealogical & Chronological Time: a Titanic Marriage. Submitted to *Kronoscope.*

Fraser, J. T., 1999. *Time, Conflict and Human Values.* Chicago: University of Illinois Press.

———, 1998. From Chaos to Conflict. In *Time, Order, Chaos: The Study of Time IX*, edited by Fraser, J. T., Soulsby, P. and Argyros, A. Madison: International University Press.

———, 1996. Time and the Origin of Life. In *Dimensions of Time and Life: The Study of Time VIII*, edited by Fraser, J. T. and Soulsby, M. Madison: International University Press.

———, 1987. *Time: the Familiar Stranger.* Redmond, WA: Tempus Books.

———, 1975. *Of Time, Passion and Knowledge.* Princeton, NJ: Princeton University Press.

Freeman, Walter, 2000. Perception of Time and Causation through Kinaesthesia of Intentional Action. Journal of Cog. Proc. 1.

Goodman, Nelson, 1977. *The Structure of Appearance, 3rd edition.* Boston: Dordrecht Reidel.

Hameroff, Stuart, 2001. Time, Consciousness and Quantum Events in Fundamental Space-time Geometry. In *The Nature of Time: Geometry, Physics and Perception*, edited by Buccheri, R., Saniga, M. and Stuckey, W. Amsterdam: Kluwer Academic Publishers.

Hawking, Stephen, R. Laflamme and G. W. Lyons. *The Origin of Time Asymmetry. Phys. Rev. D* 47, 5342–5356 (1993).

Hume, David, 2000. *A Treatise of Human Nature.* Edited by Norton, D. F. and Norton, M. Oxford: Oxford University Press.

Jackson, Frank, 1986. What Mary Didn't Know. The Journal of Philosophy LXXXIII, 5 (May 1986): 291–95.

Kant, Immanuel, 1998. *Critique of Pure Reason.* Translated and edited by Guyer, Paul and Wood, Allen W. Cambridge, Cambridge University Press.

Leibniz, Gottfried, 1714; Monadology. Reprinted in *Philosophical Writings: Leibniz*, edited by Dent, J. M., 1995. Everyman's Library, Rutland VT: Orion Publishing Group.

Lucas, J., 1999. A Century of Time. In *The Arguments of Time*, edited by Butterfield, Jeremy and Isham, Chris. Oxford: Oxford University Press.

McTaggart, John, 1908. The Unreality of Time. Mind 17: 457–74.

Modell, A., 2002. Intentionality and the Experience of Time. Kronoscope 2(1):21–39.

Newton-Smith, W. H., 1980. *The Structure of Time.* New York: Routledge.

Oaklander, Nathan, 1984. *Temporal Relations and Temporal Becoming: A Defense of a Russellian Theory of Time.* Lanham, MD: University Press of America.

Øhrstrøm, Peter and Hasle, P., 1995. *Temporal Logic: From Ancient Ideas to Artificial Intelligence.* Amsterdam: Kluwer Academic Publsihers.

Park, David, 1996. Consciousness and the Individual Event in Scientific Theory. In *Dimensions of Time and Life: The Study of Time VIII*, edited by Fraser, J. T. and Soulsby, M. Madison: International University Press.

———, 1972. The Myth of the Passage of Time. In *The Study of Time*, edited by Fraser, J. T., Haber, F. C. and Müller, G. H. Berlin: Springer Verlag.

Penrose, Roger, 1994. *Shadows of the Mind.* Oxford: Oxford University Press.

Priest, Graham, 2004. What's So Bad About Contradictions? In *The Law of Non-Contradiction:*

New Philosophical Essays, edited by Priest, Graham, Beall, J. C. and Armour-Garb, Bradley. Oxford: Oxford University Press.

Price, Hew, 1996. *Time's Arrow and Archimedes' Point: New Directions for the Physics of Time*. Oxford: Oxford University Press.

———, 1995. Cosmology, Time's Arrow and the Double Standard. In *Time's Arrows Today: Recent Physical and Philosophical Work on the Direction of Time*, edited by Savitt, Steve. Cambridge, Cambridge University Press.

———, 1989. A Point on the Arrow of Time. Nature (July, 1989): 181–182.

Reichenbach, H., 1958. *The Philosophy of Space and Time*. New York: Dover Publications Inc.

Rovelli, Carlo, 1995. Analysis of the Distinct Meanings of the Notion of "Time" in Different Physical Theories. Il Nuovo Cimento vol. 110B, n. 1.

Savitt, Steven, 1995. Introduction to *Time's Arrows Today: Recent Physical and Philosophical Work on the Direction of Time*, edited by Savitt, Steve. Cambridge, Cambridge University Press.

Schlesinger, George, 1980. *Aspects of Time*. Indianapolis: Hackett Publishing Company.

Schulman, L.S., 1997. *Time's Arrows and Quantum Measurement*. Cambridge: Cambridge University Press.

Sklar, Lawrence, 1974. *Space, Time, Space-time*. Berkeley: University of California Press.

Taylor, Richard, 1955. Spatial and Temporal Analogies and the Concept of Identity. Philosophy LII.

Tooley, Michael, 1999. The Metaphysics of Time. In *The Arguments of Time*, edited by Butterfield, Jeremy and Isham, Chris. Oxford: Oxford University Press.

———, 1997. *Time, Tense & Causation*. Oxford: Oxford University Press.

Webster's Third New International Dictionary, 1993. Springfield MA: Merriam-Webster Co.

Weyl, Hermann, 1963. *The Philosophy of Mathematics and Natural Science*. New York: Atheneum.

Wittgenstein, Ludwig, 1958. *Philosophical Investigations*. Translated by Anscomb, G. E. M. New York: Macmillan Publishing Co.

SECTION III

COMMEMORATION

COMMEMORATION: WHERE REMEMBERING AND FORGETTING MEET

Michael Crawford

Commemoration carries a strange resonance that is both private and public. In the private sphere it often lends a moral dimension to remembering: obligations to forebears are implicitly or explicitly acknowledged when their virtues and sacrifices are extolled as exemplars. For this reason more than perhaps any other, commemoration in the public sphere can provide either a powerful altruistic function or a convenient tool for the cynical manipulation of historical recollections and future agendas. Memory in this case becomes, as Ricoeur puts it, "instrumentalized".[1] For individuals, memory in its everyday sense is based upon direct experience and tends to put the "self as center of the past".[2] By contrast, commemoration represents an officially recognized sign post to demarcate community experiences, and to anchor them in a familiar context. This recognition necessitates the collaborative construction of histories, and in a sense commemoration weaves the experiences and recollections of individuals into a communal fabric. We consent to remember together, and we acknowledge the importance of this remembering, but we do so from within a version of history to which we can collectively subscribe. As such, commemoration is a strange creature that inhabits a world between official remembering and tacit forgetting, or as Ricoeur puts it, "abuses of memory, which are also abuses of forgetting".[3]

Few other representations and acts of remembering cover such a broad and smooth spectrum of experience: commemoration is an activity that scales from the individual, family, and group, through to an entire society. Furthermore, commemoration serves an interesting temporal role in that it is a vehicle for establishing a consensual version of the past in order to

---

[1] Ricoeur, Paul. *Memory, History, Forgetting*. Translated by Kathleen Blamey and David Pellauer. Chicago: University of Chicago Press, 2004. p. 80.
[2] Engel, Susan. *Context Is Everything: The Nature of Memory*. New York: W. H. Freeman, 1999. p. 44.
[3] Ricouer p. 80.

Jo Alyson Parker, Michael Crawford, Paul Harris (Eds), Time and Memory, pp. 225–228

lend depth of meaning to the present, and thereby set an agenda to negotiate hospitable avenues into the future. Perhaps paradoxically, it is a form of memory that is designed to project from past experience into the future. It is a re-lived or re-constituted memory that emanates from a fixed starting point but is resurrected in periodic fashion, usually with hopes for perpetual delivery ("let us never forget…"). Finally, depending upon one's perspective, commemorative events can be seen as imperfect vessels for the transmission of memory. They are politically susceptible to flexibility both with regard to the way in which the past is represented, as well as with regard to the intentions articulated for the future. Temporally speaking, past, present, and future are indeterminate even though the cultural accoutrements of commemorative ritual are designed to relay the perception of immutable inscription.

Moreover, with the passage of time and generations, commemoration is removed from the remembrance of direct experience to a recollection of learned stories. In many cases the re-animation of commemoration stories creates an empathetic category of synthetic or virtual "remembering." The empathetic experience curiously attains an arm's length distance through ritual. Indeed perhaps this is what makes commemoration of painful events endurable. By eliciting an empathetic response, commemoration not only speaks across time, but also entrains and employs emotion to rekindle the life of a historical person, group, or event. The entraining nature of ritual within an emotional context serves a potent cohesive purpose in communal events. Not incidentally, the public event is rendered intensely personal when license is granted for us to express emotion, to share thoughts, or to vicariously re-experience historical sentiments and aspirations.

Commemoration, then, is neither entirely remembering nor forgetting, and it is situated neither entirely in the past, present, nor future. It is both a product of, and a strange attractor for, societal mores and behaviours.

Trauma shares attributes with commemoration in several regards. It is sometimes large enough in scale to be recognized by a community, and frequently it can only be redressed through formal community action. Jeffrey Prager argues that traumatic humiliation and injury tends to be re-lived rather than be remembered by the afflicted. As a consequence, trauma assumes residence in the past, present, and if untreated, in the future. The reiterative timelessness imposed by the engendered sense of helplessness that comes with traumatic injury can best be broken by empowerment in a forum that restores timeliness. Like commemoration, the distance imposed by a formal structure of people willing to stand in for, speak to, apologize for, or forgive past transgressions can help to sever reiterative timelessness and thereby restore a modicum of normality. The Truth and Reconciliation

Committee in South Africa was a commemorative event in the sense that it provided a forum to recognize past events while re-establishing the dignity of the injured. It offered a future by observing, in a civilized manner, what had been left behind.

Similarly Ann Marie Bush looks at how different traditions—the cultures of Africa, the culture of slaves in America, and the personal histories of their descendants—are combined in interesting ways though art to permit self-empowerment and pride. The artists that are described (re-)take ownership of their histories by physically mapping them out in artworks that combine traditions: the descendants of African tribes and slaves overcome humiliation and degradation and supplant them with pride of survival, pride of cultural richness, and pride of spirit. By contrast, Efrat Biberman takes a fascinating look at how commemorative art has dealt with traumas that can foster no pride and that seared nations. Whether considering the assassination of Abraham Lincoln or of Yitzhak Rabin, for many the trauma of the event went beyond the capacity of words to describe, or of state funerals to dignify. Using the examples of Peto and Berest's paintings, we see how "commemorative" art can not only cause us to remember a moment, but also to relive it by actively reconstituting it as if for the first time.

Carmen Leccardi analyzes the cultural and familial context of young adults in the South of Italy who have been educated far beyond their parents. Together, remembering and forgetting are employed as social tools to provide a history that fosters a sense of belonging, of social cohesion, and of future possibilities. Stories and histories of parental sacrifice are combined with the deployment of family connections to provide offspring with unprecedented access to education and social mobility. Having heard and acknowledged the stories of their parents, the children articulate an obligation to avoid physical labor, and to succeed in new realms of activity and financial success. In a real sense, the successes of a community's children both represent and are seen to commemorate the labor of love and sacrifice endowed by their antecedents.

Dawna Ballard takes a refreshing look at the style and meaning of meetings and minutes in the corporate grammar of temporality. As a ritualized forum for discourse, the meeting enshrines its values and aspirations by inscribing them in its minutes. Meetings and minutes serve to provide consensus-derived signposts to demarcate institutional progress and individual perceptions of corporate history. Minutes articulate a collaboratively constructed past that informs the structure of the present and helps to set an agenda for the future. While some might argue that minutes are worth "inscribing and forgetting," they are nevertheless "co-memories", and when duly noted and accepted, they enter into the realm of minor commemorations.

Commemoration bears a burden of responsibility for descendents who "owe" their progenitors a debt of gratitude and honored remembrance. This form of memory is explicitly imbued with duties: duties to honor, to respect, and to resurrect the memorial occasion on a regular basis. Furthermore commemorations are embedded within a rigorous framework of cultural values and have the tendency to enshrine and to valorize attributes and aspirations that may be not be appropriate to modern sensibilities and circumstances. Unlike real memories, where the good ones often weigh equally with the bad, the virtual memories that are elicited by commemorations tend to selectively focus upon what is valued and to dispense with, or denigrate, inconvenient frailties. Perhaps this is what often makes them such powerful instruments for shaping a societal sense of past, present, and future, or as Ricoeur deplores while quoting Nora, for perpetuating the "the cult of continuity".[4] By contrast, as the first four of the following essays show, commemorations that intentionally incorporate representations of frailties have the potential to offer uniquely balanced epigrams to guide the intelligent and the sensitive.

---

[4] Ricouer p. 404.

CHAPTER TWELVE

JUMP-STARTING TIMELINESS: TRAUMA, TEMPORALITY AND THE REDRESSIVE COMMUNITY[1]

Jeffrey Prager

Summary

This paper argues counter-intuitively that psychological trauma describes not an event in the past but a condition of the present. Trauma is a memory illness characterized by the collapse of timeliness, when remembering prior experiences or events intrude on a present-day being-ness. The social basis of traumatic remembering is defined: an *a posteriori* and critical remembering of those who, either because of their presence (as perpetrators) or their absence (as protectors) generate suffering. Trauma endures through time when, in the absence of a reparative community, no capacity is available to allow for closure of past events. If timelessness—the inability to demarcate past from present—is symptomatic of trauma, then trauma's cure requires the jump-starting of timeliness, and timeliness depends on the existence of a community that colludes in the illusion of an individual's current day well being. How to restore to an individual the experience of the world's timeliness? The paper considers the conditions necessary for social redress, the restoration of community, and trauma's cure. Apology and forgiveness are described both as constitutive features of trauma's redress and as dependent upon the creation of a new liminal community (of apologizers and forgivers) whose members are temporally demarcated from the past.

Trauma Defined

Psychological trauma is a condition of the present. It is a memory illness. It manifests itself in individuals, as in collectivities, as a collapse of timeliness, when remembering prior experiences or events intrude on a present-day being-ness. Some events or experiences in our lives differ from ordinary moments because of their capacity to forever conflate when these experiences occurred with the after-the-fact remembering of them. The present is distorted to incorporate the memory of an un-metabolized, or unprocessed,

---

[1] A version of this paper was presented to the International Society for the Study of Time, *Time and Memory Conference*, Clare College, Cambridge, July 26–29, 2004. Original research completed while in residence in the Redress in Law, Literature and Social Thought Seminar of the University of California Humanities Research Institute, Irvine, Winter-Spring, 2003. Special thanks to Stephen Best, Cheryl Harris and Saidiya Hartman, seminar conveners, as well as to all members and participants.

past; a *then* folds in upon the *now* largely without awareness or distinction. Certain powerfully affect-laden experiences or events from before (that can include mental disturbances) yield the for-ever collapsing of time and the experiencing of the past *as if* it were the present, even when, in other respects, the demarcation of a then and a now remains clear. Traumatic remembering can preclude the possibility of a being's unencumbered movement into the future, as it can impede the potentiality of a group creatively responding to a changing world.

Trauma, because of its timelessness, cannot be specified exclusively in terms of properties of the past. While a prior overwhelming experience or horrific event—a moment described as inflicting upon the sufferer a *wound* (Van der Kolk, et al.)—is a defining condition for trauma, even that depends on its *post-hoc* remembering. Nonetheless, contemporary trauma research and theory tend not to emphasize trauma's negotiated relation between subsequent re-visits and prior experience, but give primacy to the events or experiences of the past, seeing them as driving all subsequent effects. In this spirit, trauma is described as an experience so overpowering as to defy representation and symbolization. The result, it is argued, is an inability to achieve a healthy distancing from the shock (Caruth, 152–3). The failure of language to soothe and contain, portrayed often as a typical feature of the precipitating occurrence, here is misconstrued as constitutive of psychological trauma itself. Similarly, trauma is defined by the symptoms it yields: photographic-like, veridical reproductions in memory, what have been characterized as intrusive *flashbacks*, in which what happened becomes belatedly recalled (Van der Kolk et al.; Caruth, 5). Among these scientists, little interest has been shown in discovering the ways in which a present-day "return to the past"—remembering—and the past itself may significantly differ. They accept at face value the subjective experience of those who describe this uncanny return to the past *as if* no time has elapsed.[2]

These characterizations of trauma, what Ruth Leys critically describes as "the science of the literal," (Leys, 229–297) insist upon the determinativeness of the past on the present and assert that trauma's meaning lies intrinsically in the character and nature of the material event that sets traumatic recall (and suffering) in motion. The victim becomes condemned to repeat, through performance, the meaning and significance of the trauma since the experience defies the capacity for representation or articulation (Leys, 266–7). By the same logic, the power of the past, that which cannot be represented, has no hope of mitigation. It is an experience without the possibility of closure. It

---

[2] For an excellent discussion of the relation of trauma to timelessness, see Leys (229–65).

represents an understanding of a past that is invulnerable to redress: history is destiny.

Contrary to those accounts that eviscerate the *post-hoc* and here-and-now in explaining the persistence of the past on the present, trauma can only occur *after* the event, as a memory of an experience that becomes inscribed on the individual.[3] Memory, of course, can assume a narrative form, a conscious story about prior events or experiences. Yet "representational memory" (Loewald, 164–5; Prager) is itself inevitably subject to distortion. When, for example, an earlier experience is recalled, even one remembered as traumatic, a visual image of the experience is typically constructed which includes the rememberer but now as a third-party. By seeing oneself as an actor on stage, significant alterations have already occurred, surely colored by the emotion attached to the experience. Memory inevitably refashions the experience itself, now shot through with affect. This is as true for the memories of "a people," i.e. a collectivity, as it is for a person. A narrative recounting of a traumatic past, in every instance, is deeply imbued with multiple layers of post-hoc affectively-charged constructions of those events, reflecting an effort, to be sure, to reconstruct a veridical memory of the past but inflected by after-the-fact affects, interests and impulses.[4]

But, in addition, memory may also be expressed unconsciously, as embodied knowledge (Prager, 178),[5] or as Hans Loewald describes it as

---

[3] See Thoma and Cheshire, for an extensive discussion of Freud's interest in challenging traditional understandings of psychological causality when he suggests that retrospective memory makes the "pure" retrieval of experience impossible. To that end, Freud, from an early point in his writings, defines trauma in temporal terms, always mixing experience with the re-working of it through memory. These authors argue that Strachey, Freud's English translator, often sought, perhaps unwittingly, to reinstate a more conventional understanding of the primacy of the material event to psychological thought. Thoma and Cheshire cite Freud's original German in a passage from *The Scientific Project* (p. 410), along with a pre-Standard Edition Strachey translation of it to read: "as a memory...which becomes a trauma only after the event." Strachey alters his own translation for later publication in *The Standard Edition* to reflect both a consistent terminology for Freud—in this case, the term *nachtraglichkeit* that he defines as "deferred action," but that results in reinstating the primacy of past experience on psychological understanding. In contrast to a view of deferred action as implying a kind of latent festering that, some time later, manifests itself, Freud suggests rather that prior (external) experience effects a person's inner world when it later becomes re-worked in terms of feelings of helplessness separate from the experience itself. See, too, Mather and Marsden.

[4] To the comparability between individual and collective memory, Freud (1954a, 206) writes in his "Notes upon a Case of Obsessional Neurosis," i.e., the Rat Man case: "If we do not wish to go astray in our judgment of their historical reality, we must above all bear in mind that people's 'childhood memories' are only consolidated at a later period; and that this involves a complicated process of remodeling, analogous in every way to the process by which a nation constructs legends about its early history."

[5] While the idea of unconscious memory derives from Freudian thought, it is also worth

"enactive memory," seemingly automatic, without benefit of conscious reflection and revision, in which memory and perception itself are inextricably interconnected. Loewald (164, emphasis in original) writes, "From the point of view of representational memory, which is our ordinary yardstick, we would say that the patient, instead of *having* a past, *is* his past; he does not distinguish himself as rememberer from the content of his memory." Just as perception is inescapably filtered through motive and desire, so too is memory. Because of the unreliability of memory along the lines just described, trauma resides in the special processing of moments that produce psychological rupture, or breaks in a sense of the on-goingness of life and the collapse of life's timeliness, rather than in some external event— outside of time—whose meaning and significance for the individual is taken *post-hoc* as self-evident. Moreover, if trauma is to be in any way undone, it is by jointly working on the experience of it, i.e., understanding the ways in which the experience has become memorialized by those who suffer from it, not simply by acknowledging an event's reality, specifying in detail its horror, remembering it graphically, etc.[6]

---

noting that non-Freudians, speaking from within different disciplines and intellectual tradition, also posit knowledge as contained within individuals, acted upon, without a self-consciousness of it, that while embodied within the person express his or her embeddedness in a broader social universe. See, for examples, Pierre Bourdieu on "Bodily Knowledge," Charles Taylor, "To Follow a Rule," Antonio Damasio, *Descartes' Error: Emotion, Reason and the Human Brain*, and Daniel Schacter, "Emotional Memories."

[6] Contrast this view, with that of Judith Herman, a clinician and theorist who emphasize the ways in which traumatic moments determine the present. Herman, responding to those who focused on a pathological predisposition to traumatic responses rather than to trauma's perpetrators, seeks to assert the centrality of the "crime" to trauma, not a "victims" propensity to become traumatized. Laudable in itself, Herman nonetheless "overcorrects" by isolating the event itself as the psychological source of trauma. For her, the "cure" is to recapture the moment in as much detail as possible. She (175) writes, "In the second stage of recovery, the survivor tells the story of the trauma. She tells it completely, in depth and in detail. This work of reconstruction actually transforms the traumatic memory, so that it can be integrated into the survivor's life story." Here, therapeutic cure requires a return to the traumatic event, "telling the story," to enable its psychological integration. While my emphasis on the environment that sustains traumatic memory does not preclude an exploration of the past and trauma's perpetrators, it does insist that the multiple ways in which a trauma of disillusionment has been sustained in memory need to be the objects of enquiry, not the detailing of the horror of any single event or experience. "Telling the story" of trauma risks elision of the multiple sources of suffering, including the social context in which suffering occurred.

The Past Defined Traumatically and the Price it Exacts

Traumatic harm is understood, as Freud sought to make clear from his early writings onward, not by a description of the external injury alone but by the ways it becomes internally processed and remembered. He writes, "Man seems not to have been endowed, or to have been endowed to only a very small degree, with an instinctive recognition of the dangers that threaten him from without... The external (real) danger must also have managed to become internalized if it is to be significant for the ego. It must have been recognized as related to some situation of helplessness that has been experienced" (1954c, 168) For Freud "the idea of trauma is not to be conceived so much as a discrete causal *event* [but] as part of a *process-in-system*;" a system comprised of drives, events, precipitating events, "all playing out in the context of a continuing struggle between an instinctive apparatus versus a defensive apparatus" (Smelser, 35). The context of helplessness registers experience as traumatic and on-going reminders of it keep trauma alive.

Trauma endures through time, it will be argued, when no capacity is currently available to allow for closure, to enable the understanding of the past as past, to permit the distinguishing between present-day acts of remembering from the memory itself. Said differently, the context of contemporary experience, in its deficiency, keeps alive in memory an earlier moment of psychic rupture. Trauma is a function of the present failure of the environment to provide safety and security and wholeness—what Freud (1954c) adumbrates as "helplessness"—to buffer the person against the intrusive reminder of a world neither safe, secure nor whole.[7] Trauma is the intrusion of memory, an occurrence that affectively, i.e., with emotion, describes the failure of members of the community to contain against disappointment the memorial experience of the person. In this sense, psychological trauma is both a disease of the contemporary moment as well as a social one, when an individual's capacity to engage the world presently and orient herself autonomously to the future is insufficiently enabled by the environment. When these conditions prevail, memory intrudes and a traumatic past dominates.

Thus, psychological trauma describes not a moment occurring in the past, i.e., the experience of an instant of terror or horror suffered alone by the

---

[7] On the experience of helplessness in infancy, see, especially, D. W. Winnicott, who I discuss below; also, Martha C. Nussbaum, *Upheavals of Thought, The Intelligence of Emotions* esp. "Emotions and Infancy," pgs. 181–190.

rememberer. It is not even only the memory of such an instant. Danger may generate instinctive reactions, whereby involuntarily instinctual responses to danger link human beings to all animals. But autonomic response, at least for humans, becomes meaningful belatedly, i.e., traumatically, because it is re-experienced in memory and it is social-ized. When it manifests, trauma is never an asocial encounter with the past; in fact, it is replete with an *a posteriori* and critical remembering of those who, either because of their presence (as perpetrators) or their absence (as protectors), generate suffering. It is always, therefore, an egocentric experience of the profound failure of particular members of one's own community to provide and protect. It is never impersonal and abstract (though it may become defensively understood impersonally and abstractly, as experience-distant).[8]

Psychological trauma is characterized, on the one hand, by the memory of a person or people who profoundly exploit the victim's vulnerability and, on the other, by the memory of those who disappoint by failing to offer necessary protections, who fail to defend against suffering. Firstly, it conjoins the sufferer with the memory of the perpetrator; i.e., the guilty party responsible for the breaching of innocence, for the shattering of expectations, and for the harsh intrusion of ugly reality against whole-some fantasy. Secondly, trauma indicts in memory the victim's intimate community—principally mother, father or other caregivers—who, at the time of such overwhelming experience, is felt to have failed to protect the victim. The wish, however irrational, to imagine the world as complete and good, with oneself as safe and secure within it, has been thwarted.

Disillusionment and harm are inextricably intertwined. When the perpetrators and the community members who fail to shield the individual are one and the same, disillusionment, of course, is likely to be more devastating and the trauma more persistent.[9] Physical or sexual abuse by a parent is one

---

[8] For this argument, I am following the lead of others who distinguish between traumatic memories whose origins are "person-made", and not a result of natural disasters, like hurricanes or earthquakes. It is likely that the experience of betrayal by others is a universal one to trauma; nonetheless, the psychodynamics of abuse, abandonment or loss undoubtedly differ in appreciable ways from "acts of God." Especially because of my focus in this paper on the intersubjective sources both of trauma and its repair, I limit my discussion of trauma to those that result from intra-human interaction.

[9] Think, perhaps, of those children of the perpetrators who have to endure both the love of their parents and the guilt felt by their transgressions. Here is another difficult traumatic "processing challenge," now being explored extensively in terms of the German experience during World War II. See, for example, Gunter Grass, *Crabwalk* and W. G. Sebald, *On the Natural History of Destruction*.

instance of such a confluence; so, too, is the premature death of a mother or father. By the same token, overwhelming experiences become processed differently depending on one's place in the life cycle. Generally speaking, the younger one is the more devastating the confrontation with disillusionment. But each age generates its own particular response, expressing some calculus between the nature and intensity of threat, the meaning given to the experience of helplessness, and the trajectory of omnipotent dependency in one's psychic life.

The memory illness' onset can immediately follow the determining, or disruptive, experience and it can, arguably, persist over long periods of time—even across generations—transmitted from parents to children (Prager 2003). Its onset can also be delayed, manifest only after defensive strategies have proven futile, like a false self or pseudo-independence; when various ego-driven efforts to ward off disillusionment and vulnerability have failed. But latent or manifest, immediate or delayed, it can come to shape reality in its own image, as trauma encourages action in the world that conforms to individuals' time-distorted experience of it. Paradoxically, in its collapse of present with past, when the past is lived *as if* it were current, trauma prevents the creation of the sense of a person moving through time. Time instead is experienced as frozen and unyielding, even as threatening the attachment of the person, or the collectivity, to a defining sense-of-oneself. Only memory is left, effectively closing off all the gateways to the senses.

Never Again, a reference to the determinative memory of the past and the wish never to re-experience it, becomes the recipe for life. Examples, of course, are all too plentiful but one might think of the parent of a child in the *antebellum* South whose family, in an instant, was shattered due to the slave market, or the descendant of a Holocaust survivor, overwhelmed by the knowledge of his parents' death-defying experience, or the woman whose child or husband suddenly disappears at the hands of a dictatorial regime. Memory then occupies the place of duty, an obligation to preserve in the present the past. And ironically, *Never Again* (either as representational or enactive memory) can effectively insure that the world conforms to precisely the moment meant never to be repeated. The replication of the imagined parents' traumatic reality, for example, can come to occupy the fantasy life of the child; psychological trauma can result in the child's identification more powerfully with his parents' harrowing past than with his own separate and distinctive present. Identity, here understood as a connection to one's past, to one's people, to one's history, as a resource for the present and as an orientation to the future suppresses, is overcome by, or gives way to, identification.[10] Identification constitutes an inability to extricate oneself from the burden of the past; when one, in effect, is determined to re-create and repeat the traumatic conflicts that now define oneself. Life becomes only meaningful in reference to that past, and the present becomes experienced and acted upon as if it were then. Not surprisingly, when such convictions prevail, the world indeed can be transformed to conform to the timeless past, a living testimony to past disappointments.

As Winnicott (1965, 37) describes, trauma shatters a fantasy of omnipotence: the destruction of the victim's sense that because of the perfection of the world, all is possible and anything can be achieved. Omnipotence depends on an environment that encourages the person to believe in his dependence on a benign world-in-place to provide for his or her needs. It is a seamless world that, as Winnicott (1971, 12) puts it, never asks 'did you conceive of this or was it presented to you from without?' In place of posing the question, the environment sustains the illusion that the individual omnipotently creates the world that provides for him.

Independence is a life-long process in which omnipotence is "tamed," though never fully eliminated. Through the life-course, the world ever remains an expression of one's own centrality and pre-eminence, though maturation typically mutes the fantasy on the pathway toward the world's disenchantment. Aging, and perhaps the experience of those close to us

---

[10] Nicholas Abraham and Maria Torok, in essays included in their *The Shell and the Kernel, Vol. 1*, Nicholas Rand (ed.), (Chicago, University of Chicago Press, 1994) develop a similar distinction to the one being drawn here between identity and identification. In "The Illness of Mourning" (p. 114), Torok distinguishes between *introjection* (identity) and *incorporation* (identification). "Like a commemorative monument," she writes, "the incorporated object betokens the place, the date, and the circumstances in which desires were banished from introjection: they stand like tombs in the life of the ego." Introjection is a gradual process of taking-in objects, including their drives and desires, a process that both broadens and enriches the ego, while incorporation is a secret, all-at-once moment, marking the (traumatic) instant in which the process of introjection has ceased. "The prohibited object is settled in the ego in order to compensate for the lost pleasure and the failed introjection." Torok's formulation of incorporation describes, in her words, the origins of traumatic memory, a description that corresponds to my own. My concept of identification also has an affinity to Arendt's description of fraternity, though hers without the backward-in-time dimension that I emphasize. Fraternity, for Arendt, is a formulation intended to capture the experience of Jews in the face of persecution. As Schaap (*Political Reconciliation* p. 3) describes Arendt's position, "fraternity becomes a bulwark against a hostile environment as people huddle together for mutual support against the pressure of persecution. While fraternity often produces genuine warmth of human relationships, however, it dissolves the 'interspace' between persons. In this situation, what is shared in common is no longer a world perceived from diverse perspectives but an identity predicated on a common situation." See, too, Hannah Arendt, *Men in Dark Times*, 1968.

dying amongst us, often yields a more sober understanding that the world, indeed, can (and will) exist without us. But traumatic ruptures promote the *premature* destruction of omnipotent dependency. They yield, in memory, an experience of the community's failure to indulge the illusion that the world is there to gratify me. The living of life in the shadow of this failure means that trauma cannot be placed in the past tense: the fear of its present-day return, as Winnicott (1974) describes, shapes the person's relationship to the future.[11]

In place of omnipotence, trauma can generate a precocious compliance to the external world, a premature abandonment of the illusion of omnipotence. With the loss of a sense of the world's provision of safety and security, the individual may attempt to present herself (to herself, and/or to others) as without dependent needs, as an adult (of whatever age) without a link to her child-like feelings, as no longer needing a world outside herself to provide a sense of safety and containment. These are defensive maneuvers that seek either to preserve a sense of cohesion and capacity in oneself that the memory is attacking, a "fear of breakdown" (Winnicott 1974) and/or to protect those loved ones from the anger felt by having been, at that moment, forsaken. By prematurely destroying the fantasy of omnipotence that accompanies dependence, trauma interferes with the process, occurring through the life-course, of the slow weaning from dependence and the movement toward independence. This life-long enterprise is accomplished through social relationships in collaboration, a community that only slowly gives up the collusion with the person that the world is present because of her making-it-so.

A *scar*, a permanent reminder that memory has broken through, marks psychological trauma: a breach in the social "skin" has occurred, a registration that wholesomeness has been violated (Margalit, 125). The scar constitutes the record of a past remembered that, while never fully healing, nonetheless,

---

[11] In this instance, Winnicott is describing a traumatic rupture that occurred so early in a child's life that "this thing of the past has not happened yet because the patient was not there for it to happen to. Only in the transference, Winnicott (105) argues, is it possible for the patient to regain omnipotent control over the fear because "the only way to 'remember' in this case is for the patient to experience this past thing for the first time in the present, that is to say, in the transference." But as I argue below, re-remembering trauma in the context of a redressive community similarly holds the promise that the victim can regain omnipotent control over memory so that the past can be put in the past tense. A community whose members—the victims and the guilty—jointly acknowledge the existence of the rupture—to-that-point only remembered by the victim—serves to restart a process of living in the present for an unencumbered future.

over time can ever better blend into the surrounding tissue. Its capacity to be re-opened, memory revived, as a result of traumatic triggers, however, remains ever-present; these triggers can instantaneously return the person to his past, and disrupt, once more, the timeliness of the present en route to the future (Stolorow)[12] Repair, or healing, then, is not about the return to "the scene of the crime," a revisiting of the literal or veridical event or events signified by the scar. It is rather the jump-starting of timeliness, the overcoming of a pervasive and entrenched psychic commitment to the stoppage of time.

Unlike those who suggest that traumatic relief depends on a person's return in memory to his or her unassimilated past in the form of representing and speaking it in an affect-laden language,[13] it is, rather, the restoration of a community that has disappeared and a re-engagement with an experience of a providing-world that enables moving-on. Relief derives not monologically by reclaiming one's past through its representation, but dialogically by presently describing to a listener or to a community of listeners who are willing and capable of understanding both the breach that is now occurring and its likely origin in prior disillusionment. The signifier, i.e. the memory, while a reference to the past, cannot be undone by redoing the signified; rather, its efficacy as an organizing principle for living diminishes when the social world, by listening, presently reconstitutes itself on behalf of the sufferer. The experience of "falling on deaf ears" results in the perhaps increasingly strident insistence that someone pay for the crime or crimes of the past. When there are no listeners, one begins to shout. Only when the conviction develops that significant others "know the trouble I've seen" (or, obversely, that significant others are no longer willfully denying either a traumatic past or its enduring efficacy) does it become possible to appropriate past experiences on behalf of the future. The burden of holding on to the past, sequestered in private experience, for the first time, has been lifted. Now, past events are capable of becoming integrated and mobilized to realize potentiality. But for this to

---

[12] On triggers, see Robert Pynoos.

[13] The renewed interest in dissociation, and its relation to trauma, expresses this particular formulation of the historical origin of trauma. An unassimilated, unrepresented event, inaccessible to consciousness, remains part of the mind's latent structure. It manifests itself, however, in dissociated fugue-like states, co-existing with conscious awareness but inaccessible to it. In this rendering, trauma's cure is the integration of dual mental states into one, making experience that is now dissociated part of one's conscious awareness. In identifying the problem of dissociation, it was claimed, post-traumatic stress disorder and its relation to other mental diseases like Multiple Personality Disorder could be better understood and more effectively treated. For critical considerations of this prevailing model of treatment, and the history of its origins, see Allan Young, Ian Hacking, and also, Jeffrey Prager (1998).

happen, a community disposed toward redress must be restored or, maybe for the first time, created. At the same time, the forces inhibiting its creation cannot be underestimated: resentment and cynicism on the part of the victims and of acknowledging a desire toward illusion, on the one side, and, on the other, the transgressors' defensive fear and unwillingness of losing power and authority, of having to face themselves as culpable individuals.

TRAUMA'S REDRESS

Its timelessness imposes its own demands, and challenges, for the possibility of trauma's redress, or repair. How to restore timeliness to a condition defined by a psychic investment in preserving the past? How to have the past acknowledged for its continuing efficacy, so as to reclaim from it an unburdened present?

Trauma requires community for its repair. The scar's healing-over cannot be accomplished alone. Since the preservation of memory involves the re-visiting of the experience of the world's disillusionment for its failure to offer protection, the social world presently, is responsible for redress. The reconstitution of social relationships to enable repair is never a foregone conclusion; the fabric of trust, security and protection is so exquisitely delicate, especially when confronting one's enemies, or substitutes for them, that its restoration requires an equally fine re-stitching. In the same way that psychological trauma is a function of a social community that failed, trauma's repair requires the social recuperation of omnipotence after its premature destruction, in the face of those who originally contributed to the failure, or of those whom all of mistrust and violation has become "entrusted."

Put differently, the present-day community, invested in the work of repairing a tear in the social fabric, is the receptacle of possibility where an adversarial relationship characterized by enmity might become transformed into one of civic friendship (Schaap, 5). "The commonness of the world," Andrew Schaap (2) writes, "is not merely revealed..... but constituted through politics since each perspective brought to bear on the world comes to form part of the inter-subjective reality we inhabit. Friendship thrives on the 'intensified awareness of reality' that arises from such political inter-action"[14] Precisely because the aim of civic friendship, in part, depends upon the jump-starting of timelessness, this outcome cannot be foreordained: neither the

---

[14] Schaap here is quoting Hannah Arendt on friendship, *Men in Dark Times*, p. 15.

according of forgiveness, on the one side, nor apology, on the other, can be effected independent of an "agonistic process" in which the words and deeds both of forgiveness and apology can meaningfully, at the end, be uttered and enacted. The reparation of community cannot be achieved if the process begins as pre-ordained by a presumption that it will succeed; to fore-ordain the outcome precludes the possibility of achieving a new horizon of shared understanding (Schaap, 4). Paradoxically, redress is achievable only when the shadow of absolute failure constitutes real possibility, when the potential for an even more permanent alienation between members is not foreclosed.

The community, in order for it to achieve its aim, must be comprised both of the victim(s) and perpetrator(s) who meaningfully confront one another's different perspectives. If the guilty offenders are unavailable (or unwilling) to present themselves in an effort to reconstitute community (and to restart omnipotent dependency), others, with authority to do so, must stand in for them. The redressive community in-formation is comprised, on the one side, of those disposed to replace a stance of resentment or disbelief with a disposition toward forgiveness and, on the other side, of those willing to risk a position of defensive power and authority, non-accountability, now oriented toward apology.[15] Forgiveness, understood as part of this political engagement, is not an achievement, a *fait accompli*, but a negotiated process in which, over time, those who have been harmed develop a voluntary psychological orientation in which forgiveness becomes possible. The willingness to forgive the offenders develops not before they are confronted, encountered, and talked to. Forgiveness is performed in real-time, not simply granted.[16] The stakes, of course, could not be greater but, as Schaap (105) describes it, "the possibility of setting aside resentment, of comprehending the other as more than one's transgressor, must be allowed if there is to be a place for hope and trust in the politics of a divided society."

But forgiveness within community, if it is to occur, can only happen

---

[15] Hieronymi (546) writes, "Resentment is best understood as a protest. More specifically, resentment protests a past action that persists as a present threat... a past wrong against you, standing in your history without apology, atonement, retribution, punishment, restitution, condemnation, or anything else that might recognize it as a wrong, makes a claim. It says, in effect, that you can be treated in this way, and that such treatment is acceptable."

[16] One is reminded once again of Winnicott's formulation of the psychoanalytic encounter in "Hate in the Countertransference." Winnicott describes the confrontation between analyst and analysand that, in the beginning, may mobilize in the analyst hateful countertransferential feelings toward the analysand. These, in time, are sentiments that may become redeployed in more loving ways. Yet this redeployment constitutes the achievement of a productive analytic relationship. There is nothing foreordained in this outcome.

with a concomitant movement, by those accused, toward apology. The restoration of thick relations between victims and perpetrators cannot be achieved unilaterally.[17] All too often, apologies have been issued, as an achievement, seemingly to foreclose the process of meaningful engagement with one's accusers. Yet a genuine impulse toward apology develops only with the strengthening of the community, not with its greater fracturing. The impulse, both toward genuine forgiveness and apology, expresses the capacity to experience one another, in-the-present, in a timely fashion, less conflated with past experience. Apology, like forgiveness, reveals a psychological openness toward meaningfully demarcating past from present and past actions (or actors) from present frames-of-mind. To the extent that its aim is not a cynical one, its purpose is not to deny the occurrences of the past; now, rather than being denied or defensively defended, they are acknowledged for the harm they inflicted.

The pre-condition for redress, then, is the creation of a space for speaking and listening, a community constituted neither by victim and perpetrator *per se* but rather by those willing, for the time being, to shorn themselves of their particular pre-existing positions, now with a preparedness toward forgiveness and apology in the hope of reconstituting themselves and the social world into a common future-in-the-making, to a life in common. It is, as Winnicott (1971, 13) might describe it, a "transitional space," neither comprised of selves or others, where "the strain of relating inner reality and outer reality" is mitigated by this "intermediate area of experience." *Post-hoc* communities of redress, standing-in for past inter-subjective failures, become the sites where private harm, sequestered and alienating, might find expression not in their denial but in acknowledgement, pointing toward a more wholesome relation between individuals and the collectivity and to a different, socially-constructive, forward-looking future.

---

[17] Margalit uses the term "thick relations," as I am here, to describe relations with people with whom we have a sustained, in-depth, historical relationship. Thick relations evoke moral questions about the community while thin relations—concerns, say, about the abstract individual—impose merely ethical considerations. We might think of African-American/white relationships in the United States, and European/Jewish relations in Europe as examples of thick relations. Here, the importance, as well as the difficulties, of achieving redress is more pressing because of the interweaving—both past and presently--of communal histories. Margalit argues that thick relations impose a standard of moral behavior, more difficult to realize than an ethical standard concerning, say, the treatment by humans of animals where thick relations do not obtain. In a similar spirit that stresses the especially complex and urgent task of repair in a democratic society, see Paul Barry Clarke (118) who writes about "deep citizenship," suggesting the inextricable connection between care of the self, care of others, and care of the world.

Because of their transformative potential, however delicate, communities of redress need to be *ad-hoc* in nature. Conventional juridical bodies and existing law, standing state agencies and governmental procedures dealing with harm, or other forms of institutionalized authority are all unlikely to produce, from both those who might forgive and those who might apologize, a setting in which private experience gives-way to this intermediate sphere, neither self nor other, in which a cooperatively forged dimension of illusion succeeds in diminishing personal disillusionment. At the least, existing authorities, first, are necessarily sites of suspicion in which skepticism toward their motives in articulating a language of redress must be overcome, if indeed these agencies are not simply aspiring to dampen the redressive impulse. Nor are words alone likely to define the extent of redressive action: while vocabularies of meaning may be comprised by the words of forgiveness and apology, specific concrete measures of recompense for past wrongs, mutually settled upon—whether symbolic and/or material—become the grammar for the reconstitution of community.

In each instance of a potentially redressive community, the restoration of timeliness requires this struggle and confrontation between perspectives and "a willingness to engage in an incessant discourse in which difference and lack of consensus is understood not as an obstacle to communication but a precondition for it" (Schaap, 2). Redress becomes possible only when communication succeeds in the transmutation of different perspectives into a new one: only then can omnipotent dependency, i.e., illusion, possibly become restored. Winnicott describes what needs to occur by those who have been victimized, when in a redressive setting they are provided a second chance. Speaking of the analytic encounter, he (1965, 37) writes, "There is no trauma that is outside the individual's omnipotence. Everything eventually comes under ego-control... The patient is not helped if the analyst says: 'Your mother was not good enough'... 'your father really seduced you'... 'your aunt dropped you.' Changes come in an analysis when the traumatic factors enter the psycho-analytic material in the patient's own way, and within the patient's omnipotence.' The resuscitation of dependent omnipotence, in short, is an experience that cannot be simply supplied by the outside; nonetheless, for it to occur, it must be enabled by an affectively-resonant other (a real perpetrator or a stand-in) also invested in its occurrence. And while the "burden" of this transformation appears to rest on those who have been victimized—the sufferers—it is clear that the restoration of omnipotent dependency requires the earnest effort by those in the "facilitating environment"—here described as the redressive community—to insure that the process not fail.

Psychological trauma, as I have argued, jeopardizes an unencumbered present and the capacity to freely anticipate a future. A newly-constituted redressive community, by replacing memory as a primary source of experience with a contemporary engagement with other perspectives seeks to counteract disillusionment and to restore, if possible, omnipotent dependency and to enable continuing-on. Present-day members of a community-in-formation collude, on behalf of a common future, to move beyond memory and to enable, once more, an illusionary world of possibility for everyone. When it occurs, as Winnicott (1971) suggests, individuals are able once again, each in their own way, to engage the world freely and on their own behalf.

One should not be too sanguine about redressive possibilities. What is being negotiated, after all, from the perspective of traumatic memory, as Derrida (32) describes it, is forgiveness for the unforgivable. Trauma's repair, in the end, may remain forever out-of-reach. At the very least, its elusiveness becomes the ground upon which the search for reconciliation must tread. Nonetheless, the possibilities of resuming life in-the-present make the effort at redress, however daunting, worthwhile. Through the work of a redressive community, history is freed of its obligation to provide the basis for living presently: a melancholic history of past wrongs no longer becomes the source for timeless identifications in the present.[18] The past, now acknowledged, enables those in the present no longer to sacrifice themselves to the memory of prior trauma. Memory is restored to a more modest place in social experience, now providing a resource to inspire every person to utilize fully the full panoply of sense-experience now available for living.

REFERENCES

Abraham, Nicholas and Maria Torok. *The Shell and the Kernel, Vol. 1*. University of Chicago Press, 1994.
Arendt, Hannah. *Men in Dark Times*. Harcourt, 1968.
Bourdieu, Pierre. *Pascalian Meditations*. Stanford University Press, 1997.

---

[18] In a recent collection of essays *Loss, The Politics of Mourning*, D. Eng and D. Kazanjian (eds.), there is an effort to valorize "melancholic history," suggesting that its alternative, i.e. "mourning the past, " risks the past being forgotten. To insure that prior tragedies not simply be lost to memory, a melancholic attachment to that history ought to be preserved. Not only does this constitute a misreading of Freud's (1954b) "Mourning and Melancholia" and his definition of mourning, it also romanticizes the illness that was the focus of Freud's concern. Melancholia, it must be recalled, incapacitated its victims, denying them the possibility of engaging the world presently, and resulted in their suffering unrelentingly. Used rather as a free-floating signifier without attachment to the illness being described, the contributors to the current volume idealize melancholy's attributes. See Prager (forthcoming).

Caruth, Cathy. *Trauma: Explorations in Memory*. Johns Hopkins University Press, 1995.

Clark, Paul. *Deep Citizenship*. Pluto Press, 1996.

Damasio, Antonio. *Descartes' Error: Emotion, Reason and the Human Brain*. Harper Collins, 1994.

Derrida, Jacques. "On Forgiveness." In *On Cosmopolitanism and Forgiveness*. Routledge, 2001.

Eng, David and David Kazanjian, eds. *Loss: The Politics of Mourning*. University of California Press, 2002.

Freud, Sigmund. "Notes upon a Case of Obsessional Neurosis." In *The Standard Edition of the Complete Psychological Works of Sigmund Freud, X*, The Hogarth Press, 1954a [1909].

Freud, Sigmund. "Mourning and Melancholia." In *The Standard Edition of the Complete Psychological Works of Sigmund Freud, XIV*: 237–260, The Hogarth Press, 1954b [1917].

Freud, Sigmund. "Inhibitions, Symptoms and Anxiety." In *The Standard Edition of the Complete Psychological Works of Sigmund Freud, XX*. 1954c [1926].

Hacking, Ian. "Trauma." In *Rewriting the Soul, Multiple Personality and the Sciences of Memory*, 183–97. Princeton University Press, 1995.

Herman, Judith. *Trauma and Recovery, The Aftermath of Violence from Domestic Abuse to Political Terror*. Basic Books, 1992.

Hieronymi, Pamela. "Articulating an Uncompromising Forgiveness." *Philosophy and Phenomenological Research* LXII (2001): 529–555.

Grass, Gunter. *Crabwalk*. Harcourt, 2003.

Leys, Ruth. *Trauma: A Genealogy*. University of Chicago Press, 2000.

Loewald, Hans. "Perspectives on Memory." In *Papers on Psychoanalysis*. Yale University Press, 1980.

Margalit, Avishai. *The Ethics of Memory*. Harvard University Press, 2002.

Mather, Ronald and Jill Marsden. "Trauma and Temporality, On the Origins of Post-Traumatic Stress." *Theory and Psychology* 14 (2004): 205–219.

Nussbaum, Martha. *Upheavals of Thought: The Intelligence of Emotions*. Cambridge University Press, 2001.

Prager, Jeffrey. *Presenting the Past, Psychoanalysis and the Sociology of Misremembering*. Harvard University Press, 1998.

———. "Lost Childhood, Lost Generations: The Intergenerational Transmission of Trauma." *Journal of Human Rights* 2 (2003): 173–181.

———. "Melancholic Baby, Unrequited Love as Identity Formation, A Critique." In *Identity in Question*, eds. A. Elliott and P. du Gay, Sage Publications, Forthcoming.

Pynoos, Robert. "Traumatic Stress and Developmental Psychopathology in Children and Adolescents." *American Psychiatric Press Review of Psychiatry* eds. Oldham, Riba, Tasman. 12 (1993): 205–238.

Schaap, Andrew. *Political Reconciliation*. Routledge, 2005.

Schacter, Daniel. *Searching for Memory, the Brain, the Mind and the Past*. Basic Books, 1996.

Sebald, W. G. *On the Natural History of Destruction*. Random House, 2003.

Smelser, Neil. "Psychological Trauma and Cultural Trauma." In *Cultural Trauma and Collective Identity*, ed. Alexander et al., 31–59. University of California Press, 2004.

Stolorow, Robert. "Trauma and Temporality." *Psychoanalytic Psychology* 20 (2003): 158–161.

Taylor, Charles. *Philosophical Arguments*. Harvard University Press, 1995.

Thoma, H. and N. Cheshire. "Freud's *Nachtraglichkeit* and Strachey's 'Deferred Action': Trauma, Constructions and the Direction of Causality." *International Review of Psycho-Analysis* 18 (1991): 407–427.

Van der Kolk, B., A. McFarlane and L. Weisath, eds. *Traumatic Stress: The Effects of Overwhelming Experience on Mind, Body and Society*. The Guilford Press, 1996.

Winnicott, D. "The Theory of the Parent-Infant Relationship." In *The Maturational Processes and the Facilitating Environment*. International Universities Presses, 1965.

————. "Transitional Objects and Transitional Phenomena." In *Playing and Reality*, 1–25. Tavistock Publications, 1971.

————. "The Fear of Breakdown." *International Review of Psychoanalysis* 1 (1974): 103–106.

————. "Hate in the Countertransference." In *Through Pediatrics to Psycho-Analysis*, 194–203. Basic Books, 1975.

Young, Allan. "Our Traumatic Neurosis and its Brain." *Science in Context* 14 (2001): 661–683.

CHAPTER THIRTEEN

BLACK IN BLACK: TIME, MEMORY, AND THE AFRICAN-AMERICAN IDENTITY

Ann Marie Bush

Summary

African Americans have two distinctly different past heritages: their African heritage and their American slave heritage. They must recall and accept both in order to portray an open, truthful African-American experience and an accurate cultural identity that they can honor. Painter John Biggers and story-quilter Faith Ringgold, both African-American artists, weave together contemporary moments of their respective eras with glorious and traumatic cultural memories of the past to celebrate and pay tribute to the African-American experience and cultural identity and to allow African Americans to embrace fully and with pride who they really are. Each artist fashions a unique structure that assembles pieces of the here and there and the now and then, that restores the flow of linear time from past to present to future, and that generates a trustworthy and notable portrait of the African-American experience and African-American cultural identity.

Time and collective memory have an influence on how African Americans identify and portray themselves.[1] As an ethnic group, African Americans have collective memories of two distinctly different heritages that help make up their shared identity: their African heritage and their American slave heritage. Exposing and integrating both heritages is vital in generating a more accurate group identity. However, the memories of their slave heritage have had a greater impact on the identity of African Americans, for those memories, along with the distress and trauma they carry, have also affected the memories of their African heritage and their current experiences in the United States.[2]

---

[1] See Howard Schuman and Jacqueline Scott, "Generations and Collective Memory," *American Sociological Review*, vol. 54: 359–81, pp. 361–362, 1989, who define collective memory as recollections of a shared past "that are retained by members of a group, large or small, that experienced it."

[2] See Michael Fultz. *Pride & Prejudice: A History of Black Culture in America*, videorecording, Knowledge Unlimited, 1994, who notes that "African-American identity comes out of the context of slavery" and who calls that context "traumatic." See Ron

Time also plays a part in African Americans' formulation of their collective identity. In some ways, a cessation in the flow of linear time from past to present to future has occurred. The present does not really exist in its own right for African Americans if they have replaced it with collective memories of a life of oppression lived in the past by slaves. African Americans may then be fixed in the past context of slavery, still suffering the emotional and psychological wounds of that era. By exploring their murals and story quilts respectively, we can see how John Biggers and Faith Ringgold have worked to heal those past wounds and restore the flow of linear time. In the context of the present, these artists are among those who have assumed the voices silenced in the past by slave owners.[3] Through their art, they have lifted both the trauma of the slave heritage and the richness of the African heritage to the conscious level and integrated the two legacies with their own life experiences. The bridge they have created across time between the present and the past allows African Americans to discover and embrace fully and with pride who they are as an ethnic group and to move toward an optimistic, self-determined future. They have looked back in order to move forward.

John Biggers, born in 1924 in a cotton-mill town near Charlotte, North Carolina, was an educator and a muralist. When he was a student himself, he said, "I began to see art not primarily as an individual expression of talent, but as a responsibility to reflect the spirit and style of Negro people" (Biggers and Simms 7–8), and later in his life, he noted that he traveled to Africa because he was "searching for [his] roots" (Biggers, *Ananse* 4). As an educator at a black university, he wanted to help his students attain racial and ethnic pride and "substitute a feeling of self-respect for their then-current feelings of self-contempt by developing an appreciation for their own art and heritage" (Theisen 19–20). So he "asked his students to focus on the question of self-identification [in their art]" (Bearden and Henderson 427) and stirred

---

Eyerman, author of *Cultural Trauma: Slavery and the Formation of African American Identity*, Cambridge: Cambridge University Press, 2001, p. 14, who indicates that "whether or not they directly experienced slavery or even had ancestors who did, blacks in the United States were identified with and came to identify themselves through the memory and representation of slavery." See Kenneth V. Hardy, an African-American psychologist, who says in *The Psychological Residuals of Slavery*, videorecording, Guilford Publications, Inc., 1995, that "slavery shades all contemporary experiences of African-American people."

[3] See Kenneth V. Hardy, *The Psychological Residuals of Slavery*, videorecording, producer Steve Lerner, Guilford Publications, Inc., 1993, who describes the "silencing" of African-American slaves as forbidding slaves to use their native languages and demanding they use only English and adopt Anglo names and who says that slaveowners employed "silencing" to "irradicate all sense of Africanness from the psyches of black people so they could be trained to think of themselves as slaves."

his students "to look to their own [southern rural] African-American and African heritages for inspiration in their work" (Wardlaw 71). The mural, a form of public art, proved a perfect means for his students and Biggers himself to "inform an impoverished people of their history and to build their self-esteem" (Bearden and Henderson 431).

If we look at one of Biggers' murals, *Shotguns*, we can find these three important motifs: the quilt pattern, the shotgun house, and strong, dignified women. Each of these motifs brings together collective memories of the African and the southern, rural slave heritages of African Americans and works to join the past and the present.

Shotguns shows us rows of little houses arranged in a pattern reminiscent of a quilt, and unquestionably, Biggers may be personally thinking about the quilts his grandmother and mother made with quilt patterns passed down from generation to generation of women who themselves or whose relatives came from Africa and lived as slaves in the southern, rural United States. However, the pattern is also reflective of Kuba cloth and similar African textile designs that Biggers may have seen on his visits to Africa or in his study of African art. Quilts originated in Europe, not Africa, but slaves in the United States "adopted the craft and many...applied a different aesthetic to their design" (Visona, et al. 503). African weavers of many tribes employed a textile tradition in which "abstract, figurative, and geometric designs are used separately and in combination" (Tobin and Dobard 41–42). Quilt patterns created by slave women were surely those of the African textiles woven in their homelands and lodged in their minds as memories which they brought with them from Africa and which, having been passed down from generation to generation of their families, have become an important part of the collective memory of all African Americans. The quilt was an important part of a slave woman's work. She could make warm covers for her family out of scraps of material. And although this study does not deal, as does the study of Jacqueline Tobin and Raymond Dobard, with the quilt as a secret communication of American slaves traveling on the underground railroad to freedom, we can agree with Tobin and Dobard's speculation that designs found in African-American quilts "have been passed on from one generation to the next...through cultural memories" (8). The quilt patterns that appear in many of Biggers' murals, certainly in his *Shotguns*, bring together the African-American ancestral heritage of Africans and slaves of the rural, southern United States through shared memory.

The second motif that runs through Biggers' *Shotguns* is the shotgun house, defined by John Vlach as "a one-room wide, one-story high building with two or more rooms, oriented perpendicular to the road with its front door in the gable end" (qtd. in Thompson 118). After considerable research, Vlach notes that "the shotgun house derived ultimately from the narrow one-room unit of the Yoruba...in West Africa" and that the design of the house came to the United States by way of Haiti, a common path of the slave trade (qtd. in Thompson 119). We can certainly see the similarities among houses in Africa, slave quarters in the United States, and a 19th-century shotgun house preserved in New Orleans, Louisiana, and can agree with Vlach that the shotgun houses of the United States seem traceable to architecture found in West Africa. Perhaps, slaves in the United States, responsible for building their own quarters, drew from their cultural memory of houses they inhabited in Africa when they constructed their living quarters in the United States, and perhaps, freed slaves remembered the architecture of their living quarters when they constructed their own homes after their emancipation. In many parts of the southern, rural United States, African Americans can still be found living in shotgun houses. This type of dwelling is a part of the African-American experience of life and surely something that also connects African Americans to their slave heritage and to their African heritage through collective memories. Biggers can also make a personal connection to the shotgun house, having been born in one that was built by his father in Gastonia, North Carolina (Thompson 121).

The last visual motif we will focus on in *Shotguns* is the five women who stand on the porches of the first row of shotgun houses. Each is holding a small shotgun house and standing next to a pot for washing and for cooking, and some stand next to a washboard. Their faces resemble African masks (Visona 524), and their strong, straight bodies stand with dignity, as though they are guarding the small homes. Surely, Biggers is indicating that women's devotion to family and their ordinary domestic work, keeping all clean and nourished, holds the home together. The woman on the far right stands next to an open door, through which we can see a bed, which may symbolize procreation and continuation of the people and their heritage. Through the doorway next to the woman to the right of center, we see a table with a cup and bowl, perhaps indications that the woman provides sustenance for her family and, more broadly, for her race. And on another level, Biggers has said that women are "wisdom bearers" and that the shotgun houses are their "temples," so the women holding the homes are lifting up the people, lifting up their pride in the lives they lead, the places they live, and the heritage from which they have come (*Kindred Spirits* videorecording). Women of dignity and strength project a significant presence in many of Biggers' murals and call to mind the African tradition of looking upon women as stately and holding them in the highest regard.

Starry Crown, another of John Biggers' murals, portrays three regal-looking women wearing crowns, who, according to Alvia Wardlaw, "represent the three cultures of African antiquity: Egypt, Benin, and Dogon of Mali" (192), and of course, we again see the quilt, associated with the southern, rural life of African Americans during and after slavery. But another facet of shared memory of African Americans surfaces in this mural, the re-collection of the rich oral tradition common to Africans and to African-American slaves. The title, *Starry Crown*, may come from this line, "Gonna put on my starry crown," found in one of the many versions of "Down by the Riverside," a Negro spiritual. The spirituals were religious songs sung by slaves during times of worship, but they were also sung during the day to provide the slaves "a sense of personal self-worth as children of a mighty God…and a much-needed psychic escape from the workaday world of slavery's restrictions and cruelties" (Gates and McKay 5). Another association with the oral tradition, the spinning of tales, is suggested by the spider that appears on the necklace of the woman on the right. In African folklore, the spider, named Ananse, is a popular heroic character who is said to have been given the "meaning of order in life" by God and who outwits all other creatures in tales that teach lessons of truth (Biggers, *Ananse v*). And if we look closely at the three women, we see a string flowing through the teeth of the woman in the middle to the hands of the other two and the teeth of the woman on the left. Reference has been made to the woman in the middle as "the Dogon weaver of the 'word'" (Wardlaw 192) and the string running through her teeth as a sign of "the transferal of knowledge across generations and continents through the spoken word in folk tales, proverbs and divine teachings" (Wardlaw 192).

Surely, the African and the American slave tradition of passing down stories and songs and teachings by word of mouth is not only present in *Starry Crown*, but also appears sanctified by the heavens. The flowing quilt of stars that surrounds and illuminates the three women makes them queens of the three ancient African cultures, who connect themselves and their cultures with the string they weave into a star. Because the star they create is also connected to a delicate, transparent quilt that lays over their laps, they have linked their African culture to their American slave culture.

With *Starry Crown*, Biggers has certainly shed light on the collective identity and collective memories of African Americans by voicing their rich heritage of oral tradition, and he has also integrated memories of his own personal life experiences into the mural. His father was a part-time preacher who spun African-American and African folk tales to teach and to entertain, and his mother and grandmother wove quilts for the Biggers family with the same pattern seen in the mural. Though we have only looked at two murals, nearly all of Biggers' murals tend to integrate both the traumatic slave heritage and the rich African heritage of the African American and to project positive, uplifting, beautiful images of the African American by drawing elements from their collective memory.

Like Biggers, Faith Ringgold blends her personal life memories as well as African Americans' collective memories of their African origins and American slave heritage into her work. Unlike Biggers, Ringgold, a feminist and Civil Rights activist, exhibits a more confrontational disposition in her art and blatantly exposes sexism and racism directed at African Americans. She does not entirely exclude males in her art, but concentrates on capturing the strength and success of African-American women, their courage, their values, and their dreams.

During her career, Ringgold has created paintings, soft sculpture, masks, and quilts—all leading to her most distinct and unique art form, the story quilt. The story quilt is mixed-media art comprised of pieces of fabric sewn together as a quilt on which she paints images and writes original narratives in her own hand. The stories are imaginary, but based on events, some historical, and dealing with issues that could actually take place in people's lives (Ringgold, *We Flew* 254). In her memoirs, she says, "My stories may include real-life experiences that I have had—that I know about or can imagine happening to me or to other people—but they are almost always imaginary. None of them can be read literally" (Ringgold, *We Flew* 254), and she goes on to explain that:

> Most of my stories are about women and all of my narrators are female. These narrators are fashioned after the women I heard tell stories…[when I was a child and] sat quietly, so as not to be sent off to bed, listening intently to the often tragic details of the lives of family members and friends told in that way that black women had in my childhood of expressing themselves. (*We Flew*, 257)

By creating narrators akin to those story tellers from her childhood, Ringgold has, in her own unique way, kept alive the oral tradition of Africans and African Americans.

If we focus on the woman at the head of the table in Ringgold's quilt, *Harlem Renaissance Party*, we immediately notice that the design of the woman's dress and the mask she holds in her right hand reveal an obvious African influence. Further, these traits of African art appear as well: the African textile design of squares and triangles; the bright colors; the overall symmetrical look of the piece, with the narrative running down the sides and complete squares framing the dinner table; and the polyrhythms evident in the diagonal, horizontal, vertical, and curved movements throughout the quilt—all evidence of the resurfacing of the collective memory established by African Americans of their past African culture.

Focusing again on the woman, we see she is dancing for the guests, expressing her spirit of freedom and creativity, embracing her African heritage, displaying self-pride and self-acceptance as an African American—the very behavior that writers, intellectuals, and artists of the Harlem Renaissance were advocating in the 1920's. And the distinguished dinner guests include the painter Aaron Douglass, the poet Langston Hughes, the intellectual W. E. B. DuBois, the philosopher Alain Locke, and the novelist-folklorist Zora Neale Hurston, all great figures of the Harlem Renaissance, whose words and art left a rich cultural legacy for African Americans and advocated bringing to mind the African-American collective memory of their African heritage. Tobin and Dobard describe well the melding of African and African-American collective memory in Ringgold's *Harlem Renaissance Party*:

> Stylistically, Ringgold frames her narrative with "strips" of text and a vibrant Hourglass/Bow Tie quilt pattern. Like the Dahomey artists, Ringgold tells the story of the struggle for achievement, but has portraits of prominent twentieth-century African American history makers instead of Dahomian symbolic representations of kings.... Ringgold combines the geometric patterns with her central narrative as if she were framing the action with history. (154–155)

By blending African textile designs into a quilt, a prominent vestige of the creativity of women slaves, Ringgold has, indeed, raised to consciousness and made concrete African Americans' shared memories of their African culture and their American slave culture.

In another story quilt, *Matisse's Chapel*, Ringgold has again voiced African Americans' collective memories and emphasized the importance of heritage. Though the figures portrayed in the story quilt are presented as the family of Willia Marie, the narrator of the story, they are really five generations of Ringgold's own relatives from her great-great grandmother to her brother and sister, all deceased and dressed in black and white (Cameron 9; Ringgold, *We Flew* 80). The narrative is composed of stories within a story, and the recounted stories have been handed down from generation to generation. The narrator has written a letter to her Aunt Melissa to tell the story of a dream she recently had in which her deceased relatives had come to Matisse's Chapel in Vence. They were listening to Great-grandma Betsy, who is telling a story about slavery that was originally told to her by her mother, Susie, who was born into slavery. Great-grandma Betsy refers to her mother's story, saying, "She ain' never talk much 'bout slavery, so when this white man ask her how she feel 'bout being descendant from slaves? She come back at him. 'How you feel descendant from SLAVERS'" (Ringgold, *Matisse's Chapel* 135). The white man answers by recounting a story told to him by his father about his grandfather's trip from Europe, during which the ocean liner he is traveling on encounters a ship filled with a cargo of slaves. After ending his story, the white man says that every time he thinks of the story his father passed down to him, he cannot seem to rid himself of the smell of the slaves packed on the ship. And Great-great-grandmother Susie smiles with pride at the generations of her black family who are seated around her now and who were born and have died free. The narrative indicates that both whites and blacks have to live their respective stories as descendants of slave-owners and descendants of slaves.

Michele Wallace, Ringgold's daughter, correctly perceives *Matisse's Chapel* as a story quilt "about mourning and death, at the same time that it is clearly a celebration of a tradition of resistance and a legacy of hope" (24). Ringgold exposes the trauma of her own family's historical roots in slavery, but generations of her family have lived in freedom, so we also see a death of their bondage. They feel anger and bitterness, but they also feel a sense of pride in who they are and where they came from. The hope is that they and all African Americans will draw from their individual and their collective memories, embrace their African and American slave heritages, and continue to honor themselves and their collective identity as they live their story.

Concerning the art, we see that Ringgold has provided a unique voice through her work. She has come a long way from her student days when she was taught to copy the style of the white, male European masters (*Faith Ringgold* videorecording). Just as African Americans, when slaves, blended some cultural aspects of the dominant white race under which they lived into their own unique mixed culture of slave and African heritage, Ringgold has brought together Matisse's art and the bright colors, patterned designs, and polyrhythms of African art. The blend shows her own unique voice as an African-American artist and the noteworthy contribution she has made with her story quilt to the world of art. The hope is that African Americans and the world recognize that African Americans have a unique culture and a distinct voice generated by their shared memories, collected over time, of their African traditions and slave heritage and that they have made and continue to make significant contributions to humanity.

In *Who's Afraid of Aunt Jemima*, Ringgold has taken the negative stereotype of Jemima as the very dark, obese, uneducated woman always pictured with the bandana around her head, always placed in the kitchen, always speaking in a slave dialect, and always relegated to a position subservient to a white family, and made a positive image of her. The image of Jemima, just under the title of the quilt, shows a beautiful woman wearing lipstick, jewelry, and a multi-colored, pattered dress with matching turban and complementary blue hat. She is a woman of presence, befitting her success as a restauranteur first in New York and then in New Orleans. The actual design of the quilt— created with pieces of fabric that show portraits of Jemima and members of her family, segments of the hand-written narrative, and traditional colorful cloth sections—reflects the African textile design of triangles and squares. Through the repetition of the squares and vibrant colors, Ringgold has created a lively rhythm for the quilt reminiscent of polyrhythms found in traditional African art.

Like the story quilt's design, the narrative also demonstrates the re-collection of shared memories of both the slave and the African heritages from which the African Americans' collective identity stems. Written in traditional black dialect, the saga speaks about Jemima's family—her grandparents, who bought themselves out of slavery, her parents, her husband, her children, her grandchildren—and recounts Jemima's path to becoming a wealthy, successful businesswoman (Ringgold, *We Flew* 252–254). Near the end of the story, Jemima and her husband, Big Rufus, die in an automobile accident, and their son takes them back to Harlem for an African funeral: "Praise God! Dressed Jemima in an African gown and braided her hair with cowery shells. Put Big Rufus in a gold Dashiki. They looked nice though, peaceful, like they was home" (Ringgold, *We Flew* 254). Considering the context and the language about the funeral, "home" not only refers to the heaven of the Christian religion adopted by slaves, but also to a reunion with African origins.

In picturing Jemima as a strong, confident, prosperous woman, Ringgold has called upon African-American women, and actually all African Americans, to reject existing negative stereotypes dredged up from the slave identity imposed on them by others, particularly members of the dominant, white society, and to define their group identity through the re-collection of more positive traditions lodged in their collective memories of their slave heritage and African legacy.

Though their art forms differ, both Faith Ringgold and John Biggers have created portraits of African Americans as strong, dignified people who, as individuals and as a group, can embrace and honor themselves and can feel racial pride. These artists, however, were not alone in their efforts to weave together contemporary moments of their respective personal memories with the collective memories of a past slave heritage and an African legacy common to all African Americans. Biggers' murals and Ringgold's story quilts bring to mind two important periods in American and African-American history—the 1920s, often referred to as the Harlem Renaissance or the New Negro Movement, and the 1950s to the 1970s, known for the rise of the Black Muslims and the Civil Rights, Black Power, and Black Arts movements. During both those eras, African Americans were focusing on who they were and how to portray themselves. They were accepting and rising above the humiliation, brutality, and degradation they had suffered under the trauma of slavery, they were emphasizing dignity and racial pride, and they were advocating self-empowerment and self-determinism. During those eras, African Americans not only re-collected shared memories of their slave heritage, but they also summoned up collected memories of their more distant African heritage and used both legacies with their own individual personal histories to define and portray themselves accurately as a group.

During the Harlem-Renaissance, African-American philosopher Alain Locke fostered racial pride by urging African Americans artists to study the richness of African art and culture. Intellectuals like W. E. B. DuBois and Carter G. Woodson, also associated with the Harlem Renaissance, "encouraged pride by researching Black history in the United States and in Africa to refute the allegations that the African race had bred only slaves and savages incapable of contributing to civilization" (Turner *xviii*). And according to Sterling Brown, Harlem-Renaissance artists and writers incorporated into their works new ideas, among them "Africa as a source for race pride...[and] the treatment of Afro-American masses...[and rural black] folk, with more understanding and less apology" (qtd. in Turner *xix–xx*).

Harlem Renaissance writers, artists, and intellectuals promoted the idea of African Americans recalling their collective memory, recognizing their collective identity, accepting themselves for who they were, and establishing a strong sense of self-worth that would carry into the future and prove them esteemed contributors to humanity. John Biggers and Faith Ringgold are among the artists, writers, and intellectuals of the 1950s through the 1970s and beyond who have continued what began in the 1920s.

As we have seen, in *Starry Crown*, Biggers portrays three regal-looking African women engaged in making a quilt and surrounded by a quilt of stars, and in *Shotguns*, he creates a quilt pattern from shotgun houses, a type of abode resembling American slave quarters, traceable to West African dwellings, and found even today in the American South. Faith Ringgold, on the other hand, has not used the quilt as a motif in her works of art, but has actually created quilts as art pieces. Whether as motif or as work of art, both have used quilt patterns associated with the patterns found in African textiles and slave quilts. Biggers has also exposed the rich oral tradition in his murals, most significantly in *Starry Crown*, and Ringgold has printed narratives by hand on her quilts to keep alive the African and the American slave tradition of telling stories. By depicting women as strong and self-assured, both artists have identified African Americans as a proud and dignified ethnic group.

These two African-American artists, each in his or her unique way, have specifically formed a bridge to the past by using the quilt and the oral tradition so much a part of African-American history and identity. By raising the collective memory of their people's African legacy and American-slave heritage to the conscious level in combination with memories from their own personal lives, Biggers and Ringgold have spawned a trustworthy and notable image of the African-American collective identity and have re-established the flow of linear time from the far-distant past to the present, which will continue to project into the future.

FIGURE SOURCES

John Biggers, *Shotguns*, oil and acrylic on canvas, 1987. 40 × 56 inches. Photo is by Dallas Museum of Art and is found in *Black Art Ancestral Legacy: The Impulse in African-American Art* by the Dallas Museum of Art, New York: Harry N. Abrams, Inc., 1989: 200.

Woollen Textile, Kabyle, Algeria, width 41 inches. Photo is by The British Museum and is found in *African Textiles* by John Picton and John Mack, New York: Harper & Row, Publishers, 1989: 62–63.

Traditional Beembe House, Northern Kongo, Musonda Village. Photo is by Robert Farris Thompson and is found in *Black Art Ancestral Legacy: The Impulse in African-American Art* by the Dallas Museum of Art, New York: Harry N. Abrams, Inc., 1989: 119.

Closeup of Traditional Beembe House. Detail of door and decorated façade of a Beembe House, Musonda village. Photo is by Robert Farris Thompson and is found in *Black Art Ancestral Legacy: The Impulse in African-American Art* by the Dallas Museum of Art, New York: Harry N. Abrams, Inc., 1989: 119.

View of Mulberry, painting by Thomas Coram. Oil on Canvas. Owned by the Gibbs Museum of Art in Charleston, South Carolina. Photo is by Gibbs Museum and is found in *A History of Art in Africa* by Monica Blackmun Visona, Robin Poynor, Herbert M. Cole and Michael D. Harris, New York: Harry N. Abrams, Inc., 2001: 502.

Restored 19th-Century Shotgun House. Photo is by John Michael Vlach and is found in *A History of Art in Africa* by Monica Blackmun Visona, Robin Poynor, Herbert M. Cole and Michael D. Harris, New York: Harry N. Abrams, Inc., 2001: 502.

John Biggers, *Starry Crown*, acrylic on canvas, 1987. 59 ½ inches × 47 ½ inches. Photo by the Dallas Museum of Art and found in *Black Art: Ancestral Legacy: The African Impulse in African-American Art* by the Dallas Museum of Art, New York: Harry N. Abrams, Inc., 1989: 190.

Faith Ringgold, *The Bitter Nest Part 2: Harlem Renaissance Party*, 1988. Acrylic on canvas;

printed tie-dyed and pieced fabric. 94 × 82 inches. Photo is by Gamma One Conversions and is found in *We Flew Over the Bridge: The Memoirs of Faith Ringgold* by Faith Ringgold, New York: Little, Brown and Company, 1995: 112.

Faith Ringgold, *The French Collection Part I #6: Matisse's Chapel,* 1991. Acrylic on canvas, printed and tie-dyed fabric. 74 × 79 ½ inches. Photo is by Gamma One Conversions and is found in *Dancing at the Louvre: Faith Ringgold's French Collection and Other Story Quilts,* edited by Dan Cameron, Los Angeles: University of California Press, 1998: 103.

Faith Ringgold, *Who's Afraid of Aunt Jemima?* 1983. Acrylic on canvas, painted and pieced fabric. 90 × 80 inches. Photo is by Studio Museum in Harlem and is found in *Dancing at the Louvre: Faith Ringgold's French Collection and Other Story Quilts,* edited by Dan Cameron, Los Angeles: University of California Press, 1998: 81.

REFERENCES

Beardon, Romare and Harry Henderson. *A History of African-American Artists from 1792 to the Present.* New York: Pantheon Books, 1993.
Biggers, John. *Ananse: The Web of Life in Africa.* Austin, Texas: University of Texas Press, 1962.
Biggers, John and Carroll Simms with John Edward Weems. *Black Art in Houston: The Texas Southern University Experience.* College Station, Texas: Texas A & M University Press, 1978.
Cameron, Dan, Ed. *Dancing at the Louvre: Faith Ringgold's French Collection and Other Story Quilts.* Los Angeles: University of California Press, 1998.
———. "Living History: Faith Ringgold's Rendezvous with the Twentieth Century." In *Dancing at the Louvre: Faith Ringgold's French Collection and Other Story Quilts,* 5–13. Los Angeles: University of California Press, 1998.
"Conversations with John Biggers." *Web of Life: The Art of John Biggers.* On ArtsEdNet: The Getty's Art Education Web Site. http://www.getty.edu/artsednet/resources/biggers. The J. Paul Getty Trust, 2000.
Dallas Museum of Art. *Black Art Ancestral Legacy: The African Impulse in African-American Art.* New York: Harry N. Abrams, Inc: 1991.
Eyerman, Ron. *Cultural Trauma: Slavery and the Formation of African American Identity.* Cambridge: Cambridge University Press, 2001.
Faith Ringgold: The Last Story Quilt. Videorecording. Producer Linda Freeman. Editor Alan McCormick. Writer David Irving. Home Vision, 1991: 28 minutes.
Gates, Henry Louis, Jr. and Nellie Y. McKay, Eds. *The Norton Anthology of African-American Literature.* New York: W. W. Norton & Company, 1997.
Hardy, Kenneth V. *The Psychological Residuals of Slavery.* Videorecording. Producer Steve Lerner. Guilford Publications, Inc, 1995: 18 minutes.
Kindred Spirits: Contemporary African-American Artists. Videorecording. Editor/Photographer Christine McConnell. Executive Producer Sylvia Komatsu. North Texas Public Broadcasting, Inc, 1992: 30 minutes.
Pride & Prejudice: A History of Black Culture in America. Videorecording. Producer Golden Communications Association. Project Director Jonathan Burack. Knowledge Unlimited, 1994: 28 minutes.
Ringgold, Faith. "Matisse's Chapel." Reprinted in *Dancing at the Louvre: Faith Ringgold's French Collection and Other Story Quilts,* 135–136. Los Angeles: University of California Press, 1998.
———. *We Flew Over the Bridge: The Memoirs of Faith Ringgold.* New York: Little, Brown and Company, 1995.

Schuman, Howard and Jacqueline Scott. "Generations and Collective Memory." *American Sociological Review*, vol. 54: 359–381. 1989.

Theisen, Olive Jensen. *The Murals of John Thomas Biggers: American Muralist, African American Artist*. Hampton, Virginia: Hampton University Press, 1996.

Thompson, Robert Farris. "The Song That Named the Land: The Visionary Presence of African-American Art." In *Black Art: Ancestral Legacy: The African Impulse in African-American Art, 97–141*. Dallas Museum of Art. New York: Harry N. Abrams, Inc, 1989.

Tobin, Jacqueline L. and Raymond O. Dobard. *Hidden in Plain View: A Secret Story of Quilts and the Underground Railroad*. New York: Anchor Books, 2000.

Turner, Darwin T. Introduction. In *Cane*, ix–xxv. By Jean Toomer. Reprinted. New York: Liveright, 1975.

Visona, Monica Blackmun, Robin Poynor, Herbert M. Cole and Michael D. Harris. *A History of Art in Africa*. New York: Harry N. Abrams, Inc, 2001.

Wallace, Michele. "The French Collection: Momma Jones, Mommy Fay, and Me." In *Dancing at the Louvre: Faith Ringgold's French Collection and Other Story Quilts, 14–25*. Los Angeles: University of California Press, 1998.

Wardlaw, Alvia J. "A Spiritual Libation: Promoting an African Heritage in the Black College." In *Black Art: Ancestral Legacy: The African Impulse in African-American Art, 53–74*. Dallas Museum of Art. New York: Harry N. Abrams, Inc, 1989.

CHAPTER FOURTEEN

REMEMBERING THE FUTURE: ON THE RETURN
OF MEMORIES IN THE VISUAL FIELD

EFRAT BIBERMAN*

SUMMARY

Memory, it seems, is constituted under a temporal sequence. First, some event occurs. Then the event is inscribed in our memory, while being represented as a former event. Any stimulus like a particular smell, sound, word, or a visual image, can trigger the memory of something that took place some time ago. Following the psychoanalytic discourse of Freud and Lacan, I would like to suggest another way of thinking about memories in which they function in an opposite direction, where an act that already took place draws its meaning from the future. This situation can be fruitful in understanding both works of art and the collective memory they evoke. I will demonstrate this claim by analyzing a work of art by Dganit Berest, an Israeli artist, which deals with the murder and the memory of the Israeli Prime Minister, Yitzhak Rabin.

The late Israeli Prime Minister, Yitzhak Rabin, was assassinated on Saturday evening, November 4th 1995, as he was walking towards his car after a large peace demonstration that was intended to increase public support of the peace process Rabin had been promoting. Three bullets, shot at Rabin's back by a right-wing Jewish terrorist, wounded him severely and led to his death a few minutes later. Rabin's murder was like no other political event that had ever occurred in Israel and could be described as a trauma. This event constituted a turning point which ultimately led to a comprehensive change of Israeli politics in general and specifically with regard to the peace process and relations with the Palestinian Authority.

On November 4th, 1996, the date of the first anniversary after Rabin's assassination, Deganit Berest, an Israeli artist, published a work of art on the front cover of a weekly magazine. Berest's work showed a daily newspaper dated the day before the assassination over which she had superimposed a text in which she quoted the surgeon who had examined Rabin's body,

* I would like to thank the artists Deganit Berest and David Ginton for their willing cooperation in the writing of this paper.

Deganit Berest, *Untitled*, 1996.
The Israel Museum Jerusalem.

described him before realizing who he was. The physician had written: "An old man, wearing a suit… a very old man, his face is as white as snow". The inscription was traced by hand onto the editorial section of the newspaper, which, in addition to several articles and a cartoon, included an advertisement announcing the demonstration to be held the following day. This demonstration was the very same demonstration in which Rabin was assassinated.

In what ways can visual works of art represent or preserve traumatic events that seem too traumatic to be embodied? In what manners can memories be evoked and rendered by visual means? The subject of this paper is the ways in which works of art capture the elusive nature of memories and specify their unique temporality.

Berest's work refers to several artworks from the history of art which visualize political assassinations, and hence address the issue of memory. One of the most famous paintings that come to mind is Jacques Louis David's *Death of Marat* of 1793, in which David depicts the last minutes of Marat's life. Jean-Paul Marat was one of the political leaders of the French revolution and an associate of Robespierre. He had been stabbed to death by Charlotte Corday while taking a bath. The dying Marat is holding a petition given to him by Corday. David depicts Marat as a young suffering man, in a manner that shows his great sympathy and admiration towards him. Thus, the painting is obviously a memorial for Marat by David, his friend and admirer.

Death of Marat can be regarded as a narrative painting, describing a historical event, which began several minutes before the represented scene occurred. The painting thus pertains to the connection between time and memory in two ways, as a memorial for a historical event and as a painting representing a temporal sequence by means of one image. Besides, the painting directly represents the remembered event and hence can be regarded as a visual repetition of that event. In this sense, the painting connects memory and repetition. The act of repeating is a way of remembering in which certain aspects of the event are stressed. In the case of Marat, David stresses Marat's appearance as a martyr, which, according to various interpretations alludes to other pictorial representations of martyrdom.[1]

A radically different state of affairs pertains to Berest's work. This artwork does not, at first glance at least, seem to repeat any part of the traumatic event to which it refers. Furthermore, there seems to be no narrative, or explicit description of the act of assassination, or any evident linear time sequence.

---

[1] Among the interpretations some refer to visualizations of the decent from the cross, while others suggest allusions to Francisco de Zurbaran's *St. Serapion*, of 1628.

Jacques-Louis David, *Death of Marat*, 1793.
Royal Museum of Fine Arts of Belgium, Brussels.

Yet, this work, which at first sight seems very enigmatic and puzzling, has a tremendous effect upon the viewer and clearly deals with memory, time and temporal ordering. What kind of memory does it evoke? In what way, if any, does it capture the elusive nature of specific memories and their relation to time and temporal ordering?

Memory seems to be determined by means of a temporal sequence. First, some event occurs. Then the event is inscribed in our memory, represented as an event in the past. Any stimulus, a particular smell, sound or word, can trigger the memory of something that took place some time ago. The same is true with regard to images: seeing an old photograph, or encountering a familiar landscape may awaken a memory from the past.

Do memories necessarily function only in one direction, evoking something of the past? In this paper, which focuses on the visual arts, I would like to suggest another way of thinking about memories where they function in the opposite direction. An act that took place some time ago receives its meaning at a future point in time. In this sense, the future constructs the past.[2] This temporality could facilitate an understanding of both the works of art and the memory they evoke. Moreover, reversed time ordering assumes a kind of temporality which is not necessarily progressive and accumulative and thus coheres with the visual field. That is, the traditional discourse on the temporality of visual images has stressed the restricted temporality of paintings. Reversed time ordering may subvert the fundamental differentiation between space and time, and hence suggests an alternative way of thinking about time in relation to visual images.

The theoretical bases for my argument are Freud's concepts of "screen memories" and "deferred action" and Lacan's development of these concepts. Although Freud used these terms to explain how a trauma functions, I suggest that a similar concept of temporality may shed light on works of art.

Before addressing the psychoanalytical discourse, I would like to consider another painting, which refers to Rabin's assassination. In 2001, the Israeli artist David Ginton painted a picture called *Back of a painting: Date, Rabin and the Blindram stretcher.* The painting depicts the back of a canvas, showing the oil stains, which permeated from the front of the canvas, the wooden stretcher, and the folded edges of the cloth. Although the surface of the painting is hidden, the painting provides the viewer with two hints about its subject matter: the assassination date of November 4th is written

---

[2] Slavoj Zizek describes that state of affairs in his paper "The Truth Arises from Misrecognition". In *Lacan and the Subject of Language*. Edited by Ellie Ragland-Sullivan and Mark Bracher. London: Routledge, 1990.

sloppily on the back of the canvas.[3] Another hint is the title of the painting, which alludes to a painting by the nineteenth century American painter John Frederick Peto. In 1898, Peto, known for his *trompe-l'oeil* paintings, painted a picture entitled *Lincoln and the Phleger Stretcher*, referring to the assassination of Abraham Lincoln. Like Ginton in 2001, Peto painted the back of a canvas, with a painted portrait of Lincoln attached to its frame. In addition to Lincoln, the title of the painting refers to the craftsman Phleger, who made the stretcher in Peto's painting. Similarly, Ginton's title refers to Blindram, the manufacturer of his stretcher.

At first sight, there is no apparent connection between Berest's and Ginton's artworks, except for the fact that they both allude to the same historical event. Formally, technically and stylistically, they are completely different. Yet, despite the different styles and media, both works share some common features: they both feature a minimum of information and they both say little explicitly. In both works the allusion to the assassination is implicit. Both works contain inscriptions, and relate to Rabin's assassination by means of an object with the date of the event: the newspaper page that appeared the day before the event and the painting that was allegedly painted on that day. As their background, both works use an object treated in a way that deviates from the way this object is usually used. Specifically, the canvas is reversed, and the newspaper has writing all over it. Both works convey something that is enigmatic and minimal. In a way, the act of conveying meaning is in effect also an act of obscuring meaning: in Ginton's case, he shows us something hidden; in Berest's case, the seemingly casual text uttered by the physician is presented as a riddle to be deciphered. But in both cases, the veiling is multiplied. In addition to the act of reversing, Ginton plays the old *trompe-l'oeil* trick, letting the viewer suppose he is encountering an actual reversed painting, that was mistakenly hung backwards. Berest reveals a fragment of a fairy tale to the beholder. Hence, both Berest and Ginton pacify the viewer, as if saying to her that "we are hiding something, but it is actually a joke, or a naïve fairy tale". Yet, the effect of these two artworks on the viewer is tremendously shocking. The act of hiding seems more obscure when we take the historical context of these artworks into account: the trigger of these works was, no doubt, Rabin's assassination. Their manifest content, nevertheless, does not cherish the memory of Rabin but effaces it. However, is this really the case? What is the function of these multiple layers of veiling, and what is there to be hidden? Moreover, why is such a traumatic

---

[3] According to Ginton, this inscription also alludes to a sticker, noting the date 4.11.95, which was distributed after Rabin's murder and was attached to back of cars.

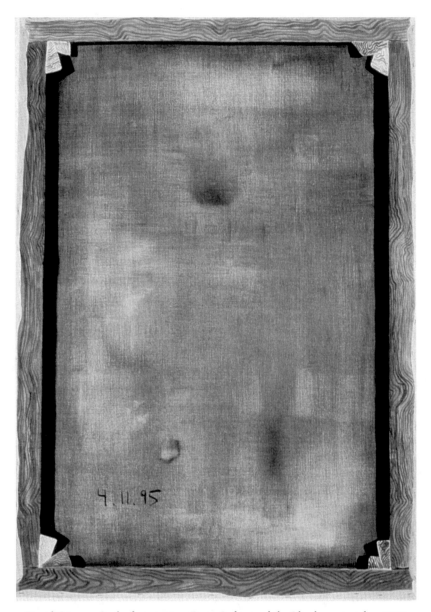

David Ginton, *Back of a painting: Date, Rabin and the Blindram stretcher*, 2001.
Private collection.

and significant event disguised and thereby denied? In many cases, the act of interpreting art amounts to deciphering a puzzle presented by the painter. The interpreter aims to reveal some hidden information that was overlooked by the spectator at first glance. In Berest's and Ginton's respective cases, the artists themselves seem to be exposing the act of disguising, as if blocking the possibility of elucidating any meaning.

I would like to argue that this multi-layered hiding cannot be distinguished from another crucial aspect of these two art works, which is the temporal order they present. In *Death of Marat*, the temporality is constructed by means of the narrative partially represented in the painting, which the viewer is required to complete. Berest and Ginton unfold temporality in a different manner. In a personal communication about this work, Berest noted the following: "The thing that was most important for me was recreating the moment of "not knowing", the moment that precedes knowing, which coexists and yet conflict with the doctor's retroactive realization. This impossible moment is allegedly hung in space, happens simultaneously on the newspaper page in which the advertisement which announces the demonstration is published, and in the deferred moment of looking at the corpse without recognizing its identity". Berest's work hence simultaneously presents several complex time sequences in which time goes back and forth: the newspaper from the day before the event, the announcement of a future demonstration, the doctor's testimony that precedes identification. Besides these temporal sequences, the artwork presents an "impossible moment" of not knowing, a moment excluded from any historical time duration. That impossibility is heightened since the newspaper page from the day Rabin was assassinated represents a day that allegedly did not occur, that is, a day without any journalistic representation, since newspapers do not appear on Saturdays in Israel.

Ginton's work also seems to present a complex temporal sequence. As in any *trompe-l'oeil* painting, the viewer, who at first sight believes that she is encountering an actual object, does not see the painting and its dazzling effect. She believes that she is standing in front of an actual picture that was hung wrongly. Only when she becomes aware of the deceit does the viewer see the painting *per se*. However, she now sees it apart from its pictorial essence. But Ginton's work has its complex time ordering for another reason: In a way, the work attains its meaning from Peto's earlier painting dated 1898, and hence conveys a linear temporality. But Peto's painting itself deals with memory. Ginton makes use of Peto's recollections in order to remember an event which takes place 130 years later.

This temporal complexity and its relation to effaced meanings which Berest's and Ginton's paintings both convey could be elucidated by referring to the psychoanalytic discourse and its views on time, memory and visual imagery. Psychoanalysis assumes that a visual image results from a specific encounter of a subject and an object, which therefore involves a specific time sequence. This time sequence is not accumulative or positive, and can only be reconstructed retroactively as the outcome of some previous events, which never really occurred.

In a paper from 1899, Freud coined the term "screen memories", which relates to memories and the temporal order of the events they represent. According to Freud, childhood memories can be characterized as displacements instead of direct testimonies of the subject's history. That is, the meaning of these memories derives from the fact that they are associated with other repressed memories and not from the fragmentary content they allegedly express. Thus Freud termed memories from early childhood "screen memories" because they conceal other memories. He claimed that these memories, in many cases visual in nature, could be considered in theatrical terms.

Having noticed that memory is selective in nature, he tried to discover the precise mechanism of this selective procedure. He discovered that significant events in his patients' biographies were often suppressed and forgotten, while marginal and seemingly insignificant memories were retained.[4] He realized that two contradictory psychical forces are involved in bringing about memories; one of theses forces seeks to remember things, while the other strives to repress and conceal events and facts from the past. These two forces do not cancel each other out, but come to a compromise: "What is recorded as a mnemic image is not the relevant experience itself—in this case resistance gets its way; what is recorded is another psychical element closely associated with the objectionable one—and in this respect the first principle shows its strength, the principle which endeavors to fix important impressions… The result of the conflict is therefore that, instead of the mnemic images which would have been justified by the original event, another is produced which has been to some degree associatively *displaced* from the former one".[5]

---

[4] Freud, Sigmund, "Screen memories," in *The Standard Edition of the Complete Psychological Works of Sigmund Freud*, edited by James Strachey (London: Hogarth Press, 1899), Vol. III, p. 306.

[5] Ibid., p. 307.

The logic of screen memories suggests that the remembered content stands for something that cannot be represented in memory. A similar logical structure characterized Freud's definition of *Vorstellungsrepräsentanz* or "representative of representation". Assuming that there is no direct access to repressed events in the unconscious, Freud argued that repressed events could be represented only by delegates that represent unconscious entities. These representatives cannot lead to the repressed content but can only represent something that represents it, while its meaning remains forever unknown. In this sense, a symptom first occurs without any particular meaning. Its meaning emerges only later in the course of analysis.[6] Berest and Ginton's respective works can be regarded as *Vorstellungsrepräsentanz*. That is, the artists do not seem to reveal meaning directly, but rather expose the veiling of meaning.

In an essay from 1901, Freud specifically relates to the temporal ordering of screen memories, arguing that they are characterized by two different time sequences. In the first sequence, a later event remains hidden and the subject only recollects an earlier one, while in the second sequence, which is more common, a reversed time sequence is presented; that is, an earlier event is hidden behind the memory of a later period.

I would like to claim that the first time sequence can be seen in Ginton's work: an earlier event, in this case Lincoln's assassination conceals a later event, Rabin's assassination. Lincoln's case, which is a well-known but distant historical event that belongs to American culture, it used by Ginton to deal with the trauma of Rabin's assassination. The term "screen memory" is hence embodied in this artwork both literally and metaphorically.

How can Berest's work be interpreted in similar terms? Here, the second time sequence, which is reversed, is relevant for an interpretation of the painting. The paradigmatic example for such a sequence is Freud's famous analysand, the wolf-man. Freud analyzes the patient's dream as a young boy. In this dream, the sleeping boy is awakened by a window suddenly opening, whereupon he sees a bunch of wolves, sitting at the top of a tree and staring at him. In analyzing the dream, Freud reveals a number of reversals: the window does not open by itself but the dreamer's eyes are opened, and the penetrating gaze[7] which the dreamer attributes to the wolves is actually his own look

---

[6] Jean Laplanche and Jean-Bertrand Pontalis, *The Language of Psychoanalysis*, translated by D. Nicholson-Smith (London: Karnac Books, 1988 [1973]), pp. 203–204.

[7] Lacan distinguishes between the look as an act of the subject's eye and the gaze, which is a lost object that can never be represented. According to Lacan, there is a fundamental split

staring at them. Freud connects these reversals to the interpretation of the dream, according to which the oneiric scene is a repetition of a scene from the patient's early childhood, when while sleeping in his parent's bedroom he suddenly woke up and witnessed them engaging in sexual intercourse. This anxiety-provoking scene, which is too traumatic for the child to cope with, was kept in the unconscious, emerging at a later stage when a present scene, which is not traumatic in nature, evoked the anxious effect of the previous one. The wolves-tree dream can thus be said to be a screen memory. The subject did not recall any of the early significant events that never actually occurred since they remained in the unconscious. The dream's screen memory is allegedly insignificant, and gains its horrific effect only retroactively through an event that anticipated it. Freud hence characterizes this temporal ordering of screen memories under the term *nachträglich*, that is, deferred action.

Following Freud, Jacques Lacan designates the term *futur antérieur* to refer to a symptom that occurs as a trace of something which has been repressed, while the cause of the occurrence of this symptom is currently unknown, only to emerge at a later point. Thus, when the repressed returns as a symptom or as a dream, it returns from the "future". The meanings of the repressed traces are retroactively reconstructed. The temporal ordering of David's *Death of Marat* refers to an image containing an implicit temporal sequence that the skilled viewer uncovers. The psychoanalytic image, however, is a fixation on a seemingly meaningless scene that can be understood only in retrospect. The temporality underlying this image thus moves backwards; the cause appears after the result. In this sense, this psychoanalytic time ordering subverts the traditional space-time differentiation according to which paintings are spatial in nature and have a restricted temporality. In contrast, the temporality of the psychoanalytic image is imposed onto the painting retroactively.

In his 1953 seminar, Lacan claims that "…Freud initially explains repression as a fixation. But at the moment of the fixation, there is nothing that can be called repression—that of the wolf man happens a long time after the fixation… how then should one explain the return of the repressed? As paradoxical as it may seem, there is only one way to do it—it doesn't come from the past, but from the future".[8] In other words, by the time the wolf

---

between the eye and the gaze. Lacan, Jacques, *The Four Fundamental Concepts of Psychoanalysis: The Seminar of Jacques Lacan, Book XI*, translated by Alan Sheridan (New York and London: W.W. Norton and Company, 1998), pp. 67–119.

[8] Lacan, Jacques, *The Seminar of Jacques Lacan Book I: Freud's Papers on technique* 1953–1954. Translated by John Forrester (New York and London: W.W. Norton and Company, 1988), p. 158.

man's look is fixed on the wolves he remembers nothing of the primal scene. Only after the analytic session does he relate this gaze to his own look years earlier as he watched his parents having sexual intercourse. In this sense, the repressed returns from the future, and not from its position in the past. Moreover, Lacan notes that "…Wiener posits two beings each of whose temporal dimension moves in the opposite direction from the other. To be sure, that means nothing, and that is how things, which mean nothing all of a sudden, signify something, but in a quite different domain. If one of them sends a message to the other, for example a square, the being going in the opposite direction will first of all see the square vanishing, before seeing the square. That is what we see as well. The symptom initially appears to us as a trace, which will only ever be a trace, one which will continue not to be understood until the analysis has got quite a long way, and until we have discovered its meaning".[9]

While analyzing the implications of such temporality, Slovoj Zizek stresses the inverted temporal order suggested by Lacan, which in a way resembles science fiction stories. He argues that "… the Lacanian answer to the question, from where does the repressed return, is paradoxically: from the future. Symptoms are meaningless traces; their meaning is not discovered, excavated from the hidden depth of the past, but constructed retroactively. The analysis produces the truth, i.e., the signifying frame which gives to the symptoms their symbolic place and meaning… [T]he meaning of these traces is not given; it changes continually with the transformations of the signifier's network. Every historical rupture, every advent of a new master signifier, changes retroactively the meaning of all tradition, restructures the narration of the past, makes it readable in another, new way. Thus things which don't make any sense suddenly mean something, but in an entirely other domain".[10]

In what sense could this notion of inverted temporality be applied to Berest's work? Berest's work challenges any way of considering time ordering. The background of the work, which is the page from the newspaper from the day before the murder, and the words uttered by the physician minutes after the assassination are both visible, while the traumatic event itself is concealed. But this cannot exhaust the complex temporality of the artwork, since every

[9] Ibid., ibid.
[10] Slavoj Zizek, "The Truth Arises from Misrecognition," in *Lacan and the Subject of Language*, edited by Ellie Ragland-Sullivan and Mark Bracher (London: Routledge, 1991), pp. 188–189.

detail in it moves backwards and forwards in the dimension of time: the newspaper page includes the advertisement announcing the demonstration to be held the next day, the physician's text is uttered after the murder but before he recognizes the identity of the victim, the work itself was published on a magazine cover a year after the assassination, and so forth.[11] All these temporal zigzags collide in order to conceal a hole in reality torn open by the assassin. Berest's work exposes this hole, which is excluded from any time duration and representation. The significance of this hole may perhaps only be comprehended in the future. In this way, Berest's work can be seen as a vanishing trace, which only the future will allow us to decipher and provide reasons for. Furthermore, unlike narrative paintings, which represent temporal sequence in various ways, the structure of Berest's and Ginton's artworks presents temporality not merely by means of representation or metaphor, but as an inherent part of the way one can perceive them. These works do not embody memory by representing an event. The works themselves force a certain complex temporality on the viewer.

As mentioned above, David's *Death of Marat*, like many other historical paintings, allegedly repeats the event whose memory it is supposed to evoke by representing a part of its temporal sequence. Berest's and Ginton's works do not repeat the event the memory of which they are engaged with, but rather represent a cover for that something which elides any representation. Is David's *Death of Marat* radically different in nature? In other words, is psychoanalytical discourse valid only for certain types of paintings? In order to answer this question I turn to Lacan's discussion on memory and repetition in *Seminar XIV*, where, following Freud, he argues that repetition is always about a loss. He argues that the act of repetition is inseparable from the repeated object; in other words, the object being repeated is in a way the repetition itself… "there is something lost by the fact of repetition…by the effect of the repeating, what was to be repeated becomes the repeated".[12] In this sense, David's *Death of Marat*, which represents a traumatic event, is not different from Berest's and Ginton's artworks, which represent an enigmatic trace, since the object to be repeated is in both cases a lost object, the loss of which is demonstrated by the act of repeating itself.

---

[11] One can think in this respect of the newspaper article titles as they appear in the background of the work. For instance, the headline article on the newspaper page is entitled "Why was it a success", another title is "Virtual celebrity". These titles gain new meaning in retrospect, besides their original context.

[12] Lacan, Jacques, *Seminar XIV: The logic of phantasy*, translated by Cormac Gallagher (Unpublished, 1966–1967), p. 116.

In conclusion, the purpose of this paper was to define the unique way memories and their relation to time and temporal sequence can be embodied by works of visual art, and to show how this treatment of temporality subverts the traditional differentiation between spatial and temporal media. However, one cannot avoid the political implications of the specific subject matter of Berest's work. Using Lacan's terms, Berest's work sends us a message from a temporal dimension that moves in the opposite direction from ours. Hence, her work first appears to us as a vanishing trace the reason for which only appears later, and at the same time, the work's elusive temporality allows us to think of an alternative future, which could have existed but never did. The work enables us to glimpse this lost future, while the flickering and fleeting image of this lost future makes reality mean.

RESPONSE

Shirley Sharon-Zisser

'O peace,' quoth Lucrece: 'if it should be told,
The repetition cannot make it less;
For it is more than I can well express,
    And that deep torture may be called a hell
    When more is felt than one hath power to tell'
        (William Shakespeare, *The Rape of Lucrece*, lines 1284–1289)

Repetition, says Shakespeare's Lucrece, "cannot make … less" of the hell/hole of inexpressible, unrepresentable trauma, in her case, the hole torn open in her flesh by the act of rape and retraced (if it is not indeed retroactively created) in the graphic real of Shakespeare's text in the multiple O's of her repetitive speech. Efrat Biberman's work admirably follows the aesthetic and psychic logic of repetition as a making present, usually in veiled form, of an object whose loss, as Freud puts it in "Mourning and Melancholia" casts a shadow on the ego. And yet the most crucial consequence of Biberman's work is that the logic of repetition as the presencing, in the delegate form of the *Vorstellungrepresentanz* (representative of representation) is an aesthetic temporality not peculiar to art works explicitly engaging a loss, but characteristic of the aesthetic act insofar as, like the inscriptions in the psyche in Freud's metapsychology, are veils for the primordial lost object, the retroactively created hole around which psychic life revolves. If so, perhaps the very aesthetic form of repetition, as Shakespeare's Lucrece also implies, itself always already carries the paradoxical logic of time and memory Biberman discerns, and does so in verbal rhetoric no less than in the domain of painting. The work of poetry too, Shakespeare suggests in his sonnet #108, is a repetition "each day [of] the very same" (line 6) which rests on a "Nothing" (line 5), a hole in the "brain" (line 1) or psyche for which the repetitive "count" of love, "thou mine, I thine" (line 7), functions as a veiling "conceit" (line 12). Poetic writing, composed of "character[s]" drawn in "ink" (line 1) is perhaps as inherently a *Vorstellungrepresentanz* of the lost object as is the work of two-dimensional visual art, composed as it is of lines, stains, and shapes dawn in ink or paint. To what extent is this temporality of loss, which enables retroactively constructed memories to return from the future, as Biberman shows, "Where time and outward form would show [them] dead" (Sonnet # 108 line 14), made possible by the graphic space common to poetry and painting? In Seminar 14, Lacan shows that graphic space enables the representation of chains of signifiers just as much as of objects appearing as an excluded remainder with respect to those chains. What might be the logic of time and memory with regard to a non-representable object-hole be in three-dimensional art, a common medium for monumental commemoration? What form might this logic take in the spoken arts (such as theater) where vocalized signifiers necessarily appear in a chain that does not allow phonemes to fall? Such are some of the intriguing questions opened up by Efrat Biberman's ground-breaking study.

References

Lacan, Jacques. *The Seminar of Jacques Lacan Book 14: The Logic of Phantasy (1966–1967)*. Trans. Cormac Gallagher from unedited French Typescripts. (Unpublished).
Shakespeare, William. *The Complete Sonnets and Poems*. Ed. Collin Burrow. Oxford: Oxford University Press, 2002.

RESPONSE

Robert Belton

Dr. Biberman's very interesting paper describes artworks—artifacts that we usually think of as narratively inert or "timeless," in the sense that they seem unable to represent beginnings, middles and ends—as temporally expressive, articulating the ebb and flow of time and memory. Discussing several works inspired by assassinations, Dr. Biberman undertakes to show that these nominally "static" objects, through their representations and their mutual reinforcements and suppressions, can articulate rich temporal sequences and their interrelations. She then deploys psychoanalytic concepts deriving from Freud and his theoretical descendants to show that the resultant narrativity is not fully open to conscious analysis but must be analysed after the fact in such a way that hidden meanings snap into focus, as if with an element of surprise. A hidden meaning is suddenly revealed, colouring further investigations of the past with conclusions drawn "from the future," as it were, making the paintings she discusses rich with narrative possibilities that cannot be separated from the sequences of time, memory, and trauma.

There is another interesting possibility that Dr. Biberman does not address in this paper, although she has used the strategy in different ways elsewhere.[1] The assassination of Israeli Prime Minister Yitzhak Rabin functions as a symbolic castration threat, a theme that appears in the psychoanalytic writings of Iakov Levi.[2] Like castration, Rabin's assassination is a trauma of sufficient enormity that it has to be suppressed. How can such a thing be represented? Facing the reality of the political events is like fearing the female form without male genitalia, which Biberman elsewhere characterizes as "castration... embodied in the gaze as an object of lack."[3] Viewers of the illustrations in the essay are certainly aware that the works evince a gap or a lack, which Biberman describes here as "a hole in reality torn open by the

---

[1] Efrat Biberman, "You never look at me from the place from which I see you," *Imagendering* 11 (2005): http://www.genderforum.uni-koeln.de/imagendering/biberman.html.

[2] For example, Iakov Levi and Luigi Previdi, "Killing God: From the Assassination of Moses to the Murder of Rabin," *Agora* 4 (2000): http://www.geocities.com/psychohistory2001/KillingGod.html.

[3] Biberman, paragraph 11.

Jo Alyson Parker, Michael Crawford, Paul Harris (Eds), Time and Memory, pp. 277–278.
© *2006 Koninklijke Brill N.V. Printed in the Netherlands.*

assassin.... The significance of this hole may perhaps only be comprehended in the future." And it is *this* particular future that will colour *ex post facto* the creation of the traumatic meaning of the works in time and memory.

With this insight—that in instantiating loss, gap and lack, the temporality of the images generate interpretative castration anxiety, to coin a phrase—I find it deeply provocative that the essay is entitled "Remembering the Future," which I cannot help but rewrite as "Re-Membering the Future" or even "Re: Membering the Future." If the future sets the tone for interpretation of the past, then re-membering the future returns to the future its "phallus as the primary signifier," to paraphrase Lacan, making an ascending interpretative loop in which symbolic (and genital) absence constantly circles around symbolic (and genital) presence, creating a recursive slope of signification.

This essay, which offered so much new insight to this commentator, seems like the first chapter of an imaginary manuscript on the wider subject of art, time, and trauma, in the light of what I have elsewhere called the hermeneutic spiral.[4] I look forward very eagerly to more essays of this calibre by Dr. Biberman.

---

[4] Robert Belton, "Hermeneutic spiral," *Sights of Resistance*, online glossary, http://www. uofcpress.com/Sights/toc_hl.html.

CHAPTER FIFTEEN

FAMILY MEMORY, GRATITUDE AND SOCIAL BONDS

Carmen Leccardi

Summary

Memory is not only that exquisitely suggestive and secret dimension, the basis of one's identity, with which we are intimately familiar, but it is also a cultural product in the proper sense, which takes form, becomes structured and changes with time and in social space. Both remembering and forgetting can be considered social actions in the proper sense, enacted on the basis of mechanisms of selection which at one and the same time make it possible to mold a given representation of the past and to transform it into an essential tool of belonging and of identity. Like memory, also gratitude (as Simmel highlighted) can be considered, from a sociological point of view, an instrument of cohesion and continuity of social life. The aim of the paper is to analyze the characteristics and potential of collective memory and gratitude thus viewed, through the empirical data of research into young people's attitudes towards work in southern Italy.

Introduction

Memory is not only that uniquely subjective and secret dimension with which we are intimately familiar, the very basis of our identity, nor can it be considered a simple "repository" of the past (Schwartz, 1982; Schudson, 1989; Jedlowski, 2001; Zerubavel, 2004). The past, in fact, is not 'reconstructed' with perfect fidelity from memory, but rather, it is redefined and restructured on the basis of precise selection mechanisms, in function with the needs of the present that are always different.[1] In particular, if we look at memory from a sociological point of view, we can affirm that it is a cultural production in its true sense, that it takes its form, is structured and changes over social time and space.

Sociological studies of memory, that have flourished particularly over the past decades in the wake of the pioneering work performed by Halbwachs in the first half of the Twentieth century, have underscored how remembering,

---

[1] "The past structures the present through its legacy, but it is the present that selects this legacy, preserving some aspects and forgetting others, and which constantly reformulates our image of this past by repeatedly recounting the story" (Jedlowski, 2001: 41).

Jo Alyson Parker, Michael Crawford, Paul Harris (Eds), Time and Memory, pp. 279–302.
© *2006 Koninklijke Brill N.V. Printed in the Netherlands.*

if considered from a societal point of view, is the fruit of complex processes of interaction and communication. Through these processes, while the continuity of a social life is maintained, the identity of groups that are the bearers of memory is confirmed. In this common reconstruction process of the past, one learns what is to be remembered and what is to be forgotten, how and why it is to be remembered or rather, committed to forgetting. Remembering and forgetting may be considered social actions in the true sense, enacted on the basis of selection processes that allow both for a molding of a particular representation of the past—"representation" should be understood in this context in the literal meaning of "making present" of that particular past—and for transforming it into an essential tool for belonging. In this sense, one can assert that the time of memory is able to undo the fabric of the linear, irreversible time of history. It proposes itself as circular time, within which not only the events of the past live again in the present, reformulated and adapted to the needs of today, but where the future itself undergoes the normative influence of the past.

Stopping a moment to turn our attention to these strategic features of memory is much more important today, in the 'presentified' climate of western society. In these societies the feeling of a strong acceleration of time (Rosa, 2003) is accompanied by a contraction of temporal horizons that tends to erode the dimension of the past and the future together. While the objectifications of memory grow—through the technological supports, prostheses of daily use that are always more sophisticated and powerful—the social environments in which individuals and groups manifest an authentic "desire for memory" (Namer, 1987: 23) make up isolated archipelagoes (today ever more tied to ethnic or religious identity). Generally, in this frame, collective memory—understood as a "living memory", preserved and trans-mitted by a social group through interaction—ends up both more difficult to construct and difficult to transmit. Nonetheless, as the reflections proposed in the following pages intend to bring to light, the strategic force of collective memory in guaranteeing continuity of social life is maintained unaltered. One can affirm that the process of 'presentification' that has accompanied the last few decades tends paradoxically to extol its social relevance.

One of these "archipelagoes of memory" is made up of a group of young people from Southern Italy, boys and girls whose cultural orientation, in particular the culture of work, I explored through empirical research in the Nineties.[2] The strength of the collective memory of these young people, as I

---

[2] The research was performed in the first half of the Nineties in Calabria: during that

will try to prove, is the fruit of the conjunction of two different factors. On the one hand, the family is recognized as a central agency and institution, that is capable of constructing memory and of guaranteeing its intergenerational passage. On the other hand, the importance of gratitude (Simmel, 1908/1964), understood as moral and social feeling at the same time, is expressed by the young people toward the preceding generations. Analogously to memory, this appears able to resist the wear of time, reinforcing interpersonal and group bonds, here specifically the bonds between the parents' generation and that of their children.

This chapter is divided into three parts. The first contains an examination of the salient characteristics of collective memory and of family memory, which represents one of its branches. The second provides an analysis of the sociological concept of gratitude, elaborated by Simmel, with the intention of showing how this sentiment, not unlike collective memory, can be considered an instrument that serves to bridge cultural discontinuities. The third and last part is dedicated to the role played by the union of family memory and gratitude in the formation of the work culture of young people in Calabria, one of the regions of Southern Italy that is most disadvantaged from an economic point of view, but characterized by a young population that is becoming more and more educated and rich in innovative potential (Fantozzi, 2003).

1. Collective Memory and Family Memory

"We are never alone... our most individual memories are closely dependent on the group in which we live." Thus French psychologist Paul Fraisse (1957: 168) wrote in the Fifties, referring to the mechanisms of individual memory. His theoretical reference, explicitly mentioned in the text, is Maurice Halbwachs. Halbwachs' work on the social dimension of memory[3]—

time, I taught Sociology at the University of Calabria (Arcavacata di Rende, Cosenza). It used qualitative methodology and was performed in Cosenza and in five small towns of its province. In this context, forty interviews of a narrative nature (Schütze, 1983) were carried out among young people of an age between 18 and 24 years old. The results are summed up in Leccardi (1993; 1995a). See also Leccardi (1998) in German.

[3] A significant contribution to the in-depth research of this dimension has also been made by another research area, the so-called social-historical tradition of Soviet psychology (Vygotsky, Leontiev, Lurija among others). Focusing on the importance of the socially constructed nature of all psychological phenomena, this school of thought has strongly emphasized the social constitution of *all* forms of memory. According to this thesis, one cannot speak of memory (at least in adults) without referring to such concepts as community, society and culture. On this theme, see Bakhurst (1990).

developed in the period between the Twenties and the Forties (Halbwachs, 1925/1976; 1941/1971; 1951/1968)[4]—had a profound effect on the social sciences and continues to be an influence even if it is not homogeneously divided among the different areas involved.[5]

Halbwachs, a fervent follower of Durkheim's line of thought, has the great merit of having delivered the theme of memory from an analytical approach limited to an individual key—of which Bergson was considered the most authoritative spokesman at that time—making it a specific subject of sociological reflection. According to this scholar, in a nutshell, there is a collective organization of memories or collective memory[6] from which the individual memory draws amply. In his opinion, the latter is to some extent a kind of derivation of the former.[7] In fact, he writes that the act of remembering would not be conceivable (Halbwachs, 1968: 47) "except on the condition that one adopts the point of view of one or more currents of collective thought."

A characteristic trait of collective memory consists of the fact that it is a "living history": its time limits, as opposed to written history, coincide with those of the existence of the group possessing it. This leads to one of its fundamental characteristics, that of continuity. To quote Halbwachs (1968: 89), collective memory is "a continuous current of thought, of a continuity that is by no means artificial, because it conserves nothing from the past except the parts which still live, or are capable of living, in the conscience of the group." It can therefore not exist and express itself without the living support of a group. It is the members of the group who through their interaction mold that particular image of the past that is transmitted in the present.

According to this approach, the past is anything but a static dimension. Rather, it changes according to the creative play of the collective memory (or the collective memories)[8] and thus, in accordance with the requirements of

---

[4] *La mémoire collective* (1968, or ed. 1951) was published posthumously.

[5] As Luisa Passerini (1987) notes, Halbwachs' work is still relatively unknown to historians, including those who work on the subject of memory as a historical source.

[6] With regard to the relationship between social memory and collective memory understood respectively as social organization of memories and organization of memories on the part of a group, see Namer (1987).

[7] On this subject, Jedlowski (2002) underscores the role played by "the missing encounter" between Halbwachs and psychoanalysis (especially as regards the concept of the unconscious) in the construction of his sociology of memory.

[8] The theme of the plurality of collective memories is developed by Halbwachs in *La mémoire collective*. To the contrary, in *Les cadres sociaux de la mémoire* his attention is focused on three "great" collective memories: the family, the religious and class memory. Implicitly, these three memories seem to exhaust the field of the social memory. On the two different approaches of Halbwachs' thought on the subject of collective memory, see Namer (1987).

the present.[9] "Society," Halbwachs points out (1976: 279), "represents the past for itself according to the circumstances and according to the times...." therefore becomes essential on the basis of this premise to "renounce the idea that the past is preserved just as it is in individual memories" (Halbwachs, 1976: 279).[10] From this point of view, the past is merely a construction, a collective image elaborated in the present and for the present, "what one has agreed to call the past," as Halbwachs writes in *La mémoire collective* (1968: 131). Once recreated, however, its effects become very real: it molds the group's vision of the world, it becomes a driving force of its actions.

The collective frames of memory, the "instruments" of collective memory (Halbwachs, 1976: XVIII) serve as filters, selecting the aspects of the past that the group *must* remember, to keep its own identity alive in the present. Collective memory thus becomes essential to guarantee the integrity and survival of the group over time. A community may use the past to guarantee its stability and at the same time obtain a fundamental anchor of sense for the present.[11]

As mentioned above, it is on the frames of collective memory that, according to Halbwachs, the functioning of individual memory is based. The nature of these frames appears complex: they are together notions, representations and norms (Namer, 1987: 58–62; see also Assmann, 1995). The elements comprising them, as Halbwachs underscores (1976: 28), "may be considered at the same time as notions, more or less logical and linked in a more or less logical fashion, which provide motives for reflection, and as imaginary or concrete representations of events or personalities, located in time and space." Collective memory can thus be considered as a *notion*, an element of a specific knowledge, of an idea or a constellation of ideas, and as a *representation*, here in the sense of image, fantastic configuration, changing creation. While different, within the picture of collective memory notion and representation appear superimposed to the point of being hard to distinguish. But this is not, as we will see, the only conceptual alchemy featured by collective memory.

---

[9] Gérard Namer (1997: 272), in his afterword to the new French edition of *La mémoire collective* (1997) underscores how collective memory can be considered a real true "reconstruction of the present performed in function of the past."

[10] This "reconstructing" vision of the past differentiates in a clearcut manner Halbwachs from Proust, for whom the past is not reconstructed but 'found again.'

[11] We are in this case dealing with what Bellah *et al.* (1985: 153) define as "community of memory".

Moreover, memory's frames are characterized by a strong normative dimension that makes them "models, examples and lessons" (Halbwachs, 1976: 151). "As a past fact serves as a lesson," Halbwachs writes in another point (1976: 282), "as encouragement or as warning, what we call the frame of memory is also a chain of ideas and judgments." Due to this characteristic, they provide the group with indications concerning the present and the future. They for instance prescribe the paths to follow today and in the future, and which paths to avoid; they convey a package of knowledge that may also serve at a later point in time. The dimension of the plan becomes visible behind the curtain of memory: the past, reconstructed collectively, in turn builds the future. According to Halbwachs, the close bond between past and future[12] is further ratified by the normative element of memory.

But the frames of collective memory have another important characteristic: they are never, by definition, anonymous. They are vivified by names, faces, histories that we are intimately bound to, which elicit unmistakable sentiments and emotions in us. They transmit a past, testifying to an *experience*. And experience, as we know from Benjamin (1955), as sediment of accumulated, often unconscious, data which flows into memory is not, in the final analysis, distinguishable from tradition. Experience, memory and tradition are arranged along the same axis, that of continuity in time. "Experience is knowledge which has been distilled over time, in which the personal past of a person is combined with the sediments of a collective knowledge conveyed by tradition" writes Jedlowski (1991: 131). Through memory, the individual and the collective are united in the concreteness of experience. Together, collective memory and experience govern tradition: as long as they resist, tradition cannot die.[13]

As Shils points out (1981: 50) in his study on the notion of tradition, if experiences lived by others who are important to us, whether they are alive

---

[12] In the experience of time, the bond between the two dimensions is, as we know, crucial: what we are able to imagine is inseparable from what we are able to remember. As Bachelard perceived already in the Thirties (1980: 46), one of the essential components of social memory frames is the "desire for a social future". Modernity, with its tendency to dissolve continuity, nevertheless introduces quite a few complications in this scheme. On this subject, see the reflections of Koselleck (1986) concerning the gap between experiences (of the past) and expectations (for the future) in the modern age.

[13] In a way, the study of collective memory enables us to go beyond the de-traditionalization thesis, avoiding conceptualizing the contemporary condition as completely opposed to tradition. See Adam (1996) for a reflection on this strategic issue in sociological analysis.

or dead, reach us through memory, not only does the image of the past which they transmit to us live again in our present, but those experiences become an intrinsic part of our identity. Our identity, thus, also includes numerous characters borrowed from members of the group (family, political, religious etc.) we belong to, and who have preceded us in time. According to Shils, this memory mechanism contributes to keeping the force of tradition alive.

As lived history, collective memory is moreover characterized by a strong affective element (Namer, 1988), which is the result of close interaction and subsequent sharing of experiences among members of the group (Alfred Schutz, 1971, would speak here of the sharing of a "vivid present"). Through this affective dimension, on the other hand, the normative character of memory is reinforced.[14]

By providing an essentially chaotic past experience with a language and at the same time a structure and a unitary direction characterized by a highly internal coherence, collective memory diminishes the differences between those participating in group life and their individual memories, guaranteeing collective force, cohesion and identity (Jedlowski, 2002).

Of the three collective memories analyzed by Halbwachs (1925/1976)— family memory, the memory of religious groups and that of social classes— our attention will now focus briefly on the first. In fact, it is the memory of the family group that plays the most important role in the development of Southern Italian young people's approach to work: the theme, as mentioned in the introduction, that has inspired these reflections.

Halbwachs dedicates the fifth chapter of *Les cadres sociaux de la mémoire* to the collective memory of the family.[15] The main topic of his reflections is the following: underlying the family group there is a "mutual fund" of memory through which the "general attitude of the group" is expressed and from which the family members "obtain their distinctive traits" (Halbwachs, 1976: 151–52). It is as if the thoughts of every member carry a mark of this "mutual fund," a cipher that they secretly share with the other members of the group. It is significant to note that Halbwachs speaks of different "ramifications" of the same line of thought. It is thanks to this memory that the family group

---

[14] With the pluralization involving collective memories in the modern age, also the sense which each of them transmit become, on the other hand, relative. This phenomenon is associated with a weakening of their affective component.

[15] On family memory, see Bertaux-Wiame (1988) and Muxel (2002).

can survive as a unit over time and despite the changes it faces, retain the feeling of its own uniqueness. This ability to remember provides the family with, to use the author's coined expression, its own "traditional armature."

The story of every family, as seen through the eyes of those belonging to it, is unrepeatable: every family, Halbwachs writes (1976: 151), "has its own spirit, its own memories which it alone commemorates, and secrets which are not revealed except to its members." These memories, as we have already had occasion to observe, not only indicate the nature, quality, and strong and weak points of the group; they also represent a set of rules to follow, an example and model to emulate. The path followed by the family memory cannot be ignored, on pain of leaving the group.

Family memory is, on the other hand, the perfect home of tradition. If one participates in the life of a family, Halbwachs observes (1976: 147), one finds oneself belonging "to a group in which it is not our personal feelings, but rules and customs beyond our control, and which existed before us, that establish our place." These rules and customs are in the first place nourished in our everyday life by the dominance of the times and places of the family. This is one of the reasons, according to the author (1976: 154), why "most of our thoughts are mixed with family thoughts." This impossibility of separating our thoughts from the family's thoughts, or our memory from the family's memory, is indicative of the profound bond that, according to Halbwachs, ties the individual members to the family group. "When a group has permeated us with its influence for a long time," he underscores (1976: 167), "we are so saturated that if we find ourselves alone, we act and think as if we were still under its pressure." The pervasiveness of the models proposed by family memory, the lessons provided by it and in general all its messages, are in the first place to be retraced to the power of habit and the daily contact between its members—in a word, "familiarity."

This nature is, for that matter, reinforced by the specific character of family relations. "Until one leaves the family," Halbwachs (1976: 163) reminds us, "as opposed to other groups whose members may change and sometime change place with respect to one another, one remains in the same position as a relative...". New family relations can be added to those preserved in the group memory: we may in turn be not only children but also parents, but old relationships cannot be cancelled. Their mark is indelible. From this point of view, family memory seems to be perfectly uninfluenced by changes: nothing appears relative or mutable within it. The trait of certainty dominates everything.

Another striking aspect of this memory is the perfect complementarity of the two aspects of memory, the collective and the individual. Thus, if every member of the group is considered here more than elsewhere in her/his uniqueness, in other words, if she/he becomes part of the family memory by virtue of her/his singularity, her/his exclusively personal vicissitudes, there is nevertheless no figure or event of which the family preserves a memory which is not seen from the viewpoint of the group. These personalities, which we are bound to by a very close relationship, become the points of reference around which family memory centers, in which entire historical phases of the group life are condensed. The single person, as well as the events concerning her/him, becomes, in the strictest sense, the symbol of the group.

As already mentioned, the affective character of the family memory is another trait contributing to its cohesive ability. Gérard Namer (1988), analyzing the affective characteristics of family memory, casts light on three different aspects. The first is related to the sentiment of uniqueness uniting its members. This feeling stimulates the development of a sense of complicity: together they possess a common heritage of secrets, which they are the only ones in the world to share—which, in turn, reinforces the affective bonds. The second is associated with the normative dimension of family memory, with the fact that it is "a lesson to repeat" and, at the same time, the vehicle of images, atmospheres that have to be renewed, that want to continue to live in the present because they speak to our heart. The third concerns that particular aspect of affection which Namer (1988: 10) defines as "compromise" or "reconciliation" and which alludes to the co-existence, in family time, of different "generational consciences" to borrow a term from Attias-Donfut (1988: 49). That is to say, of different consciences of the role and position that the generation of each member has in relation to the others. If it is true that collective memory embodies the continuity between generations, it is also true, on the other hand, that each of them will principally tend to recognize themselves within the family group, in a different dimension of time (and universe of significance): the oldest first and foremost in the past, the adult mainly in the present, and the youngest especially in the future. In this frame, the "affection of conciliation" of the family memory protects the unity of the group, preventing the memory from creating conflict among the members.

The affection which keeps members of the family group united is even more important if we consider that collective memory, as mentioned above, is also loaded with plans which the older generation more or less tacitly entrusts to the younger one.[16] The feeling of affection which keeps memory

---

[16] "Memories, recollections or testimonials attributed to a generation," Claudine Attias-

and plans together can then become the basis of a moral obligation: for instance, for children to honor the family memory by continuing along the paths which their parents have traced.

When what is remembered by the group is linked, as in the case of the family, to the spontaneous and daily interaction among members, the basis of the intimate relationship uniting them, the significance of the "images" and the "notions" conveyed by memory certainly increases. The particularly intense light that emanates is, in other words, the fruit of sharing both a symbolic universe and everyday times, spaces and rhythms. The "practicality" of knowledge of the social world conveyed by family memory, inseparable from the everyday dimension in which it is immersed, reinforces this luminosity.

A last element has to be understood with regard to family memory. Isabelle Bertaux-Wiame (1988: 25) observes, reflecting on topics associated with family histories, that an important role is played by the evaluation of the social path covered by the family group as a whole. The hopes, anxieties and desires of the present time, that give the reconstruction of the past the particular image and flavor which only the group members are familiar with, also has to be understood on the basis of this evaluation. If the path that has already been covered is valued collectively in positive terms, social pride penetrates the family memory. The presence of this feeling reinforces the sense of collective identity, while it also makes other possible paths of memory fade. Thus, the all-absorbing dimension of family memory is strengthened.

2. Gratitude as Sociological Concept and its Affinities with Collective Memory

In a nutshell, collective memory—and within it, in a particularly efficient manner considering the strong area of affection surrounding it, family memory—supposedly exercises, from a social standpoint, a dual function: while it guarantees integration, it provides the group with a valid instrument for continued survival over time. The same characters—even if in this case it

---

Donfut writes in this regard (1988: 48) "are not comprehensible unless they are related to the others, as they are sequences of a collective memory which incorporate them in a continuity in time endowed with significance and full of plans."

concerns a relationship between two people (*Zweierverbindung*) rather than a group—are part of the sociological concept of gratitude, as Simmel has elaborated (1908/1964).

Gratitude, similar to faithfulness, understood as an expression of relational continuity, interests Simmel by virtue of its capacity to preserve social connections and relations from the phenomena of destruction and wear caused by the passing of time. This is why the feeling of gratitude, like that of faithfulness—both considered "sociological sentiments" or "sociologically-oriented" sentiments as well as "psychic states" belong to "the *a priori* conditions of society which alone make society possible..." (Simmel, 1964: 381). Both gratitude and faithfulness provide stability in the world of relationships, which are by nature fluid and inclined to continuous transformation. If our interior life is similar to a current, if it is a process in constant evolution, in certain aspects magma, these sentiments "solidify" the shape of the relationship, rendering it constant over time. They practically cast a bridge across the two banks, of interiority and association, contributing to overcoming the barrier that divides them, despite their intimate bond.

As for gratitude—significantly defined as "moral memory of humanity" by Simmel (1964: 388)—the author observes, in the first place, its capacity to establish the bond of interaction also where it is not guaranteed by any external coercion. If, in all exchanges of an economic nature, the legal order guarantees the respect for the giving/receiving scheme, Simmel points out numerous other relationships based on the same paradigm obtain their support in the sentiment of gratitude. From a sociological viewpoint, gratitude can therefore be considered a supplement to the legal order being "one of the most powerful means of social cohesion" (1964: 389).

Gratitude, a feeling that develops from and in human interaction, is capable of surviving the conclusion of the relationship that gave rise to it. This is a characteristic it shares with other forms of association. Gratitude, writes Simmel (1964: 388–89), "is an ideal living-on of a relation which may have ended long ago, and with it, the act of giving and receiving". According to Simmel's analysis, the importance of gratitude is inseparable from its duality: on the one hand, a driving force of the spirit, it is also, on the other, an extraordinarily efficient means of social cohesion. Through gratitude, human actions can be reconnected to events that may even be very distant in time, of which traces may perhaps have been lost today.

The continuity of social life, which is, in the final analysis, the continuity of the interactive life, therefore obtains extraordinary support from gratitude: regardless of how tenuous and in some cases almost intangible the debt of gratitude is, the social relationship it creates is capable of pulverizing time. "Often the subtlest as well as firmest bonds… develop from this feeling," observes Simmel (1964: 389) in this regard.

Like in family memory, social cohesion and continuity preserved by gratitude do not appear analytically separable from the affective dimension of the relationship, from its intimate character. On the base of this affection, what was initially given acquires the character of incommensurability. Regardless of what the received gift consists of, the bond cannot be broken by a counter-gift.[17] The first gift, Simmel underscores, in fact boasts the unique characteristic of being the result of a voluntary act, a decision completely free from coercion. This primogeniture places the first giver in an ethically privileged position. "Only when we give first are we free, and this is the reason why, in the first gift, which is not occasioned by any gratitude, there lies a beauty, a spontaneous devotion to the other, an opening up and flowering from the 'virgin soil' of the soul…" (1964: 392–93). Those who receive the gift, on the contrary, are subject to a moral coercion—which according to the author is no less radical in its effects than a legal obligation.

The relationship created by gratitude possesses an "interior infinity" which prevents the bond of reciprocity it is based on from being exhausted. It is this *character indelebilis* (1964: 393)—the impossibility of completely eliminating the relational bond ratified by its presence—that gives gratitude a sociologically unique flavor according to Simmel. In fact, it easily survives the fading of the range of sentiments that it may have been accompanied by in the past: for instance love, friendship, respect, trust and so forth. Gratitude "seems to reside in a point in us which we do not allow to change; of which we demand constancy with more right, than we do of more passionate, even of deeper, feelings" (1964: 394). Inconvertible and unalterable, the atmosphere of obligation it gives rise to persists even after the initial gift has been matched with a counter-gift. Despite spatial and temporal gaps, its knots keep each element of society tied to another in a "microscopic" but very solid manner, "and thus eventually all of them together in a stable collective life" (1964: 395).

A fortunate harmony of accents therefore exists between collective memory and gratitude. While in different manners and forms, both focus

---

[17] The obligation in archaic societies to give gifts, to receive and return them is, as is known, at the basis of the reflections of the French sociologist and anthropologist Marcel Mauss, nephew and colleague of Durkheim (Mauss 1925/1950). Despite the importance of these reflections for the analysis of social life understood as a system of relationships, for reasons linked to the economy of work, attention here is concentrated in an exclusive way on the theoretical proposal of Simmel.

on what is *memorable*. Whether it is sealed by the bond of gratitude, or rather preserved and handed down in the form of memory of events that are crucial for the identity of the group, the 'memorable' in any case belongs to the sphere of the extraordinary. This fact preserves it from decadence and from falling into oblivion. The present that doesn't die constitute, in the final analysis, an epic time (Maldiney, 1975). Even if the heroes and heroines who populate these tales are more often than not personalities facing circumstances and experiences belonging to everyday life, the 'ordinary' *par excellence*, its exceptional quality is beyond discussion. It is the importance of these personalities, of those events, of those relationships for the identity of those who remember today which make the memorable unique. Whoever has worked with materials associated with collective memory, or with the 'memory of gratitude,' that is, with materials that reconstruct the genesis and development of this sentiment over time, can document the existence of this extraordinary quality.

3. Family Memory, Gratitude and Work Culture of Youths in Calabria

Thus family memory shares with gratitude, understood as a "sociological feeling", the ability to sustain forms of association, of making a barrier against social discontinuity, reinforcing the relationship between the past, present and future. If every collective memory may be considered a "living history" in agreement with Halbwachs, able to give rise to group action in the present and to reinforce its identity, familiar memory especially accentuates the unique traits of this memory. In an analogous way, gratitude tends to preserve over time the relationship that gave it its origins—the continuity of the relationship constitutes at the same time the goal of this feeling (gratitude *promotes* continuity) and its source for existence (gratitude *derives* from this continuity). "Movement of the soul" and together a secret means of social cohesion, gratitude in turn celebrates the uniqueness of the relationship that gave it its form. We may state in this light that both familiar memory and gratitude transform the past into an eternal present, and impede that its heritage of sense prove to be ephemeral. Despite the accentuated acceleration of the processes of social and cultural change, a characteristic of our historical age, this property remains unaltered. The results of the research on the culture of work among young people in Calabria that we are going to consider bring to light this ability.

In order to fully understand this process, it is however necessary in a preliminary way to focus on the centrality of the family in the Calabrese social context. The social strength of these two dimensions, collective memory and gratitude, in fact is directly linked to this centrality.

According to Shils (1975) every society has central areas, or society centers as he defines them, and more peripheral areas. The symbolic orders, the values, the beliefs and traditions which interest in different ways those who live in that particular social space are radiated from the former; desires, interests and collective goals are formed. The existence of those central areas is of great importance, according to Shils, for the legitimatization of the social and cultural order. Society centers, while taking on a sacred character, actually at the same time belong to the sphere of action, structuring social activities and roles. The symbolic and at the same time organizational aspects of social life thus find full institutionalization within them.

The existence of these focal points in the symbolic articulation of a society, capable of establishing the significance of behavior and of specifying ends and means, becomes very evident when one works on the materials of memory. Social memory is shaped on the basis of these centers, and clearly carries their mark. In particular, if one deals with a reality that is still wavering between tradition and modernity[18] as Calabria was at the beginning of the Nineties, where the process of pluralization of collective memories induced by modernity is weak, their presence is more easily retraceable within these materials. In fact, in Calabria there were, and still are, some collective memories whose voice is socially stronger, and makes more sense not only for the members of the group which expresses them, but also for society as a whole. Hegemonic collective memories, we might say, which incorporate other group memories tend to impose themselves as a social memory *tout-court*. Family memory is certainly one of these memories; likewise, in Calabria, the family may be fully entitled to be considered a "society center".

Both before and after the "great transformation" of the years after World War II (Polanyi, 1944), that is, before Calabria was incorporated into vaster pluralistic systems, integration and social cohesion were in the first place guaranteed by the family structure. In traditional Calabria, and in particular in the area around Cosenza, as Pino Arlacchi clearly showed (1980), the

---

[18] In a social reality subject to complex changes, as those brought about by the process of modernization, not only do aspects of tradition and modernity merge in a heterogeneous mix but, as Gusfield observes (1967), they also tend to reinforce one another mutually, the one supporting the other. With regard to the dynamic between tradition and modernity in Calabria, see in particular Piselli (1981) and Fantozzi (1982).

family cum agricultural enterprise could for instance be considered the fundamental pivot of economic and social life. Along with meeting essential human needs, it coordinated the economic relationships; it guaranteed specific roles, values, social functions for its members; it assured, integrating economic and non-economic spheres of existence, order and social control. Within it, productive obligations and personal relationships were inextricably welded. The main instrument of this welding was the relationship defined by Thomas and Znaniecki (1918–1920) as "family solidarity," a relationship that is manifested in the control and the help exercised on each member of the family group by every other member. By virtue of this solidarity, the life of each member develops as a declination of the family as a whole. Consequently, as Arlacchi underscores (1980: 37), in Calabria no one "can rise or fall without to some extent also taking his family group along".

Due to this complex order of motives, the family here more than elsewhere is a social institution capable of providing existential orientations, to prescribe the appropriate choices, not only with regard to models of behavior, but also to final goals. The basic questions of human existence find paths of reply within it. This normative character of the family as institution is reinforced by the individual's inability to escape its embrace: those who do not belong to a family, and thus lack social protection, as a matter of fact find themselves on the margins of social life.

But the family remains the central structure, even in a social context that is amply transformed, also in modern post-war Calabria. The analyses of Fortunata Piselli for instance (1981; see also Arrighi and Piselli, 1985) have clearly shown that in the social scenario of the last decades of the Twentieth century, the family succeeded in maintaining its centrality unchanged. Thus, in the community in the Cosenza area analyzed by the author, "the penetration of market mechanisms (…) not only did not provoke the disappearance or weakening of traditional relationships, but created life conditions which tended to perpetuate them, in different forms, as the principal factor of cohesion and social stability" (Piselli, 1981: 5).

In a region where the traditional forms of relationship fade along with the subsistence economy they served to support (Arrighi and Piselli, 1985: 462), the protection provided by the family net nevertheless does not lose even a fraction of its efficiency. It continues to guarantee order in the community and social integration, even if linked to the new political and economic interests that have recently developed.

"Family norms and relationships," write Arrighi and Piselli (1985: 466), "no longer represent a conditioning structure determining individual behavior… On the contrary, they form a flexible structure, subject to manipulation on the part of individuals for purposes of reinforcement and consolidation of their own economic, social and political position."

In launching initiatives in the field of economics as well as in politics, in the transition from school to work and in the subsequent long or very long period spent searching for a steady occupation, in guaranteeing its members the possibility to advance socially, the family structure still shows unchanged vitality. In short, it remains a true North Star in the sky of Calabrese society.

Especially for younger members, family relationships appear to be a safe harbor in the present-day instability and uncertainty. Young people are fully aware, for instance, that if they are to succeed in finding a 'steady job' or making a career in their profession after graduating, it depends to a large extent on the system of family relationships. The idea that thanks to "good connections" it is possible to pursue personal interests, obtaining favors and benefits connected with public resources, is in fact widespread. The diffusion of political clientelism[19] in Southern Italy is a well known reality, just like its ability to influence the processes of socialization of the very young.[20]

The experiences of preceding generations in relation to clientelism were transmitted to the young people in the family like in the example that follows:

> My father told me that with his family they went to work the land, then they needed money and the majority of them [my relatives] had to emigrate to the North. My father emigrated when he was 17 years old, he worked in building construction and as a miner. When he returned to Calabria in the Eighties, he began to work in a company of asphalt paving (…), he paved the roads. Now he's changed, he is a cook in a hospital, he always liked cooking (…) Here the PSI [Italian Socialist Party][21] is in control so since he's a socialist he's had more chances to get in than someone from another party… there in the hospital the majority are also socialists… you can't get away from it, this is what it means to be recommended. Even if you get out of school with the best grades, if you don't have a recommendation, you don't get in anywhere. Your family can help you in this sense. (Giorgio, university student, 21 years old, Mormanno)

---

[19] On the concept of political clientelism see Eisenstadt and Lemarchand (1981). See also Eisenstadt and Roniger (1984).

[20] Arrighi and Piselli (1985: 471) argue that "kinship is now maintained in existence as a framework of protection, as a last defence against the fluctuations of the market, a sane instrument of strength of the individual in the local labor market, as a means of climbing the social ladder to the top; naturally, by means of the exercise or support of political power". I reflected at length on the relationship between clientelism and the condition of young people in the South in Leccardi (1995b).

[21] In the second half of the Ninties the PSI disappeared, after probes of corruption.

Young people appear thus to be well aware of the importance of the family as a bridge toward the labor market, passing through politics. But, on a level that may be banal but no less important in terms of everyday life and identity, they are also aware that their chance to play the role of consumers in a society in which the manufacturing structure is traditionally weak, is inconceivable without the economic support of the family. Finally, we must not forget that the everyday existence of young people in Calabria continues to take place within the family circle. Even if the generation gap is, of course, a reality here as well—it concerns above all education, life styles and consumerism—the majority of young people in Southern as well as in Central and in Northern Italy still live with their parents.[22] And if the grandparents do not share the house of the grandchildren, they usually live in the same neighborhood or in the same small town. Even though education has made a significant contribution to breaking this tight connection—consider for instance the strong modernizing influence of the principle of residence for the students at the University of Calabria—nonetheless, attempts to bridge this 'cultural gap' are clearly visible in everyday life.[23]

The family as a reference point in the life of young people in Calabria is all the more significant if we consider the social progress which a considerable number of families in Calabria have made in recent years. From the grandparents who knew neither how to read nor write and who worked to survive, to the parents who are generally barely literate, who have had a very hard life but who have nevertheless enjoyed a better lifestyle, to the sons and daughters who at least potentially may try to plan their own future, especially thanks to new levels of education, this transition is decidedly revolutionary for Calabria.[24]

Reflecting on the lives of grandparents and parents, two young female students express themselves this way for example:

> My grandmother stayed home, she had to take care of the children, they had many children my grandparents... seven... she always stayed home she told

---

[22] The "famiglia lunga" (long family), that is the cohabitation under the same roof even if the children have already finished their studies, has been a characteristic of Mediterranean countries during the last few decades (Cavalli and Galland, 1993). In 2000, only 30% of young Italians of an age between 25 and 29 years old did not live with their parents (Buzzi, Cavalli and de Lillo, 2002).

[23] Thus, on Saturday and Sunday (as well as other holidays) the University of Calabria campus is almost completely empty. The great majority of students return to their families, mostly small towns or villages.

[24] In the span of a few generations in Southern Italy, we have gone from a situation characterized by low levels of education in which illiteracy was still widespread to a situation of education that is much closer to the European model.

me because my grandfather drove a boat, and with the boat he often had to go
to England (...) My mother worked hard for us children... she's a housewife...
plus we have the land... a vineyard... so my mother always stayed home and
worked in the field... she watched how things were done and she learned... she
found herself forced to work the fields because my father was not around, he
had emigrated to Germany. She told me that when she was my age, there was no
chance of studying, she knows how to sign her name, even how to write some
things. She tells me that she would have liked to study, she always repeats that
I have to continue with my studies, go to the university (Margherita, 19 years
old, high school student, Calopezzati).

My paternal grandfather was a hired hand in the fields for a Count, then he left
and set up his own work. He was a very strong person but at the same time very
sweet... that's what my grandmother says. Now my grandfather is dead. My
grandmother has always been a housewife and worked in the fields along with
my grandfather. She was a very energetic woman because they had six children
and had to maintain them (...) With respect to my grandparents, there is a
better tenor of life, the work has changed too. My mother works as a janitor
in a school, my brother and I study and I would like to go into the field of
computers (Anna, 17 years old, high school student, Crosia).

The grandparents and parents tell their sons and daughters these biographical
stories with an abundance of detail, about the efforts made in their precarious
work situations, without certainty for the future, marked by the experience of
emigration.[25] The young people consider the transmission of these memories
an obvious thing. In the interviews, they speak without hesitation about hard
or very hard work experiences on the part of their grandfathers and fathers,
about the life as a 'housewife'[26] of their grandmothers and the majority of
their mothers that familiar memory has preserved and transmitted. In their
eyes, the social trajectory that has been covered in a short time span—and
that has allowed them to acquire well-being and education—appears first
and foremost to be the result of their parents' long litany of denial, their

---

[25] Between 1951 and 1971, for instance, 800.000 people emigrated from Calabria, that is
one inhabitant out of three. See Congi (1988: 21).

[26] In the situation of Southern Italy, women's work has traditionally been socially invisible
(Cornelisen, 1977). Even if, as Amalia Signorelli writes (1983: 71–72) in spite of "the
stereotype according to which they were all housewives, Southern Italian women have always
worked outside the home, except for some brackets of middle-class women... However, the
productive capacity of women, in particular the work done outside the home, has been, as we
know, culturally denied, as a sign of dishonor for the man: consequently women, in particular
among farmers, have never had, I will not go as far as to say professional identity, but not even
awareness of themselves as workers".

spirit of sacrifice, rather than the result of a social and cultural evolution. The cultural and social range of this representation emerges for example clearly from this excerpt of interview:

> My maternal grandfather worked in agriculture and devoted to sheep breeding... He tells me often how he lived day to day, how he was forced to try to get by with a big family... my father is ignorant in the sense that he only went up to second grade in elementary school, then his father died young and to maintain a big family he had to accept all kinds of work... His life was full of sacrifices and wasn't a happy life like I am living now: school, home, fun (...). Now my father is a trucker ... it's a job with sacrifices because many times he has to work far from home, and he cannot have even a minute for stopping because the less time he takes to do his job, the more chance he has for doing another one when he gets back (....) Fun things, he hasn't had very much, because his life has been on a big truck ... my father and my mother and their parents before them made a ton of sacrifices to allow us children to have a better future... their sacrifices haven't been thrown away... they taught us a lot, how to live in society, how to care for a family, how life is not a walk in the park... they gave us a lot (Roberto, 22 years old, university student, Cosenza).

According to Roberto's point of view—the same as that of the great majority of the young people interviewed—it is always the individual who subsumes the social, and never the other way round. It is a viewpoint ratified by a cultural model that is by now centuries old. This also gives rise to the tacit debt of gratitude that binds children to parents: children feel profoundly grateful with regard to their family members because in the family they see the first source of the well-being and freedom they enjoy. It is also thanks to this profound and timeless bond of gratitude that, despite the profound changes that have affected lifestyles, their relationship is so strong. Young people have the impression of proceeding along the path that their forefathers have traced.

In this framework, the reasons why family memory is one of the crucial centers (if not *the* center) of social memory in Calabria become clearer. Through family memory, the past—characterized for older generations by hard manual labor, by the emigration experience, by the interminable rosary of sacrifices made to provide oneself and one's family with a dignified existence—becomes indelible, is turned into a perennial present.

Thus, family memory provides the younger members with important resources for the definition of an identity. In the first place, it indicates the starting point for the path of social ascent whose final stages they are covering. A guide for the future as much as a lesson from the past, the "living history" conveyed by family memory becomes, for the young people forced to face a vague future, a protective shield both against the unknown and interior wavering, and against any incidents faced in the struggle for status. Thanks to its vitality, young people can count on a path whose sense is socially unchallengeable. Its affective, intimate cipher blunts the feeling of difference within the family group and increases the latter's ability to unify different worlds. It is thanks to the power of family memory that the feeling of the children's gratitude can be consolidated, resisting the new uncertainties of the present. This memory in fact manages to transmit intact the profound sense of the effort made by the older generations to guarantee the young people a more acceptable existence. Gratitude and the force of family memory contribute to reinforcing emotional involvement, while underscoring the immutability of that which 'matters in life'. In turn, collective memory strengthens gratitude and the latter, for its part, increases the former. This circular development also serves to legitimate a model of work culture for young people in Calabria.

Preserved from the wear of time due to this twofold influence, archaic dimensions survive in the orientation of young people, even if with a perfect change in character associated with the present work culture. In other words, the memory of difficult work experiences of the preceding generations preserved in the family memory allows young people to relish their own new social condition that is free from the obligation of manual labor. They are in fact passing through the last phase of a path of ascending social mobility that has proceeded from the agricultural work of their grandparents (and in part still of their parents) to their new condition as students. The memory of physically hard labor, above all associated with working the land, thus becomes a tool for the young people to reject manual labor. The stories of their grandparents and parents make the effort of work real and show to their grandchildren and children what should be avoided. More precisely, that which the new access to non-manual labor guaranteed by education makes it possible to avoid. The memory of separation from one's beloved, the poisoned fruit of the emigration experience, is transformed, in the increasingly educated present of the sons and daughters, into a refusal to sever, even temporarily, one's territorial and affective ties to find work in other parts of Italy. Thus, after the hardship associated with the emigration of grandparents and parents, which often lasted many years, the fact of living and working in the same place appears, to the eyes of young people, to be an inalienable request. Finally the memory of existential precariousness, many jobs without any quality and economic security, the everyday "coping" faced by the grandfathers and fathers (for the latter, also in a relatively recent period) is changed into the primary importance assigned by young people to the stability of the job as compared to its concrete contents.[27] The dominant model of work culture among the grandparents and parents, which may be summarized in the statement "any job is fine, as long as it permits one to live and keep one's family alive" is transformed, in the changing scene of contemporary Calabria, into their offspring's dictate "any job is fine, as long as it is not manual." Together with the new conditions of the present, the family memory that is transmitted thus produces a reference framework for values and norms that is renewed by the young people, but without losing an ounce of its own strength. In short, the teaching that is imparted by collective memory manages to remain whole in substance, even if the forms with which it expresses itself are modified today, falling into line with a social situation that is increasingly well-off. In this situation—and it could not be otherwise—even the ways of working have been transformed, in accordance with new levels of education of the population.

The presence of the feeling of gratitude of the younger members with regard to those who are not as young, as has been brought to light, appears in turn doubly linked to the dedication shown by the latter with regard to the family. This dedication, constantly celebrated in the family memory, is part of the present and leaves its mark even on the work cultures. The gratitude that the research has underscored thus seals a real "pact between generations", that the transmission of the family memory in turn contributes to strengthening.

The results of the research suggest that the lesson provided by family memory has been profoundly absorbed by the dominant model of work culture among young people in Calabria. The indissoluble bond created by the obligation of gratitude uniting the sons and daughters to parents whose "sacrifices" have guaranteed the broadening of the horizon of choices through access to the educational system seals it, and reinforces its timelessness.

Instruments of cohesion, integrity and social continuity, collective memory and gratitude remind us that even in the 'society of acceleration' (Rosa, 2003) in which we find ourselves, it is illegitimate to postulate the pure and simple wiping out of the past. Despite the tendency to make "short time" a privileged point of reference in the relationship with timeframes

---

[27] This aspect tends to gather importance as the so-called 'contingent work' (jobs without guarantees for long-term employment) and the new existential precariousness associated with it become more diffuse. On this point, see Beck (2000). For young Calabrese people, this situation is aggravated by the high levels of unemployment: according to a recent study (Fantozzi, 2003: 9), 45.2% of those between 15 and 34 years old in Calabria are unemployed (as opposed to 18.4% of the national average for the same age bracket).

of life, the interviews show how the great majority of the young people are well aware of the inseparable link between the past, present and the future. By showing in this way strong sensitivity with respect to processes of social and historical order that have changed the horizons and daily lives of the generations in a radical way in an arc of time that is relatively short. No less, the movement of time in which young people perceive themselves to be inserted is not resolved simply in a fast passage between different presents. The past emerges from the interviews as a time full of meaning, in the first place on the topic of interpersonal and family relationships,[28] that place it in direct conjunction with identity. In short, this prevalent elaboration of the past as a "lived time" that has been morphed by sentimental ties and emotional bonds, safeguards the possibility of maintaining a broader temporal horizon despite the contemporary problem of building a future in the long-term.

References

Adam, B. "Detraditionalization and the Certainty of Uncertain Futures." In *Detraditionalization*, ed. P. Heelas, S. Lash and P. Morris. Oxford: Blackwell, 1996, pp. 134–148.

Arlacchi, P. *Mafia, contadini e latifondo nella Calabria tradizionale (Mafia, Peasants and Landed Estate in Traditional Calabria)*. Bologna: Il Mulino, 1980.

Arrighi, G. and F. Piselli. "Parentela, clientela e comunità (Kinship, Clientelism and Community)." In *La Calabria*, ed. P. Bevilacqua and A. Placanica. Turin: Einaudi, 1985, pp. 367–492.

Assmann, J. "Erinnern, um dazuzugehören. Kulturelles Gedächtnis, Zugehörigkeitsstruktur und normative Vergangenheit." In *Generation und Gedächtnis*, ed. K. Platt and M. Dabag. Opladen: Leske und Budrich, 1995, pp. 51–75.

Attias-Donfut, C. "La notion de génération. Usages sociaux et concept sociologique." *L'homme et la société* 4 (1988, pp. 48–60).

Bachelard, G. *La dialectique de la durée*. Paris: PUF, 1936/1980.

Bakhurst, D. "Social Memory in Soviet Thought." In *Collective Remembering*, ed. D. Middleton and D. Edwards. London: Sage, 1990, pp. 203–226.

Bauman, Z. *Liquid Love. On the Frailty of Human Bonds*. Cambridge: Polity Press, 2003.

Beck, U., ed. *Die Zukunft von Arbeit und Demokratie*. Frankfurt a. M.: Suhrkamp, 2000.

Beck, U. and E. Beck-Gernsheim. *The Normal Chaos of Love*. Cambridge: Polity Press, 1995 (or. ed. *Das ganze normal Chaos der Liebe*. Frankfurt a. M.: Suhrkamp, 1990).

Bellah, R. N., et al. *Habits of the Heart: Individualism and Commitment in American Life*. Berkeley: University of California Press, 1985.

---

[28] It is possible that the changes that have taken place over during the last decades —calling into question the stability of traditional love bonds as well as the norms of the family (Beck and Beck-Gernsheim, 1995; Bauman 2003)—made even family memory more evanescent. However, in Calabria during the first half of the Nineties a widespread de-structuring of family norms appeared to still be far off.

Benjamin, W. *Schriften*. Frankfurt a. M.: Suhrkamp, 1955.

Bertaux-Wiame, I. "Des formes et des usages. Histoires de famille." *L'Homme et la société* 4 (1988, pp. 24–35).

Buzzi, C., A. Cavalli and A. De Lillo, eds. *Giovani del nuovo secolo (Youth in the New Century)*. Bologna: Il Mulino, 2002.

Cavalli, A. *I giovani del Mezzogiorno (Youth in Southern Italy)*. Bologna: Il Mulino, 1990.

Cavalli, A. and O. Galland, eds. *L'allongement de la jeunesse*. Arles: Acted Sud, 1993.

Congi, G. *Imprenditori e impresa in Calabria (Entrepreneurs and Business in Calabria)*. Cosenza: Marra, 1988.

Cornelisen, A. *Women of the Shadows*. New York: Vintage Books, 1977.

Eisenstadt, S. N. and T. Lemarchand, eds. *Political Clientelism, Patronage and Development*. Beverly Hills- London: Sage, 1981.

Eisenstadt, S. N. and L. Roniger. *Patrons, Clients and Friends*. Cambridge: Cambridge University Press, 1984.

Fantozzi, P. "Clientelismo e mutamento. Il caso del Mezzogiorno d'Italia (Clientelism and Social Change. The Case of Southern Italy)." *Classe* 20 (1982).

———. *Politica clientela e regolazione sociale (Politics Clientelism and Social Regulation)*. Soveria Mannelli: Rubbettino, 1993.

———. "Introduzione (Introduction)." In *Giovani in Calabria (Young People in Calabria)*, ed. P. Fantozzi. Soveria Mannelli: Rubbettino, 2003.

Fraisse, P. *Psychologie du temps*. Paris: PUF, 1957.

Gusfield, J. R. "Tradition and Modernity: Misplaced Polarities in the Study of Social Change." *American Journal of Sociology* 72 (1967, pp. 109–127).

Halbwachs, M. *Les cadres sociaux de la mémoire*. Paris-La Haye: Mouton, 1925/1976.

———. *La mémoire collective*. Paris: PUF, 1951/1968 (English translation "The Collective Memory." New York: Harper and Row, 1980).

———. *La topographie légendaire des Evangiles en Terre Sainte. Étude de mémoire collective*. Paris : PUF, 1941/1971.

Jedlowski, P. "Sull'etica, la critica e la memoria (On Ethics, Critics and Memory)." In *Etica e scienze sociali (Ethics and Social Sciences)*, ed. F. Crespi. Turin: Rosenberg & Sellier, 1991, pp. 125–138.

———. "Memory and Sociology." *Time & Society* 10:1 (2001, pp. 29–44).

———. *Memoria, esperienza, modernità (Memory, Experience, Modernity)*. Milan: Angeli, 2002.

Koselleck, R. *Vergangene Zukunft. Zur Semantik geschichtlicher Zeiten*. Frankfurt a. M.: Suhrkamp, 1979.

Leccardi, C. *Giovani in Calabria fra tradizione e modernità. Le culture del lavoro (Youth in Calabria between Tradition and Modernity. Work Cultures)*. Cosenza: Marra, 1993.

———. *Crescere nel Mezzogiorno (Growing Up in Southern Italy)*. Soveria Mannelli: Rubbettino, 1995a.

———. "Growing Up in Southern Italy: Between Tradition and Modernity." In *Growing up in Europe*, ed. L. Chisholm, P. Büchner, H.-H. Krüger, and M. du Bois-Reymond. Berlin- New York: Walter de Gruyter, 1995b, pp. 95–104.

———. "Frauenarbeit und sozialer Wandel in Kalabrien. Vorstellungen junger Frauen über die Arbeit." *Diskurs* 1 (1998, pp. 40–47).

Maldiney, H. *Aîtres de la langue et demeures de la pensée*. Lausanne: L'Age d'Homme, 1975.

Mauss, M. "Essai sur le don." In *Sociologie et anthropologie*. Paris: PUF, 1925/1950 (English translation "The Gift. Forms and Functions of Exchange in Archaic Societies." New York: Norton, 1967).

Muxel, A. *Individu et mémoire familiale*. Paris: Nathan, 2002.

Namer, G. *Mémoire et société*. Paris: Méridiens Klincksieck, 1987.

———. "Affectivité et temporalité de la mémoire." *L'homme et la société* 4 (1988, pp. 8–15).

———. "Postface." In M. Halbwachs, *La mémoire collective*. Paris: Albin Michel, 1997, nouvelle édition.

Passerini, L. "*Postfazione* (Afterword)." In M. Halbwachs, *La memoria collettiva (Collective Memory)*. Milano: Unicopli, 1987.

Piselli, F. *Parentela ed emigrazione (Kinship and Emigration)*. Turin: Einaudi, 1981.

Polanyi, K. *The Great Transformation*. New York: Holt, Reinhart & Winston Inc., 1944.

Rosa, H. "Social Acceleration: Ethical and Political Consequences of a Desynchronized High-Speed Society." *Constellations* 10:1 (2003, pp. 3–33).

Schudson, M. "The Present in the Past Versus the Past in the Present." *Communication* 11 (1989, pp. 105–113).

Schutz, A. *Collected Papers*. The Hague: Martinus Nijhoff, 1971.

Schütze, F. "Biographieforschung und narratives Interview." *Neue Praxis* 3 (1983, pp. 283–293).

Schwartz, B. "The Social Context of Commemoration: A Study of Collective Memory." *Social Forces* 61 (1982, pp. 374–402).

Shils, E. *Center and Periphery*. Chicago: University of Chicago Press, 1975.

———. *Tradition*. London-Boston: Faber and Faber, 1981.

Signorelli, A. *Chi può e chi aspetta (Somebody Can, Somebody Waits For)*. Naples: Liguori, 1983.

Simmel, G. "Faithfulness and Gratitude." In *The Sociology of Georg Simmel*, ed. K. H. Wolff. New York: The Free Press, 1908/1964, pp. 379–395.

Thomas, W. I. and F. Znaniecki. *The Polish Peasant in Europe and America*. 5 vols. Boston: Richard G. Badger, 1918–1920.

Zerubavel, E. "The Social Marking of the Past: Toward a Sociosemantics of Memory." In *Matters of Culture. Cultural Sociology in Practice*, ed. R. Friedland and J. Mohr. Cambridge: Cambridge University Press, 2004.

CHAPTER SIXTEEN

TIME TO MEET: MEETINGS AS SITES OF ORGANIZATIONAL MEMORY

DAWNA BALLARD & FELIPE GÓMEZ

SUMMARY

Meetings are regularly treated as the backdrop of time-sensitive activities, but rarely considered as an important socio-temporal structure in their own right. Through positioning organizational activities in a timeframe, their decision making function links members to a socially constructed past that resides in their collective memory and simultaneously shapes their present and future. In the present examination, we illustrate our argument drawing examples from the very meeting of which this paper was a part. As central communication structures drawn on regularly to effect a variety of goals, meetings are at the heart of organizational communication and temporality which accounts for their vitality in establishing, debating, and reflecting a group's collective memory.

From the afternoon of Sunday July 25th to the morning of Saturday July 31st 2004, members of the International Society for the Study of Time (ISST) gathered at Clare College in Cambridge, England for the purpose of sharing research and ideas and developing (and maintaining) collegial ties. The conference opened on Monday morning with a brilliant and moving Founder's Address in which J. T. Fraser took great care to situate this meeting within the collective memory of the group. This, the twelfth triennial conference, was the largest such Society meeting in the history of the organization and represented a critical turning point in the "timing" of previous conference traditions. At previous gatherings, the membership was small enough so as to avoid the need for concurrent paper presentation sessions. However, the growing size of the membership and corresponding number of conference attendees in Cambridge gave way to concurrent paper sessions for the first time in the organization's history. This meeting, centered on the theme of Time and Memory, offers excellent occasion to consider its very focus.

As central communication structures drawn on regularly to effect a variety of goals, meetings are at the heart of organizational communication and temporality which accounts for their vitality in establishing, debating, and reflecting a group's collective memory (Gheradi & Strati, 1988). Following

Jo Alyson Parker, Michael Crawford, Paul Harris (Eds), Time and Memory, pp. 303–314.

Weick's (1979) notion of communication cycles, meetings are held to manage equivocality in the environment—i.e., to construct knowledge or learn information not easily gleaned from other sources. This may range from a daily staff update—designed to effect continuous process improvement— to a triennial meeting of scholars committed to the interdisciplinary study of time—designed to facilitate the exchange of new ideas via formal and informal discussions. As Cooren (2006) describes, meetings essentially "talk" an organization into being. Time is also a critical aspect of the very substance of a meeting. As one example, the frequency of a meeting shapes its character. The purpose of a triennial versus daily meeting is starkly different—conveyed, in part, by the amount of time elapsed between gatherings. While both may be equally important, a daily staff update will reflect a greater focus on the present where a triennial meeting will be more focused on the future and past, reflected each by their periodicity.

These two features—communication and temporality—underlie the centrality of memory to the business of meeting. The notion of meeting "minutes," a record of group communication bound by time (i.e., organizational memory), provides insight into this characterization. The Oxford English Dictionary defines a minute as, "an official memorandum, especially one authorizing or recommending a course of action" where *memorandum* is defined as "a note to help the memory; a record of events, or of observations on a particular subject, esp. for future consideration or use" (Simpson, 2005). Thus, Weick's (1979) notion of retention, where organizational members reflect on past actions in order to chart future directions, is a central activity of meeting. This process is facilitated through minutes, as they allow members to consider the successes and failures of the past as a guide for actions in the future based upon their understanding of the environment in the present . The common practice of approving the meeting minutes highlights the social nature of memory (Zerubavel, 2003), as it provides for member discussion about the perceived accuracy of the organization's memory. Even in the absence of a formal record, members still invoke the past as a reference for current decision-making (Oswick, 2006), and contested memories of that past are similarly common.

Despite their potential richness for understanding members' social constructions of the past, present, and future as well as the processes through which these temporal foci are shaped, meetings are overlooked in current conversations regarding organizational temporality. They are regularly treated as the backdrop of time-sensitive activities (Gersick, 1988), but rarely considered as an important socio-temporal structure in their own right (see Cooren, 2006, for a notable exception). Key aspects of the meeting make it an ideal unit of analysis to consider issues of time and memory. Meetings focus our attention at the meso level of analysis—i.e., organizational events, or routines, that link multiple units and levels of analysis (Ballard & Seibold, 2003). A key assumption guiding Ballard and Seibold's (2003) model of organizational temporality is that members' temporal experiences are best understood through their link to the practical demands of the institution (Bourdieu, 1977). Among the practical demands (referred to as communication structures) identified in the model are coordination methods and feedback cycles (Ballard & Seibold, 2003, 2004a). The use of meetings as both a coordinative tool and feedback signpost for members' tasks make them a relevant structure to explore in analyses of organizational temporality. Additionally, meetings introduce a new dimension of time, past time focus, into this framework (Ballard & Seibold, 2004b) by identifying a mechanism through which this dimension is shaped.

In the following pages we consider the ways in which meetings are important sources of information about organizational and group communication processes—specifically, the ways in which organizational memory both shapes and is shaped by these processes. We begin below with a synthesis of disparate literatures in order to develop our argument that through positioning organizational activities in a timeframe, the decision making function of meetings draws members into a past that resides in their collective memory. We turn next to Schwartzman's (1986) theoretical framework of meetings and offer a socio-temporal perspective to consider the practical implications of the role of meetings in shaping organizational memory. Finally, we conclude with a summary and point to directions for future research. Throughout we reflect on a meta-meeting, held at the twelfth triennial conference of the ISST, in which organizational members drew on a collective memory in order to make sense of their future in the present.

Meetings in the Past, Present, and Future: Organizational Memory in the Making

The open council meeting of the ISST on Thursday evening was characterized by extended sense-making (including retrospective sense-making) about the appropriate course of action to manage the present change in their environment (i.e., marked growth in membership) and its impact on past traditions and routines. Having enacted a vision of the future in which the number of conference attendees would only continue to increase, an important agenda item concerned members' feelings about the success of the concurrent session format "tried out" in Cambridge and suggestions for managing the growing membership vis-à-vis the logistics of paper sessions at future conference sites. It was not a decision-making activity per se, but initial input was sought for decisions to be made in the near future. New and old members alike were present for this sense-making activity.

While the time "of" meetings, such as their periodicity, referenced earlier is a relevant aspect of organizational temporality—such as how a triennial meeting typically has the effect of making interaction more hallowed and precious than a daily event, contributing to a stronger orientation to the past and future—it is the time "in" meetings that shapes and is shaped by members' collective memory. Time "in" meetings (i.e., interaction time) shapes the whole organization (McPhee, Corman, & Iverson, 2006). While this is the case for every activity in an extended meeting like a conference in Cambridge, the structured nature of decision making directly casts light on the interconnections among an organization's past, present, and future as it places organizational members and their activities in a timeframe (Butler, 1995).

Butler (1995) observes: "We experience time in the present, but only by relating ourselves to a past and to a future.... The present is preceded by a whole series of events and decisions which become sedimented into some kind of order codifying our experience.... Codes signify (Giddens, 1984: 31; Clark, 1990: 144) states learned from past action (Cyert and March 1992: 174) and enable communication about those states to actors in the present.... Codes contain the history of an organization, but as March (1988: 13) says, history is notoriously ambiguous" (pp. 928–929). In terms of the present discussion, collective memories constitute these codes, but they are not faithful reproductions of the past: memories are social constructions in the present (Zerubavel, 2003). For example, meetings are the communicative events wherein organizational history and knowledge becomes codified and where the meaning of those codes gets debated. Codes determine what gets attended to in discussion vis-à-vis the agenda or impromptu contributions— they are used to define and draw attention to a problem (and to ignore others) as well as signal that a decision must be made to address it. This decision is based on members' retrospective enactment of their environments (Weick, 1979).

The open council meeting illustrates these issues. The organization's "attention" codes (Butler, 1995) defined the growing size of the organization as an issue about which decisions needed to be made. While the issue was consistently described as "a good problem to have," it necessitated a decision, nonetheless, as it represented a departure in the organization's history. In order to move forward effectively, it had to be made sense of in terms of and integrated into existing codes. Members experienced this issue in the present (as they were in the midst of the largest conference in the history of the organization), by drawing on the past (i.e., previous ISST, and other scholarly, conferences) and relating it (i.e., the feasibility of this new size and corresponding format) to the future. Organizational codes, described below, were drawn upon to frame the discussion as well as to opine upon it.

The history of the organization, founded in 1966, is of a small, close knit cadre of scholars. Over the years, this size has enabled them to develop a tradition of gathering for a full week of plenary sessions at intimate, exotic locales consistent with the conference theme and in unique accommodations that afford once-in-a-lifetime opportunities. However, traditions such as this are dependent upon a small membership, and the scholarly interest concerning issues of time across a number of disciplines has increased exponentially since the turn of the century. For example, organizational and group scholars have witnessed a rise in the popularity of studies of workplace temporality. Within the past five years alone, *Academy of Management Journal*, *Academy of Management Review*, *Organizational Studies*, *Small Group Research*, *Work & Occupations*, *Culture and Organization*, and *Organization* have all held special journal issues dedicated to this very topic. This does not even take into account all of the myriad individually located articles on the subject that have been published in about the same time period. What once was an overlooked, understudied aspect of organizational and group life (Bluedorn & Denhardt, 1988), has become a fertile ground of scholarly inquiry.

While membership in the ISST is applied for and selectively granted, if organizational and group studies offer any indication, the number of people who might satisfy these criteria has grown. If this trend continues, then it is reasonable to expect the number of qualified persons applying for and gaining membership to increase as well. Extant codes indicate that an expanding membership jeopardizes critical organizational traditions, and it had already impacted at least one: the tradition of all non-overlapping, or plenary, sessions. Officers wanted to know how members—old and new—felt about this change so that they could use this feedback to plan future conferences. For old members, this sensemaking revolved around comparing their present experience to past ISST conferences. For new members, this involved comparing their present experience to other conferences they had attended and even, paradoxically, how they imagined past ISST experiences (based on what old members described during the meeting). The latter observation supports Zerubavel's (2003) claim that through shared narratives members are able to experience a past, a set of collective memories, which precedes their actual existence as part of the group.

This is consistent with Gheradi and Strati's (1988) findings that members negotiate shared truths about the group's history and social identity through meetings. They assert that meetings "may be regarded as representing the moment of *present* time that ensures coherence and continuity between *past* and *future*—whether the decision is regulative of the course of action or whether it introduces change" (p. 151, italics added). Thus, through positioning organizational activities in a timeframe, the decision making function of meetings draws members into a past that resides in their collective memory. Below we examine the practical implications of this line of argument.

INTEGRATING TEMPORALITY INTO SCHWARTZMAN'S THEORY OF MEETINGS

Schwartzman (1986) describes the various ways in which meetings are so often maligned both by group members and organizational practitioners as useless, poorly ran, and ineffective. Instead of refuting these complaints, she suggests that serious consideration of the ways in which these characterizations are true can lead scholars to a deeper understanding of what purposes meetings really do serve for social collectives. Schwartzman considers three images of meetings that suggest how scholars might reconsider the function meetings serve for organizational members—meetings as *homeostats, rituals*, and *social metaphors*. Below we apply a socio-temporal perspective to this framework and consider the implications of these three images for organizational memory.

MEETINGS AS HOMEOSTATS

One image of meetings suggested by Schwartzman (1986) is that of meetings as homeostats that validate the current social structure and maintain the status quo. Schwartzman even suggests that meetings organized in order to promote change in organizations fail due to their homeostatic function in upholding current structure. Because members have developed a collective memory, or code, through prior meetings and interactions, meetings reinforce existing codes through their "tradition-celebrating" role.

Applied from a temporal perspective, organizations and groups that are pleased with their past decisions and shared history are less likely to embrace changes in their meeting routines. For instance, they are more likely to maintain a regular set of meeting times consistent with past practices and to resist changes that are inconsistent with their history. This is consistent with Ancona, Okhuysen, and Perlow's (2001) proposition that activities, like meetings, in the category of *repeated activity mapping* help to preserve an organization's social system.

Additionally, the periodicity of recurrent scheduled meetings can serve as a source of entrainment (Ancona & Chong, 1996; McGrath & Kelly, 1986) for organizational temporality. For example, regular daily, weekly, monthly, or annual meetings can serve as *zeitgebers* (Bluedorn, 2002) that direct members' attention toward particular temporal signposts, or feedback cycles, as more central to the group than others. Schwartzman (1989) found that this pace, or cycle, is then used to construct collective memories through the kind of temporal grids (of key meetings) that groups use to interpret their shared history. While irregular meetings may punctuate a group's history in important ways, the entrainment that regular meetings permit has the unique position as a tradition-celebrating structure. In contrast, irregular or unscheduled meetings are more likely to carry the capacity to change existing organizational cultural values or policies.

During the open council meeting in Cambridge, the homeostatic function of meetings emerged most strongly as ISST members old and new bemoaned the loss of close intimacy afforded by non-overlapping sessions—despite that fact that "overlapping" meant "dual," not the standard multiples of many conference meetings. New members complimented the wonderfully deliberate, pleasant pace of a conference with no concurrent sessions and were aghast at the idea of abandoning the Wednesday "free day" reserved for touring the city (in order to assist in returning to non-overlapping sessions). So it was not simply synchronicity, but pace, that they wanted to preserve. The new membership saw both temporal features as part of the unique charm and closeness of Society meetings and wanted to maintain the original vision of the founder. Alas, a sense of collective memory had been borne all too quickly for some eager scholars, as Dr. Fraser congenially informed the group that the Wednesday "free day" was never his idea and had evolved over the years through the desires of the membership. At this point, as part of that group of "eager beavers," the first author realized that her strong sense of group identity had led her to create memories of a shared past that not only predated her existence, but did not exist for past members either! Still, the pride and love that a range of members old and new held for the Society led to clear resistance to part with the past. Nonetheless, everyone continually acknowledged that growth is a good problem to have and resigned to the inevitable change (even if it was not an individually experienced change, but existed as a new development in the shared history of the group).

MEETINGS AS RITUALS

Schwartzman (1986) asserts that, as rituals, meetings have symbolic significance that both structures and reflects members' social reality. Specifically the meeting is a:

> ...powerful and ongoing symbol for an organization because it assembles a variety of individuals and groups together and labels the assembly as "organizational action." .... [Thus]... meetings provide participants with a way to both negotiate and interpret their social reality... (p. 250)

From a temporal perspective, because memory consists of judgments about the efficacy of past decisions, meetings represent an ideal opportunity to disagree about past decisions and continue to relive these disagreements through ongoing struggles in the present. In contrast, statements that encourage "leaving the past in the past" or "letting bygones be bygones" characterize an alternate discourse and ritual. In these groups, disagreements are acknowledged as part of the groups' shared history and, in so doing, members attempt to transcend their effects in the present. Putting them "out in the open" is seen as a way of undermining their influence, yet because issues are framed relative to the (forgotten) past, this temporal discourse still serves a ritualistic function.

The image of meetings as rituals is not limited to past conflicts. Similar functions of meetings also regularly draw upon collective memories of success, including organizational heroes and heroines in addition to shared triumphs. These rituals often involve stories and myths (Beyer & Trice, 1987) that allow members' to revel in their history and discuss shared norms and values in relation to present and future concerns. Because meetings highlight the ways in which current interaction is shaped by the past and, in turn, shapes the future, they provide members with an ideal location, or opportunity, to act out on a range of temporal foci (past, present, and future) as well as an array of social issues.

During the conference, the role of meetings as temporal rituals was evident in the open council meeting through the opportunities that members took to express their feelings about the past, present, and future of the organization. This occurred not only through discussions of the presentation sessions, but also through discussions about both the location and theme of future conferences. Given the inordinately high valuation of the British pound compared with the currency of many conference attendees, discussions about the location of future conferences were made through referencing the shared values and norms found at past conferences. Members told awe-inspiring stories of the tiny Italian village they traversed during the last triennial meeting and the wonderful collegiality and familial environment it inspired as well as the ability to travel with one's family (the latter owed in good measure due to a more favorable exchange rate). While the present conference was also seen as collegial and familial in nature, the importance of these values for future meetings was underscored in members' narratives.

MEETINGS AS SOCIAL METAPHORS

Schwartzman (1986) suggests an image of meetings as social metaphors. Through meetings, she argues,

> individuals metaphorically mix their formal and informal relationships and feelings with organizational issues, problems and solutions... because in this context one thing can always be talked about in terms of something else... In this way the meeting allows individuals to engage in a variety of expressive activities while they appear to be engaged in instrumental behavior (p. 251)

The business of meetings—decision making (in a variety of forms)—can be a metaphor for a group's need to establish "memory." Following Weick's (1979) theory of organizing, organizations' evolutionary development follows three stages. In the first stage, enactment, organizations create the environment that faces them. That is, through a variety of ways, they notice certain aspects of their environment and become poised to act on these observations. In the second step, selection, organizations set about to react to their enacted environment, i.e., to make choices that increase their chances of survival. This occurs through a process of sense-making that utilizes both rules (routines, often found in organizational documents and policies) and cycles (communication, which typically require meetings). Finally, in the third stage, retention, organizations reflect on the success of various choices and remember, or retain, what worked and repeat it in the future.

The implication of this theory for considering meetings as social metaphors lays in the fact that sense-making occurs either through rules or cycles. A prerequisite for the reliance upon rules is that an organizational group must have faced the same environment in the past and formulated an appropriate response. Groups without a shared past, newer groups, will have faced fewer environments together and thus formulated fewer agreed-upon rules. Instead, new groups must rely more often on cycles, or meetings, to determine appropriate responses to their environment. In contrast, groups with more shared history, and a collective memory, will be able to rely on rules in more cases. As a result, the frequency and duration of meetings may be indicative of the "time" of the group. A greater frequency and duration of regularly scheduled meetings implies that more time needs to be spent on establishing shared norms and consensus, respectively. This characterizes both new groups and established groups undergoing a great deal of change. In particular, new groups are likely to hold regular meetings on a frequent basis in order to chart their course and establish a shared vision of the future. Older groups that lack appropriate codes (Butler, 1995) to operate in their current environment also need to meet more often.

This concept of meetings as temporal metaphors was reflected in the position of the ISST as an established group undergoing change. Although, the group has a great deal of history and tradition behind it, the need for alternate meeting practices was apparent. A range of possibilities designed to address the concurrent presentation sessions were generated that could not be easily decided upon within the confines of a single meeting. Several options were taken under advisement, but the group seemed to acknowledge that the changing constituency represented a meaningful challenge for the organization that could not be neatly disposed of within one meeting. The officers committed to further meetings and discussions about the issue in order to devise an appropriate strategy to manage this new development.

CONCLUSION

The socio-temporal aspect of meetings represents an important source of information about group and organizational communication processes. Schwartzman (1986) suggests "meetings are expressive forms that serve expressive functions much better than they serve instrumental ones" (p. 244). A temporal lens to understanding meetings offers a distinct perspective because time lies at the nexus of these (expressive and instrumental) functions. The instrumental function of meetings is explicitly about the business of time—e.g., drawing on collective memories to revisit past decisions, hash out future directions, or celebrate present achievements— and yet the social construction of these "times" is accomplished through expressive means. Stohl (2006) recognizes this tension between the topic of discussion and the social context in which it takes place. For example, in order to engage in the sense-making needed to effect instrumental goals, like strategic planning or establishing a new policy, members often invoke collective memories (whether real or imagined), such as the founder's original vision or recent developments in the group's history, in order to endorse particular paths of action. Thus, memories (i.e., social constructions of the past) serve expressive functions as they are drawn on through members' discourse to opine, while being used (as "data") to impact instrumental functions concerning the group's present and future directions.

Through positioning organizational activities in a timeframe, the decision making function of meetings links members to a socially constructed past that resides in their collective memory and simultaneously shapes their present and future. In the present examination, we illustrated our argument drawing examples from the very meeting of which this paper was a part. Given their pervasiveness in organizational life, meetings afford researchers regular access to large amounts of data that can be analyzed drawing from a number of different methods and epistemological perspectives. Depending on the scope of the data, it can point to important and practicable insights for organizational scholars, members, and practitioners. For example, in a single organization, future investigations can yield information that helps members to better understand their unique decision making dynamics and assumptions. Across several organizations, relevant analyses might allow generalizations on topics like the role of contested memories and shared memories in shaping group development and decision outcomes. Considering meetings as sites of organizational memory has the potential to inform a variety of literatures—from group communication and decision making, to organizational assimilation, to learning, to strategic planning, to team identity, to name a few. We hope that the issues and arguments raised here can be assistive in this regard.

REFERENCES

Ancona, D. and C. Chong. "Entrainment: Pace, cycle and rhythm in organizational behavior." In *Research in organizational behavior*, Volume 18, ed. B. M. Staw & L. L. Cummings, 251–284. Greenwich, CT: JAI Press, 1996.

Ancona, D. G., G. A. Okhuysen and L. A. Perlow. (2001). "Taking time to integrate temporal research." *Academy of Management Review* 26 (2001): 512–529.

Ballard, D. I. and D. R. Seibold. "Communicating and organizing in time: A meso level model of organizational temporality." *Management Communication Quarterly* 16 (2003): 380–415.

———. "Communication-related organizational structures and work group members' temporal experience: The effects of interdependence, type of technology, and feedback cycle on members' construals and enactments of time." *Communication Monographs* 71 (2004a): 1–27.

———. "Organizational members' communication and temporal experience: Scale development and validation." *Communication Research* 31 (2004b): 135–172.

Beyer, J. M. and H. M. Trice. "How an organization's rites reveal its culture." *Organizational Dynamics* 15 (1987): 5–24.

Bluedorn, A. C. *The human organization of time: Temporal realities and experience.* Stanford, CA: Stanford Business Books, 2002.

Bluedorn, A. C. and R. B. Denhardt. (1988). "Time and organizations." *Journal of Management* 14 (1988): 299–320.

Bourdieu, P. *Outline of a theory of practice.* Cambridge: Cambridge University Press, 1977.

Butler, R. "Time in organizations: Its experience, explanations, and effects." *Organization Studies* 16 (1995): 925–950.

Clark, P. A. "Chronological codes and organizational analysis." In *The theory and philosophy of organizations,* eds. J. Hassard and D. Pym, 137–166. London: Routledge, 1990.

Cooren, F. *Interacting and organizing: Analyses of a management meeting.* Mahwah, NJ: Lawrence Erlbaum, (2006).

Cyert, R. and J. G. March. *A behavioral theory of the firm,* 2nd ed. Oxford: Basil Blackwell, 1992.

Gersick, C. J. G. "Time and transition in work teams: Toward a new model of group development." *Academy of Management Journal* 31 (1988): 9–41.

Gherardi, S. and A. Strati. "The temporal dimension in organizational studies." *Organization Studies* 9 (1988): 149–164.

Giddens, A. *The constitution of society: An outline of structuration.* Cambridge: Polity Press, 1984.

March, J. G. *Decisions and organizations.* London: Blackwell, 1988.

McGrath, J. E. and J. R. Kelly. *Time in human interaction: Toward a social psychology of time.* New York: Gilford Press, 1986.

McPhee, R. D., S. Corman and J. Iverson. "We ought to have… gumption…" A CRA analysis of an excerpt from the videotape "After Mr. Sam." In *Interacting and organizing: Analyses of a management meeting,* ed. F. Cooren, 133–61. Mahwah, NJ: Lawrence Erlbaum, (2006).

Oswick, C. "Closing words (and opening discussions?): An afterword on *After Mr Sam.*" In *Interacting and organizing: Analyses of a management meeting,* ed. F. Cooren, 289–96. Mahwah, NJ: Lawrence Erlbaum, (2006).

Schwartzman, H. B. "The meeting as a neglected social form in organizational studies." In *Research in Organizational Behavior,* ed. L. L. Cummings. Greenwich, CT: JAI Press Inc, 1986.

———. *The meeting: Gatherings in organizations and communities.* New York: Plenum Press, 1989.

Simpson, J. et al., eds. *Oxford English Dictionary (OED) Online.* Oxford: Oxford University Press, 2005.

Stohl, C. "Bringing the outside in: A contextual analysis." In *Interacting and organizing: Analyses of a management meeting,* ed. F. Cooren, 185–98. Mahwah, NJ: Lawrence Erlbaum, (2006).

Weick, K. E. *The social psychology of organizing,* 2nd ed. New York: McGraw-Hill, 1979.

Zerubavel, E. *Time maps: Collective memory and the social shape of the past.* Chicago: University of Chicago Press, 2003.

INDEX

Jo Alyson Parker, Michael Crawford, Paul Harris (Eds), Time and Memory, pp. 315–321.
© *2006 Koninklijke Brill N.V. Printed in the Netherlands.*